Earth, Water, Fleece and Fabric

Through a richly detailed examination of the practices of spinning yarn from the fleece of llamas and alpacas, *Earth, Water, Fleece and Fabric* explores the relationships that herders of the present and of the past have maintained with their herd animals in the Andes. Dransart juxtaposes an ethnography of an Aymara herding community, based on more than ten years' fieldwork in Isluga in the Chilean highlands, with archaeological material from excavations in the Atacama desert. Relevant historical evidence is adduced.

This work investigates the material culture of pastoral communities at the transition from a hunting and gathering way of life over three thousand years ago, its relationship with domestic processes, and how spinning and weaving in contemporary Isluga express the values of a herding way of life. These values are intimately related to the perceived importance of the landscape, with its resources of earth and water, in the transformation of pasture into fleece. Impeccably researched, this book is the first systematic study to set the material culture of pastoral communities against an understanding of the long-term effects of herding practices. It offers original insights into understanding gender relations among the herders, who establish the working relationships with their animals that enable them to produce yarns and fabrics, while also adopting a dynamic perspective on studying technical changes that have occurred in textile production in the Andes.

Penelope Z. Dransart is Chair of the Department of Archaeology at the University of Wales, Lampeter. Previous publications include *Elemental Meanings: Symbolic Expressions in Inka Miniature Figures* (1995) and *Andean Art: Visual Expression and its Relation to Andean Beliefs and Values* (1995).

Earth, Water, Fleece and Fabric

An ethnography and archaeology of Andean camelid herding

Penelope Z. Dransart

Routledge
Taylor & Francis Group
New York London

First published 2002
by Routledge
2 Park Square, Milton Park, Abingdon, Oxfordshire OX14 4RN

Simultaneously published in the USA and Canada
by Routledge
711 Third Avenue, New York, NY 10017

First issued in paperback 2014

Routledge is an imprint of the Taylor and Francis Group, an informa business

Typeset in Goudy by
Keystroke, Jacaranda Lodge, Wolverhampton

British Library Cataloguing in Publication Data
A catalogue record for this book is available from the British Library

Library of Congress Cataloging in Publication Data
Dransart, Penny.
Earth, water, fleece, and fabric: an ethnography and archaeology of Andean camelid herding / Penelope Z. Dransart.
p. cm.
Includes bibliographical references (p.).
1. Aymara Indians—Agriculture. 2. Pastoral systems—Chile.
3. Aymara textile fabrics—Chile. 4. Camelidae—Chile. I. Title.

F2230.2.A9 D73 2002
305.5′55—dc21 2001048679

ISBN 13: 978-0-415-27959-8 (hbk)
ISBN 13: 978-0-415-75382-1 (pbk)

Contents

Plates

Figures

Tables

Preface

In writing this book, I have drawn on research that I have been doing in the Andes since 1982, when I first went to an Andean country. Why this work takes the form that it does has roots that run deeper. Little did I know it at the time, but my own upbringing in a rural community in Scotland had been shaping and informing my experience of a way of life that depended greatly on the herding of animals. It provided a grounding that helps to explain why I became interested in the research questions addressed here.

I grew up in a small community outside the market town of Inverurie, in north-eastern Scotland, in an area noted for the quality of its cattle, especially that destined for the beef market. However, it was not until I had begun research in the Andes that I fully realized the contribution of my earlier experiences to this work. Back in Scotland in about 1990 I was invited to give a talk at a meeting in a church hall in Inverurie during which I showed slides of llama and alpaca herding in the Andes. I cannot now be sure when this meeting was, but I distinctly remember the place and the comments of an elderly lady who had spent her life on a croft, caring for herd animals and drawing water from a well to supply her kitchen. On seeing the desert conditions in the Atacama, she said 'I wisna worried about the folk, but what did the animals eat?' It was her commitment to the well-being of the animals for which she had been responsible that I found striking in connection with my fieldwork in the Andes.

In Scottish literature, the responsibility that human owners have for their animals has been most powerfully evoked by Lewis Grassic Gibbon (the pen name of James Leslie Mitchell) in *Sunset Song*, the first novel in his trilogy *A Scots Quair*, originally published in 1932. The novel begins shortly before the First World War and it tells the story of Chris Guthrie, whose mother Jean had poisoned herself and the twins, the youngest children of the family, when Jean found that she was pregnant again. Chris abandoned her aspirations to become a school teacher. Her brother married and went to Argentina to work as a cattleman 'on a big Polled Angus ranch' (Gibbon 1978: 86). Chris kept house for her father until he died. His funeral was followed one night by a severe thunderstorm. Chris rose from her bed, realizing that the cattle were in the byre, and would be safe as long as the lightning did not strike it, but that the horses were out in the ley field. Earlier in the year her father had put up round that field newly introduced barbed wire, of which folk said 'it was awful for drawing the lightning' (Gibbon 1978: 105). So, taking a lantern, Chris ventured out to look for the horses:

> Then she saw that the barbed wire was alive, the lightning ran and glowed along it, a living thing, a tremulous vibrant serpent that spat and glowed and hid its head and quivered again to sight. If the horses stood anywhere near to that they were finished, she cried to them again and stopped and listened, it was deathly still in the night between the bursts of thunder, so still that she heard the grass she had pressed underfoot crawl and quiver erect again a step behind her. Then as the thunder moved away . . . something tripped her, she fell and the lantern-flame flared up and seemed almost to vanish; but she righted it, almost sick though she was because of the wet, warm thing that her body and face lay upon.
>
> It was old Bob, he lay dead, his tongue hanging out, his legs doubled under him queerly, poor brute, and she shook at his halter a minute before she realized it was useless and there were still Bess and Clyde to see to.
>
> (Gibbon 1978: 106)

In Chris Guthrie, Lewis Grassic Gibbon created a character who was keenly aware of the responsibilities that were required to care for animals and of her sense of belonging to the land, which supported both human owners and herd animals. The experience of rescuing the horses from the lightning tested her determination to follow the path that she had begun to devise for herself. Instead of selling her father's possessions and disposing of the animals in order to go and live with an uncle and aunt, she resolved to take the unusual step of running the farm on her own until she married. As the novel makes clear, the farming way of life she chose for herself was inexorably to change. Farming communities suffered enormously with the outbreak of the disease BSE in the dairy herds of England during the 1980s and the subsequent banning of beef from all of the United Kingdom for export to other countries. Since I finished writing this manuscript late in 1998, parts of the UK have suffered outbreaks of foot and mouth disease and many farmers have been forced to agree to the government-imposed cull of their herd animals. Despite the vicissitudes, animal rearing still characterizes rural life in north-eastern Scotland.

Scotland is also a country with its own textile traditions, based on the spinning of sheep wool. I do not quite know when my interest in textiles and spinning first began – presumably the skills of my mother, Betty Dransart, as a skilled knitter were influential. In addition, I was a spinner before I went to the Andes, and I wish to recognize the assistance the late Mrs Robertson of Bridge of Walls, Shetland, gave me in teaching me how to spin.

On my first visit to the Andes in 1982, I participated in an archaeological project led by Dr Frank Meddens and Beverley Meddens, based in Pampachiri, Andahuaylas, Peru. I wish to thank them for introducing me to the archaeology of the Andes. However, it was the offer of a Pirie-Reid Scholarship to undertake postgraduate study at the University of Oxford that gave me the first formal steps towards realizing my research. In Oxford I was attached to the Pitt Rivers Museum, where I read for a Master of Studies in Anthropological Archaeology (1984–5), followed by doctoral research. I wish to acknowledge the support my supervisors Dr Donald Tayler and Prof. Howard Morphy gave me. My training in Social Anthropology prepared me with an appropriate framework for channelling and directing the research questions that grew out of my interests in textiles and relationships between human and herd animals.

I began my fieldwork in Isluga, northern Chile, in 1986. In Enquelga I received generous hospitality from many people, including the late Natividad Castro Challapa, Marcos Castro Challapa, Soria Mamani Challapa, Felipe Castro Flores, the late Gabriela Mamani Challapa, Isidro Castro Castro, Silvia Challapa Castro, Pedro Mamani Flores, Isabel Challapa Castro and Domingo Castro. I wish to thank Vivan Gavilán for introducing me to the people of Enquelga. In 1986–8, there was no public transport to Isluga. I am very grateful to the officials working for CONAF (Corporación Nacional Forestal) and SAG (Sociedad Agro-ganadera) for providing me with means of transport for the arduous journey from the coast to Colchane in Isluga.

In 1986 I also established a relationship with the staff of the Museo R.P. Gustavo Le Paige S.J. in San Pedro de Atacama, Chile. I am enormously grateful to Dr Lautaro Núñez Atencio for seeing the value of my project and for providing me with the opportunity to carry it out. Dr Núñez incorporated me into his team working for the Chilean-funded project entitled 'Multidisciplinary analysis of the domestication and early herding of camelids in the north of Chile'. I wish to thank the staff of the museum for their interest in my work and for providing assistance. In particular, I would like to mention Dr Agustín Llagostera, María Antonietta Costa Junqueira, Nilda Escalante Pistán, Manuel Abán López, Felipe Abán López, Tomás Cruz López, Timoteo Cruz Salvatierra, Santiago Ramos Ramos, Héctor Luis Ramírez Ramírez and Luis Ramírez Abán. In addition, I owe thanks to Carmen Selti Pastrana and Ester Cervantes Puco for helping me with the enormous and painstaking task of sorting excavated material from the Tulan sites. Dr Timothy Holden also formed part of the team coordinated by Dr Núñez, and I wish to thank him for his advice on botanical matters. Christina Torres and Dawn Holmes helped in excavation work under hard conditions in locations of difficult access at Puripica and in the Tulan Quebrada, and I am very grateful to them.

During the course of my research I have studied many museum textiles. In the context of this book, I would like to thank the following people for giving me permission to see museum collections and for making time available for me: the late Percy Dauelsberg, Guillermo Focacci, Liliana Ulloa Torres and Dr Mario Rivera (Museo San Miguel, Universidad de Tarapaca, Arica, Chile); Claudio Castellón (Museo Escolar, María Elena, Chile); Héctor Garcés Hill (Museo R.P. Gustavo Le Paige S.J., San Pedro de Atacama, Chile); Elizabeth Rojas Toro (Museo Antropológico de la Universidad San Francisco Xavier de Chuquisaca, Sucre, Bolivia); Dale Idiens (National Museum of Scotland, Edinburgh); and Dr Jonathan King and Elizabeth Carmichael (Museum of Mankind, British Museum,

London). I would also like to thank María Antonia Benavente Aninat of the Universidad de Chile for allowing me to see Chiu Chiu 200 yarns and fabrics in spite of difficult times in Santiago in January 1988.

I wish to extend my gratitude to Dr Michael L. Ryder for explaining and demonstrating the methods he has devised for examining sheep fleece, and for discussing research problems with me. Dr Roy S. Sinclair put me in touch with Dr Jan Wouters of Brussels, Belgium, who provided the results of the dye analysis included in Chapter 8. In Chile, Dr Roberto Rojas helped me obtain samples of vicuña fleece and in Argentina, Dr Carlos Nuevo Freire provided me with samples of guanaco fleece. To all these people I am deeply grateful.

My archaeological work benefited greatly from discussions with a large number of people. Among them, I wish to mention by name Carlos Aldunate del Solar, the late Dr Martha Anders, Prof Warwick Bray, Juan Chacama Rodríguez, Dr Niki Clark, Dr Paul Goldstein, Ray Inskeep, Dr Helen Keeley, Tim Malim, Ivan Muñoz Ovalle, Dr Amy Oakland Rodman, Dr Daniel Olivera, Calogero Santoro V., Dr Constantino M. Torres and Hugo Yacobaccio.

The anthropological aspects of my work have benefited greatly from discussions with and comments supplied by a number of people who include: Nancy Alanoca Astigueta, Dr Denise Arnold, Dr Lindsey Crickmay, Olivia Harris, Felicity Nock, Juan Ossio Acuña, Dr Alison Spedding and Cassandra Torrico.

I am very grateful to the late María Astigueta of Arica who spent much time helping me with Aymara vocabulary. She introduced me to the method for writing Aymara used by Juan de Dios Yapita of ILCA (Instituto de Lengua y Cultura Aymara). Subsequently, I had the good fortune to meet Juan de Dios Yapita in La Paz. I am indebted to him for the advice he freely gave me on linguistic matters, and I wish to thank him and Dr Denise Arnold for all the hospitality they have offered me in La Paz. I first met the late Dr Lucy Briggs in the ILCA office, and I wish to acknowledge the help she gave me on the Aymara language.

Other people who have provided valuable advice include Julia Córdova González, the late Dr Lindsey J. O'Callaghan, Dr Stephen Moorbath (particularly on the problems involved in identifying rocks) and Lucila Reycart Haynes (who helped me devise the intial questionnaire I took to Isluga). Dr Pamela Wace translated Fallet (1961) from the German and helped me enormously by proof-reading my doctoral thesis. Dr Gerry Brush advised me how to set up a database using DBase III Plus, which helped me greatly in producing the results reported in Chapter 8. Then, as my access to software changed, Nicholas Bogdan helped me by converting into an Excel spreadsheet. The library staff at the University of Wales, Lampeter, have helped greatly, and I particularly wish to thank Kathy Miles and Haf James for dealing with my frequent requests for inter-library loans.

I wish to acknowledge the financial support I have received and which has made this work possible. The Pirie-Reid Scholarship not only covered my education in Oxford, but also contributed to my travel expenses to the Andes. My fieldwork has been funded also by contributions from the Emslie Horniman Anthropological Scholarship Fund, the Trustees of Linacre House Trust, the Committee for Graduate Studies (University of Oxford), the Inter-Faculty Committee for Latin

American Studies (University of Oxford), the Pantyfedwen Fund (University of Wales, Lampeter) and the British Council, Santiago, which paid my travel expenses to enable me to spend three months as an Overseas Visiting Professional based at the Museo R.P. Gustavo Le Paige S.J. in 1989.

This work focuses on the spinning of fleece from animal fibre. I finally wish to reiterate the debt that I owe to two women, who lived in different parts of the world, and both of whom taught me so much about fleece and its potential. Both taught me in their old age and, at the time, Mrs Roberston of Bridge of Walls, Shetland, had lost her sight. She spun the fleece of Shetland sheep on an upright Orkney wheel, while Doña Natividad Castro Challapa spun camelid fibre using a drop spindle. The cultural worlds they knew were so very different, yet they shared in common the skill to spin beautifully soft yarn with a very long draft, much longer than I have seen other spinners use. I hope this book provides a fitting recognition in their memory.

Note on the presentation of Radiocarbon dates

Radiocarbon dates are expressed in years before present (BP) as mean and standard deviation (where that information is provided in the source consulted), together with the code and number of the issuing laboratory. Thus radiocarbon dates take the following form: 5320 ± 90 BP (OxA 1842). Unless otherwise stated, dates are uncalibrated.

Website archive

Colour is an important component in the consideration of the fleece of South American camelids. Spinners of the present and of the past are and were sensitive to subtle nuances in natural fleece colours. This book is complemented by a website containing an archive of colour images where readers may see the plates in colour.

http://ads.ahds.ac.uk/catalogue/resources.html?dransart

1 Threads through time

This book explores the relationships maintained between human beings and their herd animals through that most fundamental of human activities, the spinning of yarn from fleece. The animals considered are the South American camelids, a group that includes the vicuña, the guanaco, the llama and the alpaca, and the place is the South-Central Andes: more precisely, the highland and desert areas that are now incorporated into Chilean national territory (figure 1.1). A long-term ethnography is evoked here that focuses on the herding of llamas and the spinning of yarn from over three thousand years ago to the present. My purpose is to demonstrate how an analysis of the making of yarn and fabric responds to the productive relations of a herding society. Yarn making is the product of a particular set of social networks that characterize the societies considered here.

Koster and Chang reject anthropological explanations of pastoral peoples living in recent times that emphasize ecological adaptation at the expense of a proper understanding of the political economy. They stress the importance of the 'historical particularity of the negotiated encounters between herders and the expanding trade and political power of capitalist firms and polities. . . . Real historical time, and not just evolutionary process, is of extreme importance to us' (1994: 14). With the time depth covered in this book, written records are unavailable. It is unrealistic to aspire to writing an account that approximates to 'real historical time'. Instead, my analysis of aspects of material culture provides the basis for the discussion of the prehistoric background, which complements the discussion of my experiences of living in a herding community in northern Chile. My method is to juxtapose different aspects of a herding way of life in order to bring into focus certain trajectories in time.

In writing this book, I am mindful of Stephen Tyler's comments about the term 'ethnography' with its suffix '-graphy' indicating that it 'is contextualized by a technology of written communication' (Tyler 1986: 122). He says the purpose of ethnography should not be that of representing other cultures (which produces idealizations of form and performance), but rather of evoking that which is not present: 'Evocation is neither presentation nor representation. It presents no objects and represents none, yet it makes available through absence what can be conceived but not presented' (ibid.: 123). Tyler's contention that a writer cannot 'represent' another society or culture, but perforce has to use writing in order to

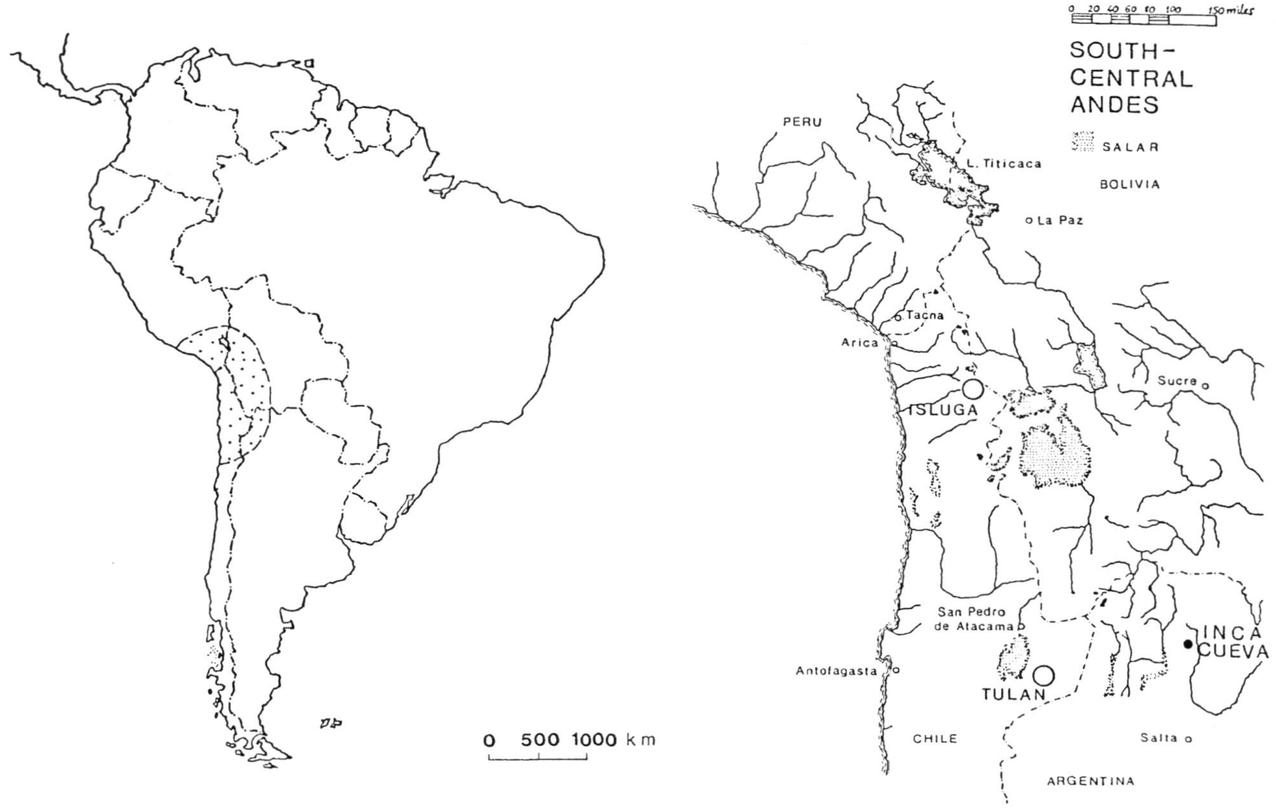

Figure 1.1 The South-Central Andes, showing the location of Isluga and the Tulan Quebrada.

connect the reader with it, has been examined by Marilyn Strathern (1991:7). She observes that by regarding ethnography as a discourse within the context of a postmodern world, Tyler brackets together reader and writer, thereby disallowing scope for comparative analysis which, she points out, has been part of anthropology's sense of its own distinctiveness that sets it apart from the other social sciences (ibid.: 8). In anthropology, generalizations about human behaviour and the ability to 'translate' cultures have traditionally relied upon cross-societal comparison (Asad 1986; Holy 1987). Indeed, my motivation to start studying the spinning of fleece and weaving in the Andes in 1982 stemmed from my personal experience of coming from a rural area in Scotland, a country with its own textile and animal rearing histories. My original strategy was therefore one of cross-cultural comparison and contrast.

If we were to produce postmodernist discourses as advocated by Tyler, Strathern comments that it would be possible to compare evocations, but she says one would have to do this for their 'resonance and effects', that is, for their aesthetic impact (Strathern 1991: 8). Yet Tyler's article is outwardly anti-form and anti-aesthetics. In her critique, Strathern ponders whether there is a hidden aesthetic form in his work: the anthropologist as tourist.[1] She asks, 'Has one killed off the fieldworker, then, only to discover the tourist?' (ibid.: 15). In her opinion, this is an impoverished way in which to think about the departure and return that is involved with intensive fieldwork, the writing that results from it, and the reading of the text. To do so would in effect reduce the fieldworker to little more than a consumer (ibid.: 15). However, Strathern takes heart from Tyler's aims for a postmodern ethnography, which include efforts 'to reassimilate, *to reintegrate the self in society*' (Tyler 1986: 135; Strathern 1991: 16, Strathern's emphasis). This means that although he thinks it is pointless to attempt to understand how 'we' understand other people's societies, he is still concerned to situate the self *in* society. Hence one of my concerns in this book is to consider the particular social and cultural contingencies within which human owners and herd animals interacted in the past, and interact in the present, from my own experience as a spinner from a country that has been heavily reliant on herd animals.

Marilyn Strathern's work alerted me to a book by Vincent Crapanzano (1985) on whites in South Africa (Strathern 1991: 13). In it he used a literary device of juxtaposition to set side by side the verbal accounts of the white residents of a community outside Cape Town along with his own comments on the effects that living in that particular political situation had on himself. He regards his book as an attempt to 're-create' what he calls the cacophony and plurivocality of his South African experience in 1980 and 1981 (Crapanzano 1985: xiii–xiv). Crapanzano juxtaposes commentaries made virtually contemporaneously within a very short space of time, although he does append a short chronology beginning *ca* AD300 at the end of his book (ibid.: 327–33).

In contrast, my use of the device of juxtaposition at times sets different contemporaneous perspectives together, and at other times it sets side by side different historical circumstances. Hence in Chapters 2 and 5, I juxtapose an 'outsider' perspective on pasture lands and camelid fleece with my own interpretation of

Isluga perceptions of these phenomena, based on my conversations with Isluga people and my observations of their cultural practices. Chapters 3 and 5 provide discussions of Isluga cultural life, again based on my conversations and observations, in order to examine the cultural context in which Isluga herders conduct their activities. These chapters highlight the important relationships Isluga people perceive between earth, water, land and fleece. Thus Chapter 3 sets a view based from within Isluga against the preceding chapter, with its outsider views and its interpretations of insider views. This is followed in Chapter 4 by an examination of a ceremony known as the *wayñu* in Isluga. The ceremony makes evident Isluga people's reliance on camelid fleece, the use of which is considered in Chapter 5.

Chapter 6 juxtaposes my interpretations of Isluga herding technology, historical material from sixteenth- and seventeenth-century documentary sources, and pre-Hispanic fabrics and textiles. My aim is to examine how fleece was used at particular historical conjunctures in order to look for technical activities that reflect specific herding practices. The chapter also serves as a link for another level of juxtaposition between Chapters 2–5 and Chapters 7–8. In the latter two, I consider archaeological material from the Tulan and Purifica Quebradas and, in so doing, I set material from the pre-Hispanic past alongside ethnographic material from Isluga from 1986 to 1997. Chapter 9 offers my conclusions, in which I take a long-term view of the relationships maintained between human herders and herd animals, with reference to the exploitation of fleece.

The strategy of juxtaposition allows me to set historically discontinuous materials alongside each other, and they are implicitly understood to be of the same order. It does not offer a real alternative to 'comparing', which Tyler regards as a rhetorical device, part of the 'whole visualist ideology of referential discourse' (Tyler 1986: 130). The manner in which Strathern follows through Tyler's arguments is instructive. Having expressed relief that Tyler is concerned to locate 'the self in society', she then goes on to look at specific theoretical developments that arose in the 1980s and the manner in which 'the problems the critiques have addressed have become attached to the critiques as their problems' (Strathern 1991: 19). Perhaps, she admits, we should not have read Tyler as making representation the converse of evocation. If evocation is 'neither presentation nor representation', then representation is 'superseded' by it. Yet Strathern sees many texts that do not conform to the realist/representational mould in 'Anthropology's past', and that Tyler's critique is over-forceful in its attitude to previous anthropological endeavours (ibid.: 19–20). She also found that fellow anthropologists considered the opposition between representation and evocation to be itself 'tendentious and rhetorical' (Strathern, citing Weiner, 1991: 124, n.21).

I would like to think that my book offers useful insights precisely through its juxtaposition of anthropology, history and archaeology. However, I also realize that anthropologists and archaeologists alike have relied extensively on comparative methodologies. This realization is implicit in Chapter 8, where I repeatedly look for correlations and contrasts between the yarns and fabrics that I have been studying and those reported in other published accounts, in order to situate those materials in a cultural context. By carefully working through analyses of the yarns and fabrics excavated from archaeological sites, it is possible to set the material in a fuller regional context. From the results of these analyses, it is possible to observe the choices and preferences people made in the past regarding their use of raw materials. Such work illuminates their appreciation of the domains from which they gathered or harvested those materials. The results of these analyses are juxtaposed with the ethnographic material in order to convey a long-term perspective on the relationships human herders maintain with their herd animals, the technical activity of spinning camelid fleece and the perceived importance of landscape for the production of fleece as a renewable resource.

However, before embarking on these chapters, I first wish to comment briefly on the concept of having herds, with particular reference to the South-Central Andes.

The concept of having herds

Pastoral societies live with and by the animals they herd. Studies of pastoral societies focus on the characteristics of the relationship established between human animals and herd animals. From an archaeological or historical perspective, authors writing about pastoralism are also concerned with the antiquity of such relationships. This is conventionally referred to as domestication, the point at which human beings are seen qualitatively to change the relationship from one of non-responsibility for non-human animal species to one of responsibility over other species. Many authors equate the notion with control, and they implicitly or explicitly attribute agency for the changing relationship to men, who are seen to exert dominion over animals. The heartlands for such studies are the fertile crescent of Eurasia and the grasslands of East Africa. A traditional picture emerged, in which anthropological and archaeological data were employed to characterize pastoral societies as patriarchal, patrilineal and a phenomenon belonging to the so-called Old World.

New approaches to pastoralism have been challenging and complementing the traditional perspective. Ingold's study (1980) of the transition from hunting to herding reindeer in the sub-arctic regions of Eurasia has been influential. More recently, Kinahan has proposed that nomadic pastoralism in the Central Namib Desert emerged from an indigenous hunting economy, and that it was characterized by autonomous household units and the social relations of property in livestock (Kinahan 1991: 12).[2]

The study of human and herd animal relationships

In anthropology, archaeology and human geography, the theme of animal domestication has been incorporated into larger-scale treatments on the exercise of human domination and control (Lumbreras 1990; Hodder 1990; Tuan 1984; Ingold 1994). In his study of Peruvian prehistory, Luis Lumbreras regards the transition from the Archaic period to the Formative as the 'Andean neolithic'. During this time, he says that human beings 'finished the task of domesticating the natural environment and subjected it to their domination' (Lumbreras 1990: 37).[3] This he regards as a key moment in a process that was to involve the adaptation of plants and animals for human use, and the adoption of wide-scale irrigation schemes, followed by the abandonment of gathering and hunting activities. The process culminated in 'modern life' as experienced by contemporary city-dwelling Peruvians, to whom he addresses his book.

Whereas Lumbreras sees domestication as an essential step in an evolutionary process inspired by the works of V. Gordon Childe, Ian Hodder presents domestication in Neolithic Europe as a step in the development of a particular symbolic order. A commitment to sedentism and intensification arose, Hodder says, through a desire to 'control the wild' (Hodder 1990: 41). The dual linkage between domestication and control is part of what he calls the structures of control in which the house serves as a metaphor for the domestication of society (ibid.: 12, 29, 30). Hodder proposes that a symbolic domestication arose and provided the social and cultural environment that enabled the domestication of plants and animals to take place (ibid.: 31). In this view, symbolic relations are seen as being internally generated within the community rather than constituting an early phase in a universal evolutionary process. However, like Lumbreras, Hodder examines the archaeological material in order to appraise the defining moment when humans first relied on domesticated plants and animals. Neither author enquires too deeply how that relationship between humans and animals was and is maintained.

In contrast, Tuan (1984: 6) adopts a descriptive, psychological approach to the 'making and maintenance of pets'. His study of the (ab)use of power is a bleak one – the affection people have for their pets 'is possible only in relationships of inequality' (ibid.: 5). While he recognizes that humans may have the power to reduce other humans and animals to the status of animate nature, an animate being does possess a will of its own, hence the exercise of ultimate power is 'the ability to overcome resistance' (ibid.: 15, 168). In this view, the converse of dominance is dependency and obedience (ibid.: 172). Tuan collated material in order to clarify relationships between humans, animals and plants. His discussion on the exercise of power centres on examples drawn mainly from historical accounts of Asian and European monarchs and potentates, and deals with the manipulation of plants as well as animate beings. However, his chapter on human cruelty and affection towards animals does incorporate ethnographic material from smaller-scale societies.

For Tuan, domestication is cognate with domination: 'the two words have the same sense of mastery over another being – of bringing it into one's house or domain' (ibid.: 99).[4] In the wider context of the exercise of power by men often characterized as ferocious despots, Tuan concludes that the keeping of pets may appear to be relatively innocuous, an activity that benefits the master that may, arguably, also benefit the dominated pet (ibid.: 6).

Ingold develops a considerably more nuanced argument. He maintains that although hunter-gatherers and pastoralists both depend on animals, the principles that characterize the relationship are 'mutually exclusive' (Ingold 1994: 16). In his view, the principle that operates among hunter-gatherers is one of trust, which, he explains, combines both autonomy and dependency and which is also expressed in the perception hunters have of their animal prey (ibid.: 12–13). According to Ingold, animals 'are not just "there" for the hunter to find and take as he will, rather they *present themselves* to him' (ibid.: 14, original emphasis). The contrastive principle is one of domination, and it involves the herder imposing his will 'whether by force or by more subtle forms of manipulation' in 'an abrogation of trust' (ibid.: 15). For Ingold, whips, spurs, harnesses and hobbles indicate mastery and control over non-human animal species. Among Eurasian reindeer herdsmen (the term used by Ingold), the transition from hunting to herding animals led to the replacement of egalitarian relations of sharing by a system of relations in which herding leaders dominated their assistants as well as their animals (ibid.: 18).

Ingold sees the principles of trust and domination as alternatives, neither being more advanced than the other. He comments that they are both ways of living with animals that offer possibilities for 're-writing the history of human–animal relations' (ibid.: 19). A problem that I have with these alternatives is that there are aspects of the herding way of life in Isluga, discussed in Chapters 3 and 4, that seem to me to have much in common with Ingold's notion of trust, which he equates with hunter-gatherer societies. In an earlier paper, Ingold himself has discussed how individual societies may combine a structural matrix of a mixed hunter-gatherer and herding economy (Ingold 1984).

Another difficulty I have is that his treatment of the alternative principle of domination is couched in terms that emphasize 'mastery' and 'control'. In common with the works cited by Lumbreras and Hodder, the responsibility that herders have for their animals is treated as human control (which I read as the exercise of a particular type of power) over other species. These views *ex*clude from their purview a consideration of gender relations among the herders who are establishing a working relationship with their animals. Tuan goes as far as to *pre*clude women from having any responsibility for introducing changes in the relationships between species. In discussing men's power and domination and 'man's role in changing the face of the earth', he says 'man is the correct word because men, not women, have brought about nearly all the major changes for good and ill' (Tuan 1984: 5). Women have as close a relationship with herd animals as men in the herding societies of northern Chile. The need to examine women and men's relationships with animals is advocated by Lita Webley in her study of changing relations among Khoe herders in Namibia (Webley 1997). In this book, I wish to characterize the long-term effects of the actions and values (as expressed in their yarns and fabrics) of both women and men in assuming responsibility for their herd animals.

Aims and strategies of this book

This book presents a long-term view of the relationship established between herd and human animals and of the exploitation of an environmental resource (camelids as fibre-producing animals). It does so in the context of particular systems of social relations that govern the appropriation of this resource. I recognize that human beings experience a tension between caring for their herd animals and using them as sources of food and raw materials. My premise is that by understanding the changing relationship between human groups and herd animals we can deepen our understanding of the development of fibre technology in the Andes. Evidence from Tulan and Isluga in northern Chile is juxtaposed, with a view to examine the 'material conditions of social life' (to borrow a phrase from Ingold 1986: ix).

Although both ethnographic and archaeological data are considered here, no attempt is made in 'archaeological ethnography', as defined by Stiles (1977). The ethnographic material was not collected in order to compare variation in artefact form, nor the details of spatial relations of the physical traces of social and economic activities, nor yet disposal practices, all of which have been listed by Stiles (ibid.: 91) as being areas of specific interest to archaeologists but which are not normally recorded in ethnographies. The chapters in this work that deal with Isluga concentrate on material practices relating to the herding of camelids and to the exploitation of camelid fleece. Very few direct analogies are made between Isluga practices and the archaeological data. The warning voiced by Payne (1968: 371), who advocates that past populations of animals should be reconstructed from internal evidence, and not from existing evidence, should be noted. Stiles (1977: 89) lists comments by authors who see 'no logical connexion' between the past and present, and hence no validity in the use of ethnographic analogies in archaeology, and Kinahan (1991: 123) reminds us that recent ethnography has critical limitations as a means for viewing the distant past.[5]

However, the juxtaposition of material from the recent past along with an interpretation of the material remains left by past societies allows certain questions to be posed. Isluga herding practices permit an examination of widely held presumptions that tend to prevail in the northern hemisphere. Did the herding of camelids by the people of the Atacama desert three thousand years ago resemble the strategies employed by the Chukchi reindeer herders (that is, the human owners respected the 'wild' behaviour and appearance of their animals, which corresponded closely with those of their wild counterparts, and used their animals as sources of raw material)?[6] Alternatively, had the people already forged a closer social bonding and recognized their animals as individuals, as demonstrated by the people of Isluga? What difference does owning one's animals have on yarn and fabric production?

I wish to prioritize neither of the two contexts considered in the course of this work. The systems of social relations that enable people to reproduce their material existence are seen as constituting responses to the particular conditions in which they live. So, although relics of the past may be observed in the present, and antecedents for future practices may be seen (retrospectively) in the past, neither can adequately explain the other.

This work aims to present qualitative rather than quantitative studies. The Isluga ethnographic account is based on participant observation in a part of Chile that was relatively isolated until the 1970s, when roads to the area were somewhat improved. There was still no public transport to the area when I did my initial fieldwork between March 1986 and January 1988, although it was gradually introduced. Since the mid-1990s I have been travelling by bus for much of the journey between Iquique and the border with Bolivia. Camelid herding is in decline in the whole of the north of Chile, and city-based Chileans tend to hold both camelid herding and the high-altitude lifestyle of the Aymara in low esteem. Herd censuses for northern Chile have been attempted by Novoa and Wheeler (1984: 117) and Guerrero (1986: 46–7, Table 12 and Figure 11), and for the community of Cariquima (Van Kessel 1980).[7] I did not consider a census of domesticated camelids in Isluga important for the aims of this work.

Transformations

In the course of this work, transformations on different levels are considered. The material base is provided by the transformation of camelid fibre into yarn, and yarn into fabric. Camelid fleece is a raw material that is perceived by the users as having inherent possibilities. Other transformations are equally important. Fleece itself is considered by the people of Isluga to be the result of the transformation of pasture and water into fibre. The increase in and abundance of camelid fleece is, and was, an important concern for the herders. Finally, the transition from hunting wild camelids to herding domesticated camelids is considered in the context of the landscape of the Purifica and Tulan Quebradas. The character of this transformation requires some comment.

With the adoption of herding as a subsistence activity, hunting and gathering yielded to a pastoral economy. Talal Asad maintains that there is no such thing as an 'essential "pastoral nomadic society"' (Asad 1978: 60), for pastoralists, if their economic organization is based on production for subsistence, may be in the same category as, for example, sedentary peasants (ibid.: 58). In reality, pastoralism is not, and was not, the only means of subsistence.[8] The Tulan archaeological record shows that after people began to herd domesticated camelids, they still continued to hunt wild rodents and birds and to gather plant foods. In contemporary Isluga, people herd llamas, alpacas and sheep, but they also grow some potatoes and *quinua*, as well as seeking waged day labour. Previously, they traded llama meat for maize and hired out their llama caravans to agriculturalists of the valleys. It is self-evident that no society lives on pastoralism alone.

Asad advocates that we consider the relation between direct producers and those who expropriate their surplus product (ibid.: 58). Deborah Caro accepts his basic premise, but she also considers that 'the process of expropriation differs according to whether the primary producers are herders or farmers' (Caro 1985: 26). The operational constraints, she says, are different for herders and agriculturalists, since animals may be consumable while still alive and while they still have the capacity to multiply (animals may provide fleece and/or milk while still alive, and meat when slaughtered). The relation between labour and land is different, too. According to Caro, land is productive only in that it transforms nutrients contained in its pasture into meat, wool and/or milk. Finally, she considers the structure of consumption to be different among herders. Herders enter into exchange relations with other producers or the market, and they are afforded a greater range of options in this respect than are agricultural peasants (ibid.: 27).

According to Asad, societies that have been studied by social anthropologists are historically determinate populations, which interact systematically. Hence it is important to study the character of 'the total systems within which nomads exist and reproduce themselves as a distinctive cultural, political and economic entity' (Asad 1978: 60). This wider analytical framework advocated by Asad is important. The relationships between the herders of Isluga and their agricultural neighbours in the valleys is an interesting topic worthy of further investigation. Isluga does not constitute a self-contained system, since I believe pastoral systems to be inherently unstable. The valleys and coastal cities to the west and valleys far to the east in Bolivia provide 'demographic reservoirs' to which people sometimes migrate for longer or shorter periods of time. Five thousand years ago, the Tulan Quebrada was also connected to peoples of other areas. Peoples using the Tulan and neighbouring *quebradas* undoubtedly established social relations with other human groups, as testified by the presence of seashells in archaeological sites throughout the period under consideration here. Given that there is no easy access from the *quebradas* to the east of the Salar de Atacama to the sea, due to the extreme aridity of the intervening desert, and the absence at this point of a river which might act as a corridor, the presence of sea shells implies the existence of a complex set of contacts with other peoples.

However, I do not propose to adopt this wider analytical viewpoint in this book. My concern here is to examine the system of social relations by which people reproduce their material existence with special reference to the cooption of camelids into systems of relationships established between people. In other words, people began to herd camelids and they came to regard certain living animals as the private property of certain individuals. The social relationships established between people through the ownership of camelids were concerned with movable property. This particular form of establishing property rights is different from that of claiming ownership over plots of land for the cultivation of crops. The transition in the valleys to the east of the Salar de Atacama was a direct one from hunting camelids to herding camelids. Whereas wild vicuña and guanaco do not belong to individual human beings and may be hunted by anyone who has the ability to do so, domesticated camelids belong to certain people, and other people become dependent on specific other individuals for access to basic requirements. If people previously lived on the principle of 'free access' to hunted animals, they relinquished that right of access to all those domesticated camelids that were not their personal property. James Woodburn describes what he calls Hadza 'egalitarianism' as tending to disengage people from property and from the 'potentiality in property rights for creating dependencies' (Woodburn 1982: 445).[9] With the adoption of pastoralism, people were now able to coopt the labour service of the camelids and to ensure access to their own supplies of food and raw materials. They also created dependencies among themselves.

Some authors (Browman 1974; Hesse 1982b) maintain that climatic changes, capable of seriously affecting the availability of game, probably caused herders to start herding camelids in order to ensure the easy availability of tame livestock as a source of meat. Reconstructions of the palaeoclimatic conditions of the area to the east of the Salar de Atacama are considered further at a later stage (Chapter 7), but they do not tend to support such claims. Of course, pastoralism, like any other subsistence activity, is subject to certain ecological constraints. However, Tim Ingold sees the inception of pastoralism as being marked by a qualitative infrastructural transformation (Ingold 1980: 94). In other words, he rejects the attribution of the origins of pastoralism both to technological advances in the science of breeding and to any underlying ecological rationality.[10] Nor does he envisage pastoralism as proceeding by imperceptible degrees. Instead, Ingold posits the source of discontinuity between hunting and pastoralism on the level of the *social* relations of production. Hence, 'the reproductive increase of the nucleus of domestic stock to substitute for wild herds as a subsistence resource simultaneously reproduces the property relations of pastoralism, but not of the original bonds of taming which gave rise to them' (ibid.: 94).

In Isluga, domesticated animals are owned by individuals and cared for by households, a situation which conforms to the usual pattern of ownership in the Andes and which is discussed further in Chapter 3. Parents devolve animate property (animals) to male and female children, and ownership is claimed for the progeny of the animals in the uterine line. Therefore camelids have known genealogies, which reveal a network of social relations between herders. The owner has power over the animal and he or she may sell it, give it to another person or keep it. If the animal is a llama, it may provide its owner with labour service as a beast of burden. In addition, once the animal has been slaughtered, the meat and the raw materials belong to the owner, who may use or dispose of these according to his or her wishes. These powers over camelids are tempered by a sense of responsibility for them. A herder in Caaquena, northern Chile, explained to me that she goes to bed thinking of her animals, and she wakes up thinking of them.

From an analytical perspective, Ingold (1980) contrasted two types of pastoralism: carnivorous and milch. This classification provides a useful counterbalance to models of pastoralism that have been proposed for the 'Old World' (Forde 1934; Zeuner 1963; Bökönyi 1969; Sherratt 1981). Ingold assigns his reindeer to the carnivorous category, as the domestication of reindeer arose from a direct transition from hunting wild reindeer (1980: 85). Milch pastoralism, according to Bökönyi, emerged at the end of the Neolithic when, he says, 'new uses' were discovered for domesticated animals, including 'milk, wool, draught etc' (1969: 222). Sherratt dubbed these 'secondary products' (1981: 261–3), and they are seen as a secondary consequence of agricultural intensification.

In the Andes, the potential for spinning camelid hair was realized long before the beginnings of domestication have been traced in the archaeological record.[11] Camelid fleece is emphatically not a 'secondary product'. The differences between the coats of the wild and domesticated camelids are not of the order as between wild and domesticated sheep (Dransart 1991b: 458–73). In addition, camelids do not produce sufficient quantities of milk to be milked. High levels of lactose are reported in llama milk (McCracken 1971: 480). This factor lessens the usefulness of milk as a source of human food, since adult Andean populations might not have the ability to metabolize it.[12] As a result, llamas and alpacas are never milked in the Andes. Nor is blood taken from them while they are still alive. Like reindeer herding in the subarctic and sheep and goats in the Namib desert, the archaeological record suggests that in the Andes camelid herding emerged as a direct transition from hunting.

Since camelid herders do not take milk and blood from their animals while still alive, Andean pastoralism cannot be assigned to milch pastoralism. However, the harvesting of fleece from live camelids means that, strictly speaking, it cannot be assigned to Ingold's category of carnivorous pastoralism, where the animal has to be slaughtered in order to provide meat and hides.

The paradox: camelids as providers of raw materials

The uses which human beings made of their herd animals in the pre-Hispanic past are attested by the remains excavated from archaeological sites. In the Atacama, the prevailing climatic conditions are arid; hence the range of animal-derived products obtained from excavations is varied, including fleece, hide, sun-dried meat as well as the bones. In less arid regions, where conditions do not favour the preservation of organic remains, animal bones provide the only direct evidence for the exploitation of animals. This study, in contrast, relies on the direct examination of a product for which the camelids were valued: the fleece itself.[13] The need for reliable supplies of fleece was, as will be demonstrated in this book, an important factor in the process of maintaining herds of camelids in the Andes. Domestication of the camelids was not a necessary step that preceded the widespread use of animal fleece in the making of yarns and fabrics. However, with the domestication of the camelids, animals are more readily available for shearing while they are still alive. Nowadays, camelids are normally shorn every other year. Since most llamas and alpacas are slaughtered at the age of seven or eight years, an animal may be shorn two or three times during its lifespan. However, much will depend on the fleece growth characteristics of the individual animal and how old it was at the first shearing. If the fleece which is pulled from the hide after the animal dies or is slaughtered is taken into account, the owner may expect to gain as much as four times the amount of fleece from one domesticated camelid as from a hunted one. This means that pastoralists, as owners of camelids, have access to greater quantities of fleece than do hunters of wild camelids. What herders did with their more abundant supplies of fleece, and how modern herders attempt to increase such supplies, are themes examined here.

From a materialistic perspective it might be said that the usefulness of vicuña, guanaco, llamas and alpacas to human beings is that they can be slaughtered to provide meat and raw materials. A superficial examination of the uses of camelid products among present-day herders might also support such a conclusion. In Isluga, most of the animal is used as a source of food, and when a llama or alpaca is slaughtered only the appendix is thrown to the cat. The long bones provide marrow, and even the lower limbs are roasted. The padded foot of the llama is considered to be a great delicacy. Some bones are used to make tools. The scapulae are used, unmodified, for harvesting potatoes, and one of the right metapodials (the metatarsal is preferred) is shaped into a pointed weaving tool, which is used to beat the weft into the fell of the fabric (see Miller 1979: 77–9). The hide may be used to cover drums, and the thick leather from the neck is cut into narrow strips for lashing beams together (for example, the rafters of a house). The neck leather may be also used to make sandals. Fleece, of course, is exploited in many different ways, which are discussed in this work. Finally, llamas may be used as beasts of burden.

However, this materialistic perspective is, at best, only a partial picture. Unfortunately, such a view has prevailed in the archaeological literature, and many authors have considered the exploitation of animals by different peoples of the past in terms of little more than a relentless quest for food. In his study of Formative period sites at Caserones and Tarapacá 40A and B, northern Chile, Lautaro Núñez lists all the animal bones registered from the site as food (Núñez 1982: tables 1 and 2). He suggests that after the maize harvest, the vegetable remains were fed to camelids along with shrubs inedible to human beings, so that plant foods with a high cellulose content were cycled to provide meat for human beings (ibid.: 106–7). This may indeed have been the case, but human–camelid relationships were undoubtedly more complex than that, as indicated by the different animal products found at these sites.

An even more serious consequence of regarding herded animals as no more than suppliers of dressed meat is to be found in an article by Gray Graffam. He argues that llamas represented a minor contribution to the 'annual caloric budget' of families living in the Titicaca Basin in the period following the collapse of the Tiwanaku state (Graffam 1992: 890). Since he fails to take into account other products afforded by llamas (such as fleece and hides at an altitude where clothing is an essential aspect of human survival), he came to the conclusion that intensive agriculture subsidized the keeping of camelids (ibid.: 887). In his scenario, herding and intensive agriculture on raised fields are seen as antagonistic activities, with the latter maintaining the people who persisted in practising the former. Nowhere does he discuss different strategies undertaken by farmers in order to minimize risk while maintaining satisfactory yields (Hastorf 1993: 29–30). Instead he repeats, but does not acknowledge, a statement originally made by C. Daryll Forde in 1934 that pastoralism never developed as a separate economy anywhere in the Andes (Graffam 1992: 890; Forde 1934: 394). Forde considered pastoralism to have arisen in the Americas with the introduction of sheep to the Navajo by the Spanish. However, Jorge Flores Ochoa conclusively disproved this position when he published in 1968 an ethnography of the pastoralist people of Paratía in southern Peru.

One of the important points emphasised by Flores Ochoa (1979a: 8) is the commitment that pastoralists make in caring for their herd animals. The quality of the bond maintained between human beings and animals is the hallmark of a pastoral way of life. This means that regarding herd animals merely as providers of food and raw materials cannot be a sufficient explanation for understanding human–herd animal relationships. At best, such an approach can seek to explore only a partial view of human motivations in the relationship. This book attempts to explore the paradoxes between caring for generation after generation of herd animal, while at the same time using those animals for their fleece and hides.

2 Camelids, land and water in the South-Central Andes

In this chapter, I examine how the term 'domestication' has been used, with reference to the archaeological literature and the South American camelids. Following this, the camelids themselves are presented to the reader, and the social organization of the wild camelids is examined. In this chapter, social organization refers to the spatial and temporal organization of individual animals within a population of guanacos or vicuñas with respect to one another. This discussion forms the necessary contrast with Chapter 3, which deals with the incorporation of herd animals into cultural life in Isluga. The present chapter ends by considering the terrain that the South American camelids inhabit. At first I attempt to present a perspective that focuses on the use the herd animals make of their land. Then I shift to a human perspective, based on a discussion of the geographical literature and on my observations derived from fieldwork in Isluga. Access to adequate supplies of pasture is of utmost importance for ensuring abundant fleece in camelids, and the theme connecting water, pasture and fleece is considered in the course of this and the following two chapters.

Attempts to define 'domestication'

Many authors have sought common criteria for defining human/animal relationships under the rubric of domestication. Tim Ingold regards the study of such relationships as a particularly human activity, but he reminds us that whether human beings account for domestication by appealing to myths of totemic origins or to the results of scientific observations, what we construct are narratives of the events that have occurred between human and non-human species (1994: 1).

The definition of domestication offered by Bökönyi involves the capture and taming by 'man' of animals of a species with particular behavioural characteristics, their removal from their natural living area and breeding community, and their maintenance under controlled breeding conditions for profit (Bökönyi 1969: 219). He proposes two main phases: (1) animal keeping, and (2) animal breeding, the basis of which is 'purposeful selective breeding and the control of both quantity and quality of feeding' (ibid.: 220). Animals under the first of these phases are implied not to be fully domesticated. In addition, fully domesticated animals are regarded, according to this definition, as having undergone morphological modifications away from the wild form of the animal. This gives rise to the use of such unsatisfactory terms as 'semi-domestication' to describe animals whose appearance or behaviour does not conform to the classic definition of domestication. Wheeler Pires-Ferreira, Pires-Ferreira and Kaulicke (1976: 488) propose a model in which human control is extended over 'herds of semidomesticated or half-tamed animals [camelids] in which breeding with wild animals regularly occurs'.

Lownie maintains that the blanket use of the word 'domestication' to describe the first tentative handling of wild camelids in the Andes several millennia ago, on the one hand, and the corralling, caravanning and breeding activities of Inka times on the other, renders the term virtually devoid of meaning, unless a breakdown of 'domestication' is attempted (Lownie 1978: 119). She recommends the adoption of the eight-point scoring system proposed by Brothwell (1975), which is designed to account for certain types of human/animal relationships.

Brothwell attributes domestication to the changing of the gene pool composition of another animal species by human manipulation, caused by different types of relatively close association (Brothwell 1975: 397). Therefore, he attempts to differentiate human-influenced, micro-evolutionary variation, basing his grades on the appearance and behaviour of the animal. Not only does this grading system imply a hierarchical approach in defining the status of domesticity, but it may also be said that the criterion of reproduction in captivity under human control does not apply to all cases of domesticated animals, witness North Ronaldsay sheep, Orkney, cited by Brothwell as 'semi-feral' (ibid.: 399), horses and bulls in the French Camargue (Ducos 1978: 54), or llamas and alpacas in Isluga. The apparent wildness of the reindeer herds belonging to the Chukchi and Koryak is not an indication of imperfect domestication, but is, as suggested by Ingold (1988: 368), a consequence of their use as sources of raw materials rather than of labour power. Ingold sees the essential difference between hunting and pastoralism in the productive relations linking human beings and animals, and not in the particular characteristics of the animals themselves (ibid.: 82).

According to Ducos, biological or behavioural features separating domestic from wild animals arise from the evolutionary dynamics that occur within human societies, and that with domestication, new types of relationships are established with herd animals:

> [D]omestication can be said to exist when living animals are integrated as objects into the socioeconomic organization of the human group, in the sense that, while living, those animals are objects for ownership, inheritance, exchange, trade, etc., as are the other objects (or persons) with which human groups have something to do. Living conditions are among the consequences of domestication, not the mark of it. . . . In hunting, man/animal relations are between populations; with domestication these relationships tend to be between individuals or between an individual and a group of individuals.
> (ibid. 1978: 54)

The problems of defining domestication have been discussed further by Bökönyi (1989) and Ducos (1989), but neither author has significantly changed his approach. Bökönyi (1989: 26) emphasizes that animal domestication is caught up in very complex relationships between animals and human beings. Ducos (1989: 30) maintains that Bökönyi's definition applies most successfully to the Neolithic and later periods (presumably in Eurasia), but that his own definition may be useful in opening up other perspectives for more ancient periods when humans appear to have deployed novel methods of working with nature.

Guillermo Mengoni Goñalons (1997/8: 425) has argued that we should distinguish the processes of 'domestication' from 'herding', so that both the biological and cultural aspects can be explored in depth. In contrast, I argue that the herding of camelids and the daily activities of caring for them resulted in 'domesticated' herd animals. This is a view that I have developed through my participant fieldwork in Isluga; it is elaborated further in Chapters 3 and 4.

Ancestry and nomenclature of the South American camelids

Camelidae, both those of South America and of Asia and Africa, have an ancestry dating from the late Eocene in North America, with their primary centre of dispersion two to three million years ago from what is now Utah (Cardozo 1975: 44). Those camelids which reached South America had already developed a characteristic fast-pacing gait as their mode of locomotion. A long stride and fast gait is suitable for long-limbed species, and it is useful for a life in open and flat habitats, although this type of pace does lessen the animal's manoeuvrability and stability (Franklin 1982: 461–2).

A form of camelid known as *Hemiauchenia* extended its range southwards early in the Pleistocene over the Panamanian land bridge. It diverged to become *Palaeolama* and *Lama*, the former being much larger than the latter (Pires Ferreira 1981/2: 206). However, the fossil record is incomplete. Fossilized remains are known from the Republics of Brazil, Argentina and Bolivia, but none are known from Peru and northern Chile. *Hemiauchenia* and *Palaeolama* became extinct by the end of the Pleistocene (Franklin 1982: 463). The four surviving Camelidae in South America are the wild guanaco and vicuña (plates 2.1 and 2.2), and the domesticated llama (plate 2.3) and alpaca (plate 2.4). The general biological characteristics of these animals are presented in table 2.1.

There is no universally accepted systematic classification of the South American Camelidae. Some authors classify the llama, alpaca, guanaco and vicuña as separate species of the genus *Lama* (Gilmore 1950; Barros Valenzuela 1963; Zeuner 1963; Gade 1977; Novoa and Wheeler 1984), while others separate the vicuña, describing it as the only species of the genus Vicugna (Koford 1957; Fernández Baca 1971; Cardozo 1975; Miller 1979; Franklin 1982; Bustinza Choque 1985).[1] Since crosses between the four forms of animal occur and produce fertile offspring (Fernández Baca 1971), and all four camelids have the same 2n=74 karyotype, Novoa and Wheeler classify these camelids under one genus, *Lama*. However, they note that there is, as yet, insufficient systematic research to ascertain whether the

Plate 2.1 Vicuña, northern subspecies, Salar de Surire, northern Chile.

Plate 2.2 Vicuña, southern or Atacama subspecies, San Pedro de Atacama.

Plate 2.3 Llamas on high-altitude, dry pasture ground above Enquelga, Isluga.

Plate 2.4 Alpacas on the wet pasture (*bofedal*) at Enquelga, Isluga.

Table 2.1 Comparison of the general biological characteristics of the South American camelids.

	Llama	*Alpaca*
Scientific name	*Lama glama*	*Lama pacos*
Domesticated or wild	Domesticated, beast of burden, meat, fleece, skins, fuel, bones used to make tools	Domesticated, meat, fleece, skins, fuel
Numbers and status	3.6 million and declining	3.2 million and increasing
Distribution	In the Andes of southern Colombia and Ecuador, from the Andes of central Peru, western Bolivia to northern Chile and NW Argentina. Two breeds: 'woolly' and 'bare'	Andes of central Peru, to western Bolivia and northern Chile. Two breeds: *wakaya* and *suri* (only in Peru)
Elevation	*ca* 2,300–4,000 m	*ca* 4,000–4,800 m
Habitat	Alpine grassland and shrubland	Alpine grassland, wet bottomlands
Feeding habits	Browser and grazer	Grazer
Social organization	Polygynous; behaviour suggests territoriality	Polygynous; behaviour suggests territoriality
Description	Uniform or multi-coloured white, brown, grey, black	Uniform or multi-coloured white, brown, grey, black
Weight (kg)	130–155	55–65
Height at withers (cm)	109–119	94–104
Gestation (day)	348–368	342 (*wakaya*); 345 (*suri*)
Birth season	December–March (if males and females are herded in continuous association)	December–March (if males and females are herded in continuous association)
Weight at birth (kg)	8–16	6–7

	Guanaco	*Vicuña*
Scientific name	*Lama guanicoe*	*Lama vicugna*
Domesticated or wild	Wild, hunted for meat and skin	Wild, hunted for meat and skin. Kept in captivity in research stations
Numbers and status	0.6 million and declining; protected in Peru and Chile	0.08 million and increasing; protected in Ecuador, Peru, Bolivia, Chile and Argentina
Distribution	The Andean foothills of Peru, Chile and Argentina, and Patagonia of Chile and Argentina. Four (?) subspecies	The high Andes of central Peru, western Bolivia, north-eastern Chile and north-western Argentina. Two subspecies: northern and southern
Elevation	*ca* 0–4,250 m	*ca* 3,700–4,800 m
Habitat	Desert grassland, savanna, shrubland, forest	Alpine puna grassland
Feeding habitats	Browser and grazer	Grazer
Social organization	Migratory or non-migratory. Resource defence polygyny with permanent to seasonal feeding territory	Non-migratory. Resource defence polygyny with year-round feeding and sleeping territories
Description	Uniform cinnamon brown, with white undersides, grey to black heads. Fibres of fleece 16–18 microns	Uniform rich cinnamon with or without long white chest bib, white undersides. Fibres of fleece 11–14 microns
Weight (kg)	100–120	45–55
Height at withers (cm)	110–115	86–96
Gestation (day)	345–360	330–350
Birth season	April, May, June (Peru)	Mid February to late April (Peru)
Weight at birth (kg)	8–15	4–6

Source: Based on Franklin (1982) and Novoa and Wheeler (1984)

correct classification of the llama, alpaca, guanaco and vicuña is at the species or subspecies level. They use the nomenclature, also adopted here, 'llama' (*Lama glama*), 'alpaca' (*L. pacos*), 'guanaco' (*L. guanicoe*), and 'vicuña' (*L. vicugna*) (Novoa and Wheeler 1984: 116). Aymara herders have developed their own taxonomies for classifying these animals and their crosses; I will consider the use of their terminology in the context of the herding cycle in the next chapter.

In addition to the controversy over whether or not the four South American camelids conform to standard definitions of species as mutually exclusive breeding populations, with the resulting differences of opinion regarding nomenclature, there is a great deal of similarity between the skeletons of these animals. To date, morphological features have not been recognized to enable the faunal analyst consistently to identify archaeological bones other than as merely 'camelid', unless there is sufficient evidence for the dentition of these camelids in the archaeological material (Miller 1979: 6). Pronounced differences between vicuña and guanaco/llama incisor closure is the basis for Franklin favouring the two-genera classification (Franklin 1982: 464). It has been proposed that incisor morphology may be used to help differentiate the camelids represented at an archaeological site; guanaco incisors are spatulate in form, with enamel on the both the lingual and labial surfaces, and have a definite crown, neck and root structure; vicuña incisors are non-spatulate, approximately square in section, have enamel only on the labial surface, and are rootless; and alpaca incisors are intermediate in form between these two, being non-spatulate and rectangular in cross-section, with enamel only on the labial surface, and root formation occurring at an advanced age (Wheeler 1984: 401,402). However, guanaco and llama incisors are indistinguishable (ibid.: 402) and, to further complicate the problem, Miller examined alpaca skulls at a research station at La Raya, Peru, and found a number which had incisors that could be easily confused with those of a vicuña (Miller 1979: 255).

Recent work on a small-sized assemblage of camelid bones from the site of Pirincay, Ecuador, has been interpreted by Miller and Gill (1990) as providing evidence for a previously undocumented form of camelid, an 'undersized llama' intermediate in size between the contemporary llama and alpaca. They point out that the existence of such a camelid, which is hypothesized as being the prevalent form north of about 10° S from the late Early Horizon until Middle Horizon times (using the Peruvian chronology devised by Rowe 1967), would cast doubt upon methods of osteometric discrimination that rely on statistical evaluation (Miller and Gill 1990: 64). Such methods use a comparative series of known skeletons as a standard to which the unknown archaeological skeletal remains are compared. Measurements of certain dimensions of each skeletal element of the comparative series are used to establish the size limits of each species, which are then used as parameters to which the unknown archaeological skeletal fragments are compared. Since there is no clear and obvious separation between the camelids (the ranges of some measurements overlap widely), statistical analyses are employed to determine the measurement(s) which seem to separate each group (Wing 1972: 329–30).[2] Potential pitfalls in the interpretation of such patterns of measurements are pointed out by Hesse and Wapnish (1985: 102). The work conducted by Miller and Gill should be taken to indicate that a previously unsuspected range of polymorphism existed in camelid populations, and they ask that increased attention be paid to regional camelid osteometrics. The existence of an undersized llama form would cast doubt on many past identifications of zooarchaeological alpacas, especially, according to Miller and Gill (1990: 64), in areas removed from the current range of *L. pacos* in the southern *altiplano*. They suggest that the 'undersized' llama may account for a small form of camelid cited by Wing (1977: 852) as prevalent in the northern part of Peru to Tarma, while a larger form predominated in the southern part of Peru as far north as Ayacucho.

The taxonomic controversy surrounding the South American camelids is also compounded by a supposed evolutionary proximity. The domesticated llama is generally regarded as derived from the guanaco, but the origin of the alpaca is less clear. Alpacas share certain morphological features and behavioural characteristics with vicuñas, and some authors suggest that this was due to hybridization with the vicuña, or that the alpaca derived from some now extinct late Pleistocene camelid. However, other authors maintain that both llama and alpaca are the direct descendants of the guanaco.[3] Obviously, the problem remains unresolved and further research is required.

Distribution of the South American camelids

The camelids may be ordered according to their decreasing requirement for succulence in food: alpaca, vicuña, llama and guanaco. Koford notes that this is also the order of increasing geographic range, which suggests that the seasonal availability of green food, and that the differing ranges of tolerance of the camelids for dry food, are important factors which limit the distribution of these animals (Koford 1957: 161). The guanaco is the widest-ranging of the camelids, occupying the western foothills of the Andes of northern Peru and the east-facing slopes of the southern Andes, including Patagonia, Tierra del Fuego and the island of Navarino. It is, therefore, adapted to dry and open habitats: for example, the extremely arid Atacama desert and clearings in the wet forests of Tierra del Fuego, which experience year-round rainfall. Present-day and prehistoric ranges of the camelids are shown in figures 2.1 and 2.2.

The coexistence of vicuña and guanaco populations is not well documented. Koford (1957: 211) observed guanacos throughout the vicuña zone. A late-eighteenth-century report from the Atacama area suggests that guanaco and vicuña occupied the same rocky terrain, since beat hunts rounded up both species (Cañete y Domínguez 1974[1791]: 249).

However, Franklin points out that there are few remaining areas where the two exist in close proximity because of the widespread decline in the numbers of these animals; he observed two instances, the first in the Pampa Galeras National Vicuña Reserve (Latitude 14° S) on the west side of the Andes, and the second in the San Guillermo National Reserve in central Argentina (Latitude 32° S) on the east side of the Andes. He reports that the vicuña and guanaco populations were adjacent, but separated by habitat and altitudinal differences; the vicuña

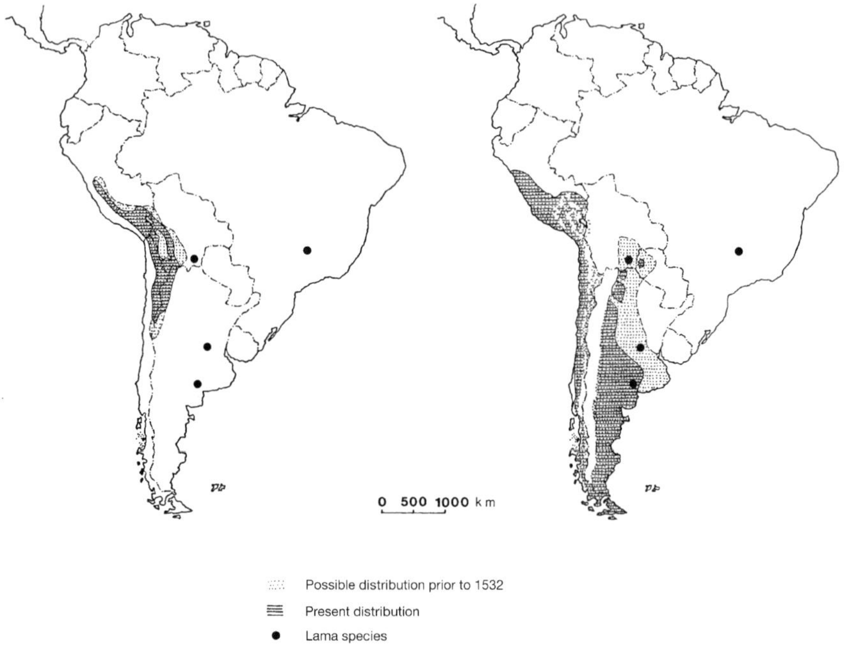

Figure 2.1 Past and present distribution of vicuña (left) and guanaco (right) (adapted from Franklin 1984 and Novoa and Wheeler 1984).

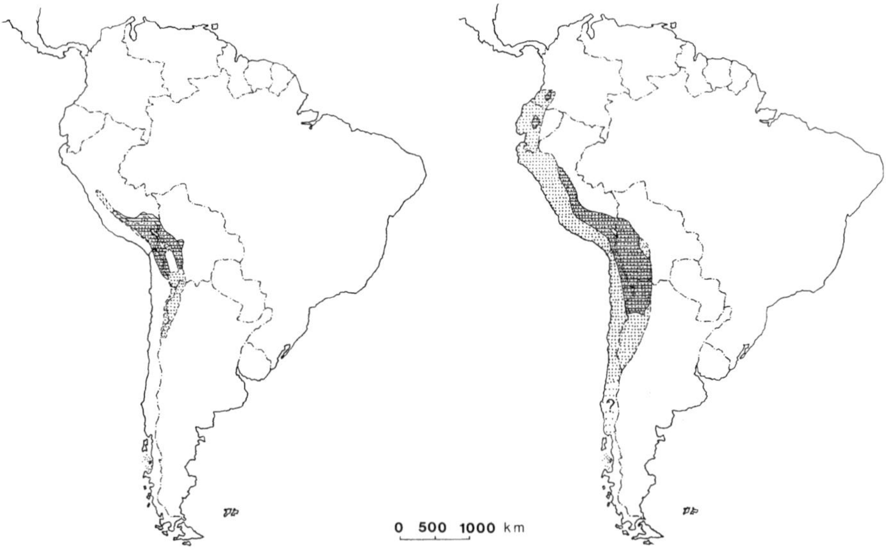

Figure 2.2 Past and present distribution of the alpaca (left) and llama (right) (adapted from Franklin 1984 and Novoa and Wheeler 1984).

occupied the upper *puna* grassland zone, above approximately 4,000 m, while the guanaco occupied the mountainous shrub land and desert-like zone below this altitude (Franklin 1982: 475). However, Wheeler identified guanaco incisors among the archaeological faunal remains from the preceramic site of Telarmachay, at an altitude of 4,420 m above sea level in the central Andean *puna* of Junín. This caused her to express concern over the current lack of knowledge concerning the guanaco. Franklin limits guanaco distribution to areas of tree and scrub vegetation at altitudes below 4,260 m, but a high-altitude subspecies (*L. glama cacsilensis*) has been described on the basis of a single specimen from Caxcile in the Department of Puno, Peru. Wheeler makes the point that the validity of this species has not been verified, and emphasizes the need for further research (Wheeler 1984: 401).

Nowadays, the respective distributions of the domesticated camelids appear to be conditioned by factors of altitude. The llama occurs in various ecological levels, while the alpaca is restricted to the highest, moist pastures of the *altiplano*. However, llamas and alpacas are hardly found in significant numbers beyond the natural living areas of the wild camelids (unless one considers recent renewed attempts to introduce llamas and llama hybrids to non-Andean countries), thus contradicting Bökönyi's definition that the 'essence' of domestication includes the removal of the animal from its natural living area (Bökönyi 1969: 219). Archaeological evidence cited below does counterbalance the somewhat distorted picture produced by the present-day distribution. The optimum geographical distribution of the llama is said to be from 2,300 to 4,000 m asl, and Cardozo (1954) mentions its tolerance to arid conditions, but only at levels above 2,000 m (cited by Flores Ochoa [1980: 64]). The llama is found in large numbers in the highlands of Peru and Bolivia, in north-west Argentina and in northern Chile, in zones parallel with the Cordillera de los Andes. Thus the distribution *appears* to be limited by climatic factors and by the specific vegetation of the Andes. The alpaca is found in greatest numbers from 4,370 to 4,800 m asl (Cardozo [1954], cited by Flores Ochoa [1980: 64]). However, alpacas are found at lower levels, where moist pasture is locally available. From my own observations, alpacas are herded from 3,700 m in the north of Chile. Besides altitude, relative humidity, rainfall, pasture, and the need to avoid certain pathological conditions such as *sarna* (llama mange) have to be taken into account (ibid.: 64). In modern times, alpacas and llamas have been virtually restricted to the high *puna*, with the former predominating in the humid part, without a noticeable presence in the wet *páramo* conditions further north of the central Peruvian highlands, which extend as far as Venezuela. Therefore, llamas and alpacas are unusual in that they are domesticated animals whose present-day distribution is more limited than that of the wild camelids.

However, this situation does not necessarily apply to pre-Hispanic times. The geographical distribution of llamas and alpacas has been modified substantially by the introduction, since the sixteenth century, of sheep, goats, bovine cattle and donkeys by the Europeans. Flores Ochoa states that the present-day distribution of llamas and alpacas is the result of an imposed political situation which favoured the expansion of sheep for the sake of non-Andean economic, political, social and cultural criteria. His study of this process explores the conditions which have intervened to create the present-day limits of high-altitude pastoralism; the marginal limits of the distribution of llamas and alpacas is reflected in the marginality to which Andean pastoralist societies have been subjected (ibid.: 65). Additionally, the archaeologists Shimada and Shimada (1985) have presented a case for the breeding and herding of llamas on the north coast of Peru in prehistoric times. Further archaeological evidence for the presence of prehistoric camelid remains in Ecuador and Colombia has been discussed by Stahl (1988). Sites which have produced such remains are few in number, but the abovementioned assemblage of camelid bones from Pirincay in the Ecuadorian highlands has been interpreted as providing evidence for the introduction of a domesticated form of camelid, the so-called 'undersized llama', in the earliest Regional Development period of Ecuador (Miller and Gill 1990: 64). This small type of llama was present at Pirincay by the late first millennium BC (Karen Olsen Bruhns, personal communication of 6 January 1991).

In more southerly latitudes, in the central valleys of Chile, Spanish-speaking travellers of the sixteenth to eighteenth centuries reported the presence of camelids known as *chilihueque* and *luan* in territories belonging to the Araucanian peoples. Pedro de Valdivia observed, in a region 'thirty leagues' south of the newly founded city of Concepción, that camelids were numerous, the people were dressed in woollen clothing and that their houses were full of wool; he saw 'an abundance of cattle [camelids] like that of Peru, with fleece which drags along the ground' (Valdivia 1953 [1550–]: 223–5). The following century, the Jesuit priest Diego de Rosales equated the *luan* with the guanaco and the *chilihueque* with llamas and alpacas (Rosales 1877 [1670]: 324; cited by Benavente 1985: 46). However, there is some dispute as to what these animals really were. María Antonia Benavente (1985) has examined various arguments which have been used in support of conflicting claims that the *chilihueque* represented a locally domesticated guanaco or that it was really the llama which had been transferred from further north, but without coming to any firm conclusions as to the identity of these animals. Striking archaeological evidence for the importance of camelids in the Valle de Elqui (which is located between the Atacama area considered in this book and Araucanian territory) comes from the site of Las Animas, some 15 km inland from La Serena. This site has given its name to a pre-Diaguitas period complex (*ca* AD 800–1000) (Ampuero Brito 1986: 24). Cemetery sites from this period in places such as La Serena and Coquimbo contain human and camelid burials, in which the corpses were laid out together. Indeed, at the site of Plaza de Coquimbo, eighteen out of a total of some thirty burials combined human beings and one or more camelids; one burial consisted of a human skeleton which was completely covered by five camelids (Castillo G. 1989: 274–5). The excavation of these burials reveals an intimate relationship between humans and animals, in which the former appear to be embraced by the latter (Ampuero Brito 1986: 24; Castillo G. 1989: 270, fig. 4).

Social behaviour of the vicuña and guanaco

Social organization, in this context, is defined by Franklin (1982: 476) as how animals in a population are spatially and temporally organized with respect to one another. Vicuñas live in family groups which consist of an adult male and four to seven females with their young. Their social organization is based on a year-round system in which polygynous groups defend permanent sleeping and feeding territories. A territory is an area that an animal defends against members of the same species (Koford 1957: 190).

Franklin (1982: 478–9) has identified five main types of vicuña social groups, based on fieldstudies conducted in the late 1960s and early 1970s at the Pampa Galeras National Vicuña Reserve in the Department of Ayacucho, Peru:

1. Permanent territorial family groups occupy the preferred habitats. Their territory consists of two parts: a feeding territory, which usually includes permanent water, and a sleeping territory on higher terrain. The size and shape of the feeding territories occasionally change slightly from season to season, but they retain their general localities throughout the year.
2. Marginal territorial family groups occupy areas marginal to the better areas. The size, shape and location of such groups are more likely to change than those of the permanent family groups. They are also less likely to have permanent water, and these groups have to move out of their territory during the dry season in order to drink, since vicuñas have to drink once or twice daily during this season.
3. Mobile family groups or temporary groups of females with a male, but without an established territory. These groups are most common during early summer, when juvenile females have been expelled from their family groups.
4. Male groups consist of non-territorial males, varying from two to one hundred individuals, but averaging twenty-two in Franklin's study area. They are chased out of the preferred habitat areas, where family groups are established. The frequent attacks on them by territorial males result in their running long distances in search of undisturbed feeding and the group size fluctuates frequently.
5. Solo males; sexually and physically mature animals that have established or are ready to establish a territory of their own. They spend much time looking for unoccupied or poorly defended sites in order to set up their own territories.

The territorial system as described by Franklin forms the foundation on which the vicuña population bases its social organization and makes use of its environment. Adult female vicuñas belong to a family group with a territorial male. The territorial male regulates the size of his family group according to the food resources found within the territorial boundaries, which are relatively constant. Franklin's fieldstudy demonstrated that in the spring, when forage conditions were poor, mean group size and total forage production were highly correlated. He observed that although habitat resources may limit group size within a feeding territory, it is the territorial male who regulates the use of that territory:

'His individuality, pugnacity, or even "perceptiveness" ultimately determines the boundaries and size of his territory and the number of adult females within his group' (Franklin 1982: 480). Successful territorial males are able to control their patch even where domesticated animals are making heavy use of the pasture.

Vicuña and guanaco social behaviour is similar, but there are some slight but significant differences. Guanaco populations are either sedentary (forming seasonally stable populations in one area) or migratory, whereas vicuña are sedentary. Migrations are movements in range which are altitudinal (vertical) or lateral (no change in altitude), caused by seasonal vegetation changes or snow covering feeding grounds. De La Tour observes that guanaco are browsers (1954: 278), which means that they are able to adjust to more varied habitats because they can both graze and browse. In addition, guanacos are more tolerant of arid conditions; in San Pedro de Atacama they are popularly believed to be able to survive several days in waterless zones. Undoubtedly, their energy and water metabolism is very efficient.

Like the vicuña, the guanaco is territorial and the male defends his territory against intruding guanacos. Guanacos observed in northern Peru were seen to form family groups (an adult male and females, with or without young) and male groups, and were also seen as solitary males (Franklin 1975: 195–6). Research in southern Chile has shown the existence of female groups in sedentary populations, consisting of females and young which spend most of the winter months in forests while the territorial male remains in their territory, and also the existence of mixed groups in migratory populations, which are formed of males and females of all ages during the winter months (Franklin 1982: 482).

The sizes of guanaco family groups seem to be variable and their social system more fluid than that of the vicuña. Guanaco males do not regulate group size according to the availability of food because family groups are more likely to change their composition from season to season. Young males and females are forcibly evicted from their group at the age of 13 to 15 months, whereas young vicuña are expelled before the birth of next season's baby vicuña. This means that during several weeks of summer following the current birth season, guanaco family groups include the previous and the current year's young (ibid.: 483).

The South American camelids and their predators

Potential predators of the wild camelids are the domestic dog, the Andean fox (*Dusicyon culpaeus*), condor (*Vultur gryphus*), puma (*Felis concolor*), wildcats (*Felis colocolo* and *jacobita*), and human beings (*Homo sapiens*). Herders accuse foxes of causing the death of the young of both wild and domesticated camelids, and they have to be particularly vigilant during the birth season of llamas and alpacas. Pumas, often called *león* in the Andes, are now very rare, but a reported sighting causes great anxiety among herders as they are capable of killing immature and mature animals (Dransart 1996: 36). It is interesting to note that Isluga people regard the wild cat as the spiritual guardian of the domesticated animals. The condor is essentially a carrion eater; Koford observed fourteen condors unsuccessfully harassing a newborn vicuña (Koford 1957: 217). In fact, the fox is seen as the greatest competitor for domesticated animals; the Isluga people call the fox a thief (*lari llunt'ata*), 'because he wants to eat llamas'. However, lightning is also a major cause of death among wild and domesticated herds. Franklin reports that of thirty-five dead vicuña discovered over a two-and-a-half-year period in the Pampa Galeras Vicuña Reserve, nine were killed by lightning (Franklin 1982: 477).

The tendency of wild animals to remain in a cohesive group in response to a threat from a predator and to act as a unit in defending themselves, is seen by Ingold as an essential element that enables human beings to control and protect animals forming the basis of domestic herds. He postulates that a direct transition from hunting to pastoralism has to involve an animal which is tolerant of crowding and, furthermore, that in its wild state it is limited in numbers by efficient predation and adapted to an open-country habitat, be it tundra, steppe or semi-desert (Ingold 1989: 45). In the case of the camelids, lightning is a non-selective agent of death, whereas the fox preys selectively on young animals. Koford (1957: 217) observed an apparently high infant mortality rate among the vicuña of Huaylarco; the ratio of juveniles to mature females at the end of the birth season was 46 to 100. At Pampa Galeras, Franklin (1982: 477) observed that 10 per cent (1969) to 30 per cent (1970) of young vicuña died before the age of four months. The case of the South American camelids therefore conforms to this aspect of Ingold's prediction regarding a direct transition from hunting to pastoralism: the animals were limited in number in the wild state, and pastoralism occurred in an open-country habitat. Significantly, in the wooded habitats in the far south of the South American continent, the Selk'nam, the Tehuelche and Yámana peoples maintained their hunting and gathering economy until the beginning of the twentieth century.

Both vicuña and guanaco, males and females, have a penetrating alarm call, which is unusual in an open-habitat ungulate. In fact, the Spanish word for the dominant male in domesticated and wild herds is *relincho* ('whinny') in Chile and Argentina (Barros Valenzuela 1963: 59). According to Koford (1957: 217), when a predator is sighted, the territorial male vicuña warns the females with an alarm trill, and positions himself between the predator and the females, allowing them to retreat. As described by Franklin, guanaco behaviour, on seeing a predator, is to maintain visual contact with the predator until it approaches too closely, then the guanacos flee rapidly (Franklin 1982: 483). Barros Valenzuela describes the dominant male guanaco as a cautious and watchful guardian of the herd, and if a potential predator is seen, the females and young run away with the male *relincho* in the rearguard (Barros Valenzuela 1963: 61–2). A comparison may be made with the behaviour of another ungulate. When an individual reindeer becomes aware of a threatening predator (in this case, a wolf), its behaviour may be similar to that of a guanaco in that it stops in its tracks, faces the predator to maintain visual contact, then the reindeer abruptly brings the stalemate to an end by fleeing in a contest of speed between the prey and predator (Ingold 1980: 50). This type of behaviour among ungulates is effective in enabling the prey to seize the advantage and possibly outrun the four-legged predator, but it is ineffective if an armed human being is involved in the contest. It has been recorded that if a human hunter lies in ambush and shoots a guanaco and remains concealed, the rest of the guanaco group will cluster round the fallen victim to sniff it, enabling the hunter to claim more prey (Barros Valenzuela 1963: 61).

The traditional methods for hunting in the Andes derive from two basic strategies: driving herd animals into an enclosure of some kind, and the interception (stalking or ambush) of wild animals. The first type of strategy is commonly known as *chaku* in the literature on the Andes; accounts exist of the *chaku* which took place in the Inka period, during which time the Inka royalty claimed ownership over wild animals. The chronicler Garcilaso, El Inga, wrote a description of the *chaku*, which he describes as a 'solemn hunt'. This paraphrase of his lengthy account is taken from Garcilaso (1609, translated by Livermore [1966: 325–6]):

> The chaku took place after the breeding season, and the Inka emperor commanded twenty or thirty thousand people to attend the beat. The people were divided into two groups, which set off in opposing lines 'until they had made a great enclosure which might consist of twenty or thirty leagues of land' (p. 325). They rounded up the animals until three or four rows of hunters were able to take the game with their hands. As many as twenty, thirty or forty thousand animals were enclosed, including pumas, wild cats, bears, foxes and genets, which were killed. Female deer of breeding age were released, as were 'such males as were necessary as sires, picking the best and largest' (p. 326). The rest were slaughtered and their meat distributed among 'the common people'. Guanacos and vicuñas were shorn, then released. Records were kept of the number of camelids on the quipus, 'the yearly accounts, noting the different species and the number of males and females' (p. 326). The other animals captured were also recorded. The fleece of the guanaco was distributed among 'the common people', while that of the vicuña was destined for the Inka emperor 'who shared it with those of the royal blood: no others were allowed to use the wool, under the pain of death' (p. 326).

Garcilaso added that *chaku* hunts took place every four years in a given place to allow the vicuña fleece to grow to its full length and to give the wild animals time to multiply, and he lamented that with the introduction of the arquebus, few guanacos and vicuñas survived. His account, even though it may be an idealized version, makes it clear that the Inka people sought to practise the selective management of wild herds, removing unwanted animals from the breeding populations. The Inka administration controlled the conservation and harvesting of wild animals; Custred describes this as Inka 'wilderness resource management' (Custred 1979: 14). The shearing strategies carried out on the wild animals, as described by Garcilaso, resembled those practised on domesticated animals, as discussed further in Chapter 5. Although the *chaku* hunt yielded meat for all the participants, its main purpose, according to the accounts of the early chroniclers, was to ensure a supply of the highly prized vicuña fleece, which was circulated among a restricted segment of the population. The royal hunt is mentioned by various authors, including Cobo, who calls it *caycu* (1956 [1653]: Bk 14, ch. 16), stating that wild animals were driven into enclosures between hills and gorges. This word is listed in an early Aymara dictionary, that of Bertonio (1984 [1612] Bk II: 39):

> *Caycu*: Walled device for hunting vicuñas.
> *Caycutha*: To hunt vicuñas in such a manner.

In Custred's opinion, the *chaku* was a more elaborate version, employing thousands of beaters over a large area, of the *kayku*, which he sees as residual *puna* hunting activity, with an ancestry dating back to perhaps remotest times (Custred 1979: 15). Undated funnel-shaped constructions of dry-stone walling, which could have been used to trap or kill wild animals using battu techniques, have been reported at an altitude of 4,600 m at Mazocruz, in the Department of Puno, Peru (ibid.: 12), and at Layo, La Raya, in the Department of Cusco, Peru (Aguilar Meza 1988: 60; Type A). A second type of trap has been described as a rectangular, stone-lined pit, which herders in La Raya are reported as saying was used by their grandfathers (ibid: 61; Type B). This is presumably not as old as the first type, but it is not clear whether it was used to trap vicuñas or condors which, as carrion eaters, would be attracted by meat.

Late-eighteenth-century camelid hunting methods have been described for the Atacama, when, according to a report made by the governor of Potosí, Cañete y Domínguez, there were still many vicuña in that area. He reported that people travelled from Salta with numerous dogs to round up thousands of vicuñas for their valued skins. He described the native method of round-up, for which large areas of ground were enclosed by means of *rodeos de hilos* (enclosures of, literally, yarns). Apparently this was the method used in Lípez, but it was not so successful in more rocky areas because if a guanaco is rounded up by chance among the vicuña, it will break through the boundaries of the enclosure, allowing the vicuña to escape (Cañete y Domínguez 1974 [1791]: 249). This is an allusion to the fact that vicuña will not cross ropes hung with streamers, a behavioural characteristic shared by llamas and alpacas. Annual round-ups of vicuña have been recorded from the Atacama in recent times. Bowman says that the people of Aguas Blancas and Toconao hunted vicuña down high-altitude valleys, where threads were strung out to impede the flight of the animals (Bowman 1924: 247–8).

There are artificially made constructions in the Tulan Quebrada which were possibly used in the hunting of wild camelids. They consist of rough stone alignments, some of which form angles, and cairns are found at intervals along the alignments (plate 2.5). Father Le Paige indicated some such alignments in a sketch map of 1957, showing the canyon between the cave site numbered TU 55 and Tulan itself, further upstream. The structures are on both banks of the canyon and run roughly parallel to the river below (Le Paige MS). Another concentration of lineal structures is spread over the slopes almost immediately upstream of Tilomonte. These structures cover an area some 500 m long, and they constitute

Plate 2.5 Stone alignment in the Tulan Quebrada.

site TU 75 listed in Núñez (1988: 233–5). As yet, these unidentified and undated structures have not been surveyed adequately or mapped. However, the cairns might have supported stakes, which could have been dressed with streamers during a beat, the function of the streamers being to encourage the hunted animals to flee towards the waiting hunters.

However, Cañete y Domínguez considered the most common eighteenth-century hunting technique of vicuña used in the Atacama to be that of ambush. The hunter and dogs lie in wait near a source of water where vicuña go down to drink during the midday heat. Specially constructed dry stone walls (*pirka*) hide them from the vicuña. They wait until the vicuña have finished drinking and start to make their return uphill. The hunter releases the dogs, which easily bring down the prey, since they are less agile when their stomachs are weighted down with water (Cañete y Domínguez 1974: 249). Wheeler Pires-Ferreira, Pires Ferreira and Kaulicke maintain that surround hunting would destroy the stability of local camelid territories, and they propose that the most likely technique would have been that of the ambush of individual animals (Wheeler Pires-Ferreira *et al.* 1976: 488).

In contrast, Custred speculates that the drive-and-surround technique was the preferential and original method of hunting camelids. He says it established a close relationship with camelids, with the eventual consequence being the domestication and herding of these animals (Custred 1979: 11). According to this theory, human control of camelid territory preceded human control over camelid breeding.

It is not clear when the *chaku* or *kayku* system was first practised, but it is evident that the Inka people sought to restrict its use and made it a ritual event. The Inka Emperor and the Empress were considered to be direct descendants of the Sun and Moon, hence their claim for the legitimacy of their expropriation, since guanaco and vicuña were *intipllama*: that is, the camelids of the Sun, according to Guaman Poma (1980 [1615]: 288). If over-used, the *chaku* would cause the social organization of the camelids to break down. Research conducted by Franklin demonstrates that vicuña have the flexibility to make short-term adjustments to a changing environment, but in 1978–9, high animal density and drought caused a large percentage of territorial males to abandon their territories and join male groups: family group density dropped, while male group density increased (Franklin 1982: 481). However, if a family group is driven from its territory, for example by domesticated animals, the group will return within several hours, even though the wild animals may have been widely displaced in the meantime (Koford 1957: 217). It seems evident that the social organization of the wild camelids would break down if subjected to frequent surround hunting, but its effects might not be so severe on herds of migrating guanaco.

Whereas the origins of Andean surround-hunting techniques are obscure, there is some evidence, collected by Custred, demonstrating its continuity until the early years of the twentieth century. Custred regards the pattern of residual hunting in the highlands, combined with herding and trading activities, as a 'symbolic complex', with a sexual division of labour (men hunted and traded, while women tended the domesticated herds)[4] and associated ritual practices (vicuña were regarded as the livestock of the hill spirits). In this analysis, hunting provided a 'secondary means of production' for meat, hides and wool, which could be processed into dried meat and woollen goods, which were then traded, thus limiting the depletion of the herders' own stocks of llamas and alpacas (Custred 1979: 16–17).

Nourishing terrains from the perspective of herd animals

Given the territorial organization of the family groups of vicuñas and guanacos that have been revealed by biologists, it is possible to posit that wild camelids have developed a strong sense of the terrain they inhabit as well as an understanding of its affordances for sustaining life.[5] It is perhaps provocative to juxtapose animals' perception of their own *habitus* alongside human understandings of *their* occupation of the land. Bourdieu (1977) developed his notion of *habitus* in order to clarify the ways in which human beings are socialized into certain practices and outlooks through their routine occupation of space, place and time. On some levels, this routinization of social practices is inarticulate.[6] Non-human animals, too, must have their own inarticulate notions of habitus.

The phrase 'nourishing terrains' is borrowed from Deborah Bird Rose's demonstration of 'country' as a place that gives and receives life among Aboriginal peoples of Australia. To Aborigines, country 'is a place that gives and receives life. . . . it is lived in and lived with' (Rose 1996: 7). She adapted the phrase from the work of the European philosopher Emmanuel Levinas, but she uses the concept shorn of Levinas's understanding of nourishment that he devised in a specific Western context that focused on maternal breast milk. From her discussions with Aboriginal peoples on their relationship with the land they inhabit, Rose suggests that their world view may be 'neither human-centred nor geared to the endless satisfaction of human wants' (ibid.: 3).

Given such possibilities for envisaging alternative human appreciations of notions of terrain, it is worth re-envisaging how wild camelids make use of their own territories. There are at least two aspects of camelid behaviour that concretize their marking of the land for their own inhabitation. All South American camelids have common wallowing places, which are conspicuous circular hollows in dusty places in their territory.[7] Second, they all void dung and urine in communal piles, although they do not use these dung piles in exactly the same way. Franklin observed that guanaco smell the pile before positioning themselves over it for defaecation–urination, but that they do not always void on existing piles (Franklin 1975: 199–200). Vicuña in Franklin's study only voided dung on already established dung piles scattered throughout their territory, and they smelled the pile first (Franklin 1982: 480). In contrast, domesticated camelids nearly always use previously existing dung piles, which are scattered throughout the zones where they are herded; they do not necessarily smell the pile first. Franklin suggests that the vicuña territorial dung piles serve to orientate members to keep within their own territory, rather than to discourage outsiders from entering. However, he points out that dung piles have a short-term function as a territorial marking system, and a long-term effect in inducing vegetation changes in the camelids' habitat. The nutrients from the dung piles help improve the plant succession around and downhill from the piles, eventually producing the preferred types of forage in dry and poor soil areas. Thus, in the long term, the vicuña improves its own environment by increasing the quantity of preferred plant species (ibid.: 481).

Although human beings are regarded as possessing abilities to make large-scale changes of the land they occupy, animals too possess the capability to modify their surroundings beyond just merely maintaining open landscapes through their browsing and grazing activities.

Nourishing terrains from an etic perspective

In this section, assessments of the highland and desert lands of the Andes are considered in the light of what has been published in the geographical literature. The classifications that have been proposed for different ecological zones in the Andes tend to be based on the observed biota and not on the fauna, which is omitted from these botanically oriented schemes (Craig 1985: 23). Tropical forest has survived in the Upper Zaña Valley of northern Peru (Dillehay, *et al.* 1989: 735). However, Craig suggests that relict tropical rainforests near Hacienda Udima in the upper Zaña Valley, at an altitude of 2,600 m, were formerly more extensive along the foothills of the western Andes, both vertically (extending to somewhat lower levels) and laterally, becoming established at the beginning of the Holocene. Markgraf (1989: 12) sets the Holocene boundary at 10,000 years ago in Peru. Steppe-like grasslands, providing a habitat for large Pleistocene herbivores, were probably characteristic above these Late Pleistocene–Holocene woodlands.

According to Craig, post-glacial desiccation did not have a significant effect on the aridity of the coastal desert, which retained its xeric character throughout most of the Late Pleistocene to the Historic period (Craig 1985: 27), although at least three previous glacial episodes, corresponding to pluvial periods during the Pleistocene, left noticeable remains in the Andes from Ecuador southwards to northern Chile and Argentina, when water accumulated in large glacial lakes (Simpson Vuilleumier 1971: 778). Old shorelines in the Puna de Atacama have been described by Vita Finzi (1959: 401), who suggested that the Salar de Atacama was formerly a deep Andean lake that later underwent a rapid drop in level. It would seem that at the last glacial stade, *puna* vegetation (steppe-like grasslands) became more extensive with the upward shrinking of the permanent snow cover on high peaks. The tropical forests became restricted between the high-altitude *puna* vegetation and the coastal desert, but they also became increasingly rare further south towards the hyperaridity of the Atacama desert in northern Chile.

An interval of approximately five to seven millennia occurred between the extinction of Pleistocene megafauna at the beginning of the Holocene and the first reported signs of camelid domestication in the Puna de Junín, in the central highlands of Peru, *ca* 6000 to 5000 BP (Wheeler 1984: 402; Lavallée *et al.* 1985: 66). Craig attributes an expansion of high-altitude grasslands at the expense of forest cover to increasing herd numbers and cultural manipulation on the part of herders. This deforestation apparently exaggerated the already existing trend of post-glacial desiccation (Craig 1985: 26–7).

Palaeoclimatic trends for *páramo* sites in Venezuela, Colombia and Peru have been postulated by Markgraf on the basis of changes in the proportions between Gramineae, Compositae, other herbaceous taxa and Andean forest taxa, compared with information on modern pollen representation (Markgraf 1989: figures 5 and 7). She notes an abrupt tree-pollen decline after about 2000 BP, coupled with increased levels of disturbance pollen (*Dodonea*), weeds (*Ambrosia, Polygonum*) and Graminae; this change is said to have been caused by human impact (ibid.: 14). Pollen records from high-altitude sites further south in Bolivia and northwestern Argentina are said to be located further from forests, hence palaeoclimatic interpretation in this area is based on proportional shifts between Gramineae and herbaceous taxa, which represent *puna* (high-altitude) vegetation versus Compositae, *Ephedra* and Chenopodiaceae, which represent *subpuna* vegetation (ibid.: 14). In this area, evidence for human impact is less clear, but Markgraf suggests that it might have begun to have an effect on the high-elevation environments after 2000 BP, when Compositae replaced Gramineae for a short time, a succession which she suggests could relate to the first episode of intense grazing which diminished the grasses but left the shrubs. Unambiguous evidence for changes induced by human activity is noticeable only in the last 500 years, and it includes the introduction of a European weed (ibid.: 15).

However, llamas and guanacos, it may be remembered, are grazers and browsers; they feed on both grasses and shrubs. Additionally, it should be stated that the making of a given biotope culturally viable by human societies is a complex affair. The first herders also lived by hunting herbivores, rodents and birds; their herds of domesticated animals were perhaps not numerous. In the Atacama area, the earliest indications of camelid domestication come from the preceramic site of

PU 1, over 4,000 years ago. The adoption of pastoralism seems to be a local development, unrelated to events in the Puna de Junín. The Junín cave site of Telarmachay had a long hunting tradition, and it witnessed a gradual change from the hunting of cervids to that of camelids, with the eventual emergence of the herding of alpacas (Wheeler 1984: 402). In contrast, the herbivore resources of the Atacama region are more restricted than those of Junín, and Hesse reports no deer among the faunal remains of the Atacama sites which he studied (Hesse 1982b: 10), although cervids are reported from coastal sites in the north of Chile. The open-air site of PU 1 apparently saw the taming and domestication of the guanaco, according to Hesse (ibid.: 10) but, to date, no evidence has been reported from the site that suggests long-range contacts with other parts of the Andes. Pastoralism in the Atacama might have been a local development, one of a number of places where communities began to herd and subsequently to domesticate llamas. Another area where this process took place is Susques in north-west Argentina (Yacobaccio *et al.* 1997/8: 427–8).

If vicuña 'improve' their environment, making it a fitter place within which their social organization takes place, then obviously the situation of human groups, who made different Andean ecological niches culturally and environmentally habitable, constitutes a far more complex state of affairs. Pastoralism is commonly regarded as having the tendency to induce and maintain areas of short grass, but it also implies human control over the exploitation of pasture. Browman regards pastoralism as the best solution for a semi-arid grassland ecosystem that supports grazing animals but which is not suitable for cultivated crops; moreover, he maintains that a distinctive characteristic of camelid pastoralism is that, 'it is integrated into and generally maintains the structure of the hunting and gathering ecosystem into which it is introduced' (Browman 1974: 188). This theoretical stance permits Browman to postulate that changing environmental conditions prompted the adoption of pastoralism, the initial effect of which was to conserve the hunting way of life (domesticated animals are available for slaughter when required), and the Andean herders and their camelid herds are seen to make a place for themselves alongside other members of the grazing succession. They are, according to this theory, in cooperation, and not in conflict, with the wild camelids. Hesse agreed that unpredictable climatic changes capable of seriously affecting game availability probably caused hunters to start herding camelids for a more secure supply of meat, but he argued that this form of carnivorous pastoralism ensured its success by destroying the potential for other systems (Hesse 1982b: 12). This is a debatable point, since the eighteenth-century report by Cañete y Domínguez cited above makes it clear that non-local people from Salta travelled to the Atacama to hunt large numbers of vicuñas in order to sell the hides to outsiders. Probably, human population levels in the Atacama were never high.

Both Browman and Hesse regard the initial effect of the adoption of pastoralism as an attempt to preserve the former ways of life against a background of adversely changing environmental conditions.[8] Yet herders must constantly seek to protect their livestock against disease and weather conditions, and also to ensure access to pasture in conveniently situated locations. Their contact with their herds is of an entirely different nature from their dealings with wild animals. Chapter 3 deals more fully with my observations on the relationship between herders and herded camelids in Isluga. Here it is my intention to discuss the management of pasture as practised by modern pastoralists in the Andes, specifically dealing with the exploitation of pasture in Isluga, where a scarcity of pasture is a limiting factor for the herders.

It is possible that hunters attempted to manage, in some way, areas of vegetation to promote plant growth and thus establish better grazing areas for wild animals. Certainly, there is ample evidence that herders did, and still do, make efforts to improve areas of suitable pasturage (plates 2.6 and 2.7). The earliest reference that I know of to the practice of irrigating pasturelands is in Garcilaso (1966 [1609]: 296), who mentioned that in the province of Poc'ra, between Parcu and Pícuy, an irrigation channel twelve feet deep and 120 leagues long served to 'irrigate the grazing land of those empty moors'. This information seems to have been disregarded by more recent commentators, for it was not until fairly recently that Palacios Ríos published a description of such a strategy for maintaining and increasing the extent of pasture by the Aymara herders of Chichillapi in the Department of Puno in Peru (Palacios Ríos 1977). The predominant vegetation types in the highlands of the South-Central Andes are perennial bunch grasses and shrubs, which are accompanied by annual grasses. In some areas, a different type of vegetation occurs; this is succulent throughout the year and it contains plant species which are high in nutritive value for camelids. A moist area of such vegetation is known in Spanish as *bofedal*.[9] It is found at high altitudes, often in seeps which form along glacial moraines and alongside rivers or around springs. It is particularly important to prevent overexploitation of this vegetational resource, since Orlove reports that native pasture does not store well in the form of hay or silage (Orlove 1982: 97). Naturally occurring *bofedales* are common in the north of Chile along rivers which have not formed steep banks, but they are not very extensive. To extend the size of these areas, and thus increase the capacity for herding greater numbers of animals, local herders construct irrigation canals and low walls of *ch'ampa* (turves) in order to retain the moisture.

Access to sufficient *bofedal*-type pasture is especially important in the herding of alpacas. Herders in highland communities in the north of Chile constantly reiterated that it is essential for alpacas to graze on moist pasture, lest their feet or, according to one woman, their noses, dry up. Palacios Ríos reported that in Chichillapi, herders stated that alpacas raised on *bofedal* vegetation had a greater body weight and that fleece yields were higher. Whereas a fleece shorn from an animal fed on moist pasture may weigh 10–12 pounds, that of an alpaca raised on dry pasture might yield only 3–4 pounds of fibre, which is 'dry' like that of a llama (Palacios Ríos 1977: 157). Research at La Raya in Peru indicates that there is a positive correlation between live body weight and weight of fleece in the two types of alpaca (Suri and Wacaya) although, statistically, the coefficients are low (Avila Felipe 1979: 36).

Herders also encourage new growth of the shrub and bunch grass vegetation of the dry land by burning areas of such pasture during the dry season. This

Plate 2.6 Canal for irrigating the *bofedal* pastures of Caquena (4,600 m asl), in the highlands east of Arica.

Plate 2.7 Mud walling to retain moisture in pasture in the Tulan Quebrada above Tchulín, opposite site Tulan 67.

encourages tender shoots to sprout when the vegetation begins to grow in about November. In some parts of Isluga, the burning of restricted areas of such *waña* pasture takes place in June or July. The patches of vegetation which have been burned have a brighter colour when the new growth begins; these patches are called *viacha*. In Chile, this practice has been criticized by the Corporación Nacional Forestal (CONAF), which states that wild animals such as guanaco, vicuña and avestruz are endangered since they lose pasture and protection, which the shrubs offer against the cold, at night (CONAF 1982: 104). The CONAF report ignores the fact that the herders' own animals spend the night on the same hillsides, and that the herders are improving the area of grazing, which is also exploited by the wild animals.

Nourishing terrains from the perspective of human herders

Isluga, with a territory covering some 200,000 hectares, or 80,000 acres, in the highlands to the east of Iquique, had a reported population of nearly 2,000 persons in the 1970s (Martínez 1975: 404; Provoste 1978: Appendix 2). This ethnic Aymara community is divided into two moieties, an upper (Araxsaya) and a lower (Manq"asaya, which literally means 'inner'); the former is located west, and the latter east, of the ceremonial centre, the town of Isluga itself (Islug marka). The people of Isluga use this focal point of the community when they congregate for the festivals of Todos Santos, Santo Tomas and Carnival, all of which take place during the rainy season (summer: October–November to the end of March). However, its pastures also constitute an important resource exploited by the people in addition to the resources available at their main place of residence in one of the villages of Isluga.

The terrain of Araxsaya is characterized by its rugged nature; high-altitude pampas alternate with rocky escarpments and narrow ravines, which suddenly broaden to form marshy bottomlands, the *bofedales*. The landscape of Manq"asaya opens up, and from Islug marka, looking towards Coipasa, one can see wide vistas more typical of the *altiplano*. They are dominated by extinct volcanoes (e.g. the mighty Kawaraya and Tata Sabaya, both in Bolivian territory, and Mama Wanapa, the focal point of Cariquima, the neighbouring ethnic Aymara group to the south of Isluga). In contrast, parts of Araxsaya are actively volcanic, and boiling waters bubble from the ground from the geysers of Puchultisa near Mawk'i. The huge eroded outline of Laram Qawani (marked on the maps as Volcán Isluga) belies its active nature, for it has two fumaroles and warm thermal water (*junt' uma*) springs from the ground 2 km east of Enquelga, before draining into the river which flows towards Coipasa. Araxsaya is, in fact, at a higher altitude above sea level than Manq"asaya, while the central town of Islug marka is at an altitude of about 3,700 m.

The mainstay of the economy in Isluga is the herding of camelids and sheep, but *quinua* (a high-altitude cereal) and potatoes are grown, especially in the villages near the Salar de Coipasa. However, as one enters Araxsaya, agriculture becomes increasingly less viable. Both potatoes and *quinua* are grown in Enquelga[10] (which is in Araxsaya), but only *quinua* is grown in Ch'awani. The villages and hamlets clustered around the *bofedales* of Parajalla have no agriculture, but the families do have access to some inherited corrals around Islug marka. In any case, severe climatic conditions and frosty nights often cause potato and *quinua* yields to fall. Families in Enquelga have not harvested enough to be self-sufficient in these crops during most of the period I have known the area. Although unreliable, *quinua* and potatoes form an important part of the diet (*quinua* is particularly rich in amino acids and vitamins), and their cultivation enables a diversification of the economic base. Pasture is much more limited in Isluga, compared with that available further north in the *altiplano* east of Arica. Before the introduction of sheep by the Spanish, herders exploiting the land in what is now Isluga would have found it difficult to increase herd sizes, a common method used by pastoralists of insuring oneself against the possible dangers of a restricted resource base. They could increase herd sizes only by exercising the careful management of pasture and by practising seasonal migrations to the *precordillera*. (Once sheep were introduced, however, it became much easier to increase herd sizes because in equatorial latitudes, unlike in northern Europe, they have two birth seasons.) There was no possible diversification into the herding of other animals, so the cultivation of *quinua* and potatoes formed an important strategy against relying on a successful outcome to only one economic activity.

Water resources in Isluga

Apart from Parajalla and Mawk'i, both of which are located within separate drainage systems, the rest of the communities of Isluga are found within easy access of a river which rises in the Chilean *cordillera* and which drains inland, following an approximate north-west to south-east course, until it eventually peters out in the Salar de Coipasa in Bolivia. This river changes its name as it proceeds, adopting the name of certain nearby villages (Arawilla, Isluga, Sitani), as marked on the official maps of the area. However, the people of Enquelga merely call it 'the river'. It is narrow and it is augmented by tributaries (many only seasonal), springs and thermal waters. Many of the communities in Isluga capture water from this river to feed canals which irrigate and maintain the *bofedales*, as described above. The river flows through the narrow Arroyo Pasiriji and is joined by a tributary before entering more open ground, where it passes through a large *bofedal*, to the north of which is located the *estancia* (one of the main settlements of Isluga) of Arawilla.[11] Once more, the river enters a ravine and emerges into another *bofedal*, to the north of which is sited Enquelga, and to the south, the villages of Ch'api Qullu and Karawanu (figure 2.3). Both Enquelga and Arawilla are sited on a dry pampa, gently sloping land between the *bofedales* and the lowest lava flows of Laram Qawani. However, the extent of moist pasture available to the people of Arawilla is much greater than that available to the people living around the other *bofedal*. This means that the people whose main residence is in one of these three communities have to exercise special care in the management of pasture.

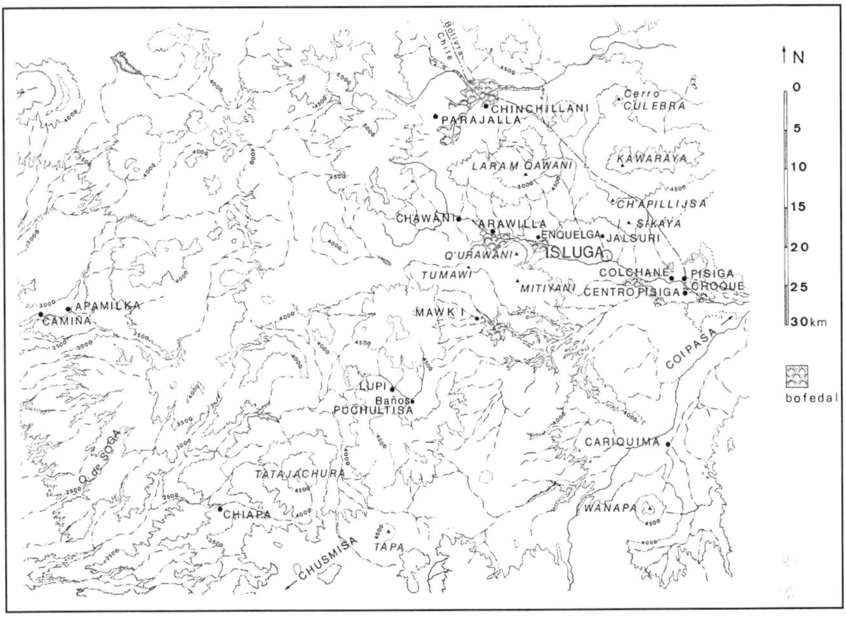

Figure 2.3 Map showing the location of Isluga and the neighbouring communities of Camiña, Chiapa and Cariquima.

At the eastern end of the Enquelga *bofedal*, the river has cut a channel that is fairly wide and steep sided. Llamas and alpacas attempting to cross the river at this point may slip and drown; consequently, unsupervised herds are not left for long here. The already mentioned thermal spring issues forth from the ground immediately north of the wet pastureland, and a small pool has been created to collect the water. An irrigation canal leads into the *bofedal* from this pool, but most of its water drains into a larger and lower pool, which feeds into the river. Yet once again, the river flows through a valley and continues on its course, south of the hills of Pukar Qullu and Najraya, and past the cemetery of Islug marka. The ceremonial centre is sited near the confluence of the river with another tributary. This tributary emerges from two large springs at a place called Jalsuri, north of Islug marka. The name Jalsuri refers to the fact that water springs (jumps) from the ground, and one of the springs is conceptually regarded as being masculine (*mallku*), while the other is considered to be feminine (*t'alla*). The theme concerning the relationship between camelids and water will be explored more fully in the following chapter. At the moment I wish to indicate the importance of this source of water for the herding economy of Isluga. The double spring is used to feed irrigation canals which create an expanse of moist ground which then narrows and becomes the so-called Irpa uq"u,[12] forming a narrow thread of lush vegetation between the *bofedal* of Jalsuri and that of Isluga.

The families of Isluga own a large house for ceremonial gatherings in Islug marka, and they may also own a small house (*paskana*) near Jalsuri. As these houses are not in constant use there are many which have the appearance of having been abandoned, where the roofs have collapsed. However, their owners at any time may reroof them, and in the meantime they may well borrow a house belonging to a relative if they wish to pasture their herds in the area. The small houses owned by members of Manq"asaya are clustered around the *bofedal* of Jalsuri (this part is known as Choque Jalsuri), and those belonging to members of Araxsaya are scattered over the dry pasture uphill and further west (this is known as Castro Jalsuri). Families using these houses have to carry water a fairly long distance uphill from the springs. One of the features of Jalsuri is a series of very straight paths which have been cut through the sparse shrub vegetation of the dry land. One such path leads from the foot of the hill called Pukar Qullu, which dominates Islug marka. It was described to me as the '*camino* Choque [the Choque path]', a surname associated especially with Pisiga in Manq"asaya. It passes the *bofedal*, without leading to it, and continues to an area of dry pasture on a slight eminence. Doña Soria Mamani told me that there is another path known as '*camino* Castro', which is associated with the Castros of Enquelga in Araxsaya. She described these tracks as an *hito*, or 'landmark', whose function appears to be to mark out areas of dry pasture (*waña*) commonly used by certain *estancias*. While members of both moieties use the *uq"u* of Jalsuri, the areas of *waña* pasture are divided between the families of Enquelga and Pisiga. The former have access to the lower slopes of Ch'apillijsa, and the latter to the slopes of Sikaya.

Families in Isluga, therefore, have a main place of residence in a settlement, with access to communally owned wet and dry pasture (*uq"u* and *waña*), for the use of which they pay taxes to the Chilean Government. In addition, they may have inherited houses in other hamlets (*caserío*). Residence after marriage is often virilocal in Isluga, thus a newly wed couple sets up home next to the main residence of the husband's parents. However, a wife never loses her right to pasture her animals, and those of her husband and children, on the communal pastures of her natal community during her lifetime. Since the *bofedal* used by the villages of Enquelga, Ch'api Qullu and Karawanu is fairly small, the families here exercise this right fully to avoid the overexploitation of the wet pasture. These families tend to have ambilocal residence in the communities of both the wife and the husband's parents. Finally, all families have the right to take their animals to Isluga, as described above.

Whereas individuals own and inherit large corrals which also serve to grow crops, especially potatoes (*corrales para sembrar*), each village controls areas of *uq"u* (wet pasture lands) and *waña* (dry pasture) which are for the usufruct of the members of a particular *estancia*. These are described as community lands. At first sight, there would seem to be no pastures which are owned by individual families. However, Provoste (1978) reported that in some cases an extended family may actually be the sole user of certain pastures and may claim ownership of the land, but non-family members may challenge this claim. Provoste suggests that communal ownership of land is a traditional right, but that individuals may, according to national law, lay claims to isolated stretches of land which are not included in community boundaries (ibid.: 29–30). The existence of communally owned pastures in Isluga is unusual. Reported accounts of land ownership from the *altiplano* to the east of Arica in Chile (Bernhardson 1985) and from Chichillapi in Peru (Palacios Ríos 1984) indicate that *bofedales* are owned and inherited by individuals in those Aymara-speaking areas. In parts of Bolivia, the situation is somewhat different. Caro (1985: 32) reports that under the Agrarian Reform titles issued in 1976, pasture lands in the Aymara-speaking area of Ulla Ulla are held in common by all persons belonging to a particular community. However, in practice, the people observe other norms in the division and distribution of lands among the families of the community. Herders born into an *estancia* inherit rights to herd their animals on the pasture lands of that *estancia*, or they may inherit usufructuary rights to pastures from parents and grandparents. Men and women who actually live and graze their animals in the *estancia* maintain active rights to the use of those lands (ibid.: 38–9).

Isluga was formerly in Peruvian territory, in an area perceived by outsiders to be a barren wasteland which, like the immediately adjacent Carangas region in Bolivia, did not see the establishment of large, colonial estates. The conditions which apply in Isluga more closely resemble those of the former Kunza-speaking Peine district (Mostny *et al.* 1954: 11 and 79), which is considered in greater detail in Chapter 6.

The seasons in Isluga

The year is marked by two seasons: *jallu pacha*, which lasts from approximately mid-October or November to March, and *t"aya pacha*. The former is the rainy season, but most of the rains fall in the months of December, January and February. Pasture begins to grow in October, and it is best between December and the end of March. From this time onward, the *bofedales* begin to wither and lose their green hue. The rains cease, and the windy season begins, characterized by strong winds which blow up in the afternoons. Extremely strong winds are especially associated with the festival of San Santiago (St James) on 25 July. At the beginning of August, these winds begin to ease. During this season, the pasture of the *bofedales* is stunted and yellow. Rivulets running through these areas often remain frozen for a greater part of the day, only thawing out for a couple of hours in the afternoon where they are exposed to the sun. Precipitation occurring in *jallu pacha* may fall as snow or sleet, but it does not usually settle. On the other hand, snow falls in late June and in July may last longer and cover large expanses of pasture. This is a serious problem, because the camelids, already weakened by the strong winds and low night-time temperatures, are unable to feed as long as the snow cover remains. However, when the heavy cloud cover disperses, the fierce sun causes the snow to disappear rapidly, and the moisture is absorbed by the parched ground.

A shortage in rainfall can have grave consequences. Cloudless skies in February, March and April bring frosty nights to Isluga, and the young *quinua* and potato plants are badly affected. In addition, reduced rainfall means that there is less pasture, and this in turn results in poor camelid fleece yields. Alpacas and llamas are shorn in November and early December in Isluga, when night-time temperatures are not so low. If the rains were good in the previous rainy season, the shorn fleeces weigh more. Luis Castro of Enquelga said that his fleece harvest in November 1986 was good because of the previous plentiful rains. In addition, seasonality in the availability of pasture causes problems for the young animals. They are born between December and March, when there is no shortage of pasture but, as lactation continues into the dry season, by the time they are weaned at six or seven months, pasture is much more scarce.

Having indicated some of the limitations in the availability of pasture in Isluga, I will now describe the normal routine of herding llamas and alpacas as practised in Enquelga. A cluster of large corrals constructed of dry-stone dykes surrounds the *estancia*. More corrals belonging to the village are located 2 km to the east, near the thermal spring, and also at a similar distance to the west, north of the road between Enquelga and Arawilla. These corrals are used every other year to grow crops. One year the villagers grow *quinua* and potatoes in the corrals immediately surrounding Enquelga. The following year, the outlying corrals are used for crops. During the intervening fallow years, the corrals are used to enclose llamas and alpacas, separately, overnight. The dung voided by the herds fertilizes the ground, which is cultivated once again the following year.

During *jallu pacha*, when pasture is freely available, herds of llamas and alpacas graze for a greater part of the day on the *bofedal* below Enquelga. Shortly before sunset, the herds are driven back to a corral for the night. However, they will only spend one night out of two in a corral (and never use the same one consecutively, so as to fertilize all of a household's corrals). For the intervening nights, the animals are sent to one of the hill slopes of Laram Qawani. Thus a typical sequence for a herd of llamas or of alpacas would be to spend the night on a hill; after dawn they descend gradually to the *bofedal*, where they stay until light fails, after which they are herded into a corral. The following morning they are driven to the *bofedal*, and so on. In fact, herds are trained to follow a sequence of spending the night on higher ground and returning to the moist pastures during daytime. A herd will maintain this routine without supervision, and it will start to leave the wet pasturelands of its own accord at the end of the day.

As the rainy season ends, and the *bofedal* begins to dry up, the llamas spend more and more time on surrounding dry *waña* pasture. They will perhaps spend two nights on the hill before returning to a corral. Alpacas continue to use the *bofedales*, but by this time of the year the one below Enquelga tends to be overgrazed. Also, camelids are said to dislike cold and frozen pasture. During *t"aya pacha*, the alpacas are more likely to be taken to Jalsuri, since the waters of the double spring are not so bitterly cold. Families who have houses in hamlets near Arawilla (Chullpa, Kullku or Kacharpaya) take their herds to the Arawilla *bofedal*. There are local differences in the types of vegetation in the various wet pasturelands of Isluga. Isabel Castro told me that she takes her llamas to Arawilla because they are used to eating *sura*, a plant which does not grow in Enquelga. Similarly, one of the common plants in the Enquelga *bofedal* is called *katari*, but this does not grow in Parajalla. However, it is said to be highly nutritious and it is considered to be favoured by llamas. Sheep, on the other hand, are said to favour a shrub called *lampaya*, which grows on dry land.

Seasonal migration within and beyond Isluga

In the higher reaches of the western valleys, pasture begins to grow in December, following the summer rains. These areas are known as the *costa*. Unlike the *altiplano*, where pastures are grazed heavily throughout the summer, those of the *precordillera* may develop in the virtual absence of herbivores grazing down the growing plants. During winters following abundant rains in the upper parts of the western valleys, the pastoralists take their herds there to avoid the effects of the strong winds in the *altiplano*. Herders who decide to make the journey to the *precordillera* (known as the *costeo*) travel between 60 and 120 km to their destination. Most of the families in Isluga do not own lands in the valleys, and they therefore have to pay rent to the agriculturalists of the valleys. This is paid in animals, calculated on the basis of the number of adult animals (that is, animals older than one year) in the herd. Lanino reports that rent up until 1974 corresponded to 1 per cent of the number of adult animals: if the herding family brought 100 adult sheep and 100 adult llamas, it would pay one sheep and one llama. In 1975, the rent was increased to 2 per cent of the total number of adult animals (Lanino 1977: 55).

Whereas most families did not go to the *costa* in 1986, many families did in 1987, presumably because of the reduced rainfall in the *altiplano* from February to April 1987. The journey is realized in stages, of 30 to 35 km every day, over three to five days. The herds are taken back to Isluga at the very end of July or the beginning of August.

Enquelga people go to the *costa* at Alto Camiña or Alto Soga. On the other hand, many of the people from Manq"asaya go to Alto Chusmisa or Alto Chiapa, although some people from Pisiga Choque and Pisiga Centro also go to Alto Soga. Citing the work of Patricia Provoste, Lanino lists the destinations in the *costa* used by the various villages and hamlets of Isluga over a decade ago (Lanino 1977: 53–4). Martínez observed that an 'ideological verticality' existed in this cycle of nomadism, in which most of the communities of Araxsaya went to graze their herds in Alto Camiña, and most of Manq"asaya went to Alto Chiapa, but with a certain crossing over, for some of the Araxsaya communities went to Alto Chiapa, while some of Manq"asaya went to Alto Camiña (Martínez 1976: 260). This 'crossing over' may not be as rigidly maintained as Martínez implies. Although many people rent land from the agriculturalists, some herders may purchase land in the *precordillera* and this land becomes heritable property.

Some families do not take their animals to the *costa*, since the advantages gained in improving the body weight of the animals during the period of the *costeo* may be lost in the course of the arduous journey back to the *altiplano*. In addition, the Chilean state is exerting external influence to encourage families to remain in their main homestead in the *altiplano* throughout the year. There is a primary school in Enquelga and children are expected to attend for the full academic year.

Parents with children of school age wishing to go to the *costa* may have to make arrangements to have their children looked after by a relative during the family's absence. CONAF also has a house in Enquelga (part of the Isluga territory is within the boundaries of a nature reserve), and local men are employed by this organization in the capacity of wardens and women as cleaners. Their herds may remain in the *altiplano* when most of the other herds belonging to Enquelga people spend the winter in the *precordillera*. Therefore, demands of Chilean national society are tending to erode this nomadic system. However, this form of nomadism is an important strategy employed by Isluga people to ensure against the vagaries of the climate when reduced rainfall limits the available pasture in the *altiplano*. It enables them to maximize herd sizes and avoid the over exploitation of pasture.

This chapter has addressed issues concerning camelids, land and water in the Andes, and the changing relationships that have occurred between human groups and groups of herd animals. In the Andean highlands of northern Chile, outsiders have regarded the sparse vegetation as marginal grazing land. From an agronomer's point of view, the land is considered to be 'unproductive', while from a traveller's perspective, it might constitute a romantic 'wilderness'. The Aymara, in contrast, have other ways of perceiving and utilizing their pasture lands as places for human activity (Dransart 1996). Their land is a nourishing terrain for both human and animal inhabitants. Chapter 3 will examine how people perceive the landscape which they inhabit, along with their animals, and in which they go about their daily herding activities. It will be seen that they do not consider their environment as an inert or neutral backdrop against which they conduct their herding activities, as the analysis of land use presented in this chapter may have implied.

3 Caring for herd animals in Isluga

Having enquired into the character of the South American camelids and of Andean pasture lands as nourishing terrain, I now wish to examine the particular relationship established between human beings and their herd animals in Isluga. My concern in this chapter is to understand how herders maintain a relationship with their camelids through the constant re-enactment of certain practices. Instead of 'domesticating' their animals, the owners continually tame new generations of camelids into an appropriate form of Isluga animal society. The fullest cultural expression of this process of caring for animals, the elaborate *wayñu* ceremony, is considered subsequently in Chapter 4. My aim here is to observe how animals are cared for during the annual cycle of events in Isluga and how these routine events unfold in place, space and time. The terminology of 'domesticating animals' and 'domestication' as used in the writings of anthropologists and archaeologists in Chapter 2 yields to one that focuses on 'caring' for herd animals within specific terrains that nourish both herds and their herders.

An important aspect of herding in Isluga has to do with how people identify their own camelids. Their classificatory schemes include cross-cutting terminologies that incorporate words for birds and wild animals. Such terminology allies pastoral practice and ideology. These herding practices can therefore be seen to serve more than just managerial strategies. They also provide a means for Isluga people to perceive the interdependence of human beings, herd animals and the land, with all its affordances, as discussed in the previous chapter.

In Isluga, people's relationship with their territory is complex and their own understanding of that relationship is deeply layered. It might be argued that people regard features of the landscape as personifications of supernatural beings and of ancestors. It has been said that this anthropomorphization of geographical formations in the Andes is to treat ancestors as though they were the 'animated dead' (Cook 1992: 356–7). However, unlike the humanity of the ancestors, other personifications of the land are much more ambiguous. Although the supernatural beings associated with such landforms are gendered, it is not always easy to assign them human or animal status. This ambiguity is lacking in how Isluga people conceptualize their relationship with their herd animals, where norms of appropriate human and animal behaviour are generally more clear cut and explicitly stated.

In common with communities elsewhere in the Andes, Isluga people recognize a cult of the saints, whom they honour in a series of festivities. The recognition of different Christian saints in the various places of Isluga also has a temporal dimension that is integrated with the annual cycle of the herding year. Fiestas present opportunities for sacrificing llamas and alpacas and convey an expression of the relationship between herders and herd animals that underpins cultural life in Isluga. This chapter therefore begins with a consideration of the annual herding cycle in relation to the annual round of events and to the places frequented by herds and their herders. The temporal and spatial use of pasture lands provides the grounding for the subsequent examination of Isluga people's classifications of their llamas and alpacas.

Cycles of time and space in Isluga

Families who have spent the winter months in the *precordillera*, as described in the previous chapter, generally return to Isluga in August. On one occasion in 1987, Doña Catalina Castro Castro and her husband Basilio reached Enquelga as early as 27 July. In this case, Enquelga was Catalina's natal village, and she had been caring for the combined herds of llamas and sheep which belonged to her own family (herself, her husband and their young children) and also those belonging to her elderly parents. The family stayed in Enquelga for a month, before the herds were separated and Catalina and her family and animals went back to her husband's village of Pisiga, in Manq"asaya, this being their main place of residence. The reason they gave for staying in Enquelga was that pasture was more plentiful there. It is generally less plentiful in Manq"asaya and, besides, the pasture had not begun to recuperate after the windy season, which was as yet unfinished.

When these herds arrived in Enquelga after the three-day journey from Apamilka, Alto Camiña, it was evening. The animals were enclosed in corrals: the sheep in a small *uywij uyu*, and the llamas in a rectangular *kancha*. A female llama and her young were missing, and subsequent searches over the next few days failed to locate them. The successful return of the remaining animals was marked by a simple ceremony as soon as they were safely in their corrals. Catalina's mother, Doña Natividad Castro Challapa, burned incense and she wafted the smoke round the corrals. Then Doña Nati and her husband Don Marcos Castro Challapa offered a few coca leaves to the Wirjin Tayka (the name by which the 'time/earth mother' is revered in Isluga), and they sprinkled a few drops of neat cane alcohol on the ground of each corral.

In the past, llama caravans were taken from Isluga to the western valleys. The herders from the highlands worked for the agriculturists during the harvest, and their llamas were hired by the farmers to transport agricultural produce. The herders were paid in kind, and they also slaughtered some of their animals in order to exchange meat for maize and fruit to take back to the *altiplano*. According to Doña Natividad, the herders did not exchange unspun fleece, only meat and *charqui* (salted, sun-dried meat). However, she did weave for people in the *precordillera*, receiving on one occasion peaches as payment for weaving a dress (*urk"u*). Older people in herding communities of the *altiplano* say that they formerly only ate fresh fruit once a year. This classic form of relationship between herders and farmers is no longer operative. Nowadays, goods and produce are usually transported by truck. In the late 1980s, only three families owned a vehicle in Enquelga. One of these families has been building up commercial trading activities, since there is no village shop in Enquelga. This type of economic specialization did not exist previously (Provoste 1978: 25). As Provoste remarked, all the people of Isluga did the same kind of activities at the same time. In fact, a yearly calendar of events is still followed in Isluga, which I will now attempt to outline.

As we have seen, residence patterns tend to be dispersed during the windy season (*t"aya pacha*), whether horizontally (in which case the inhabitants of Enquelga pasture their herds elsewhere in Isluga) or vertically (by going down to the valleys). The year is characterized by a cyclical expansion followed by a contraction when the rainy season begins. Families spend the rainy season (*jallu pacha*) in Enquelga, with visits to the ritual centre of Islug marka when the whole of Isluga congregates in order to participate in certain ceremonies. In common with most other Andean communities, a series of saints' days are observed throughout Isluga, this being an important feature of the community's definition of itself as a distinct ethnic group that is different from its neighbours (see, for example, Urton 1988: 27).

When I first went to Enquelga in 1986, all the families were at least nominally Roman Catholic. Members of one family already had attempted to convert to a Pentecostal sect, but their fellow community members exerted social pressure on them and persuaded them to revert to Catholicism. Enquelga was unusual among the *estancias* of Isluga, as the Evangelical Pentecostal Church had long gained firm support in other settlements. By 1989, some families in Enquelga decided to weather the opprobrium and they too converted to the Pentecostal sect. There are now at least three different forms of religious experience in Isluga – including membership of the Roman Catholic and Evangelical Pentecostal churches, and a third way, which is agnostic or atheist. However, in my discussions with different people I have found that the festivals associated with Andean and Catholic religious practices are still important for their understanding of place, space and time in the annual cycle in Isluga.[1]

The three important events which mark the rainy season and which are observed in Islug marka are the Christian festivals of All Souls Day at the beginning of November; the formerly traditional month of festivals leading up to the celebration of Saint Thomas, the patron of Isluga, on 21 December; and the celebration of Carnival, which begins on the Saturday immediately before the first day of Lent. It is frequently stated that Catholic festivals are never selected arbitrarily for observance in the Andes, and it is significant that 21 December is the summer solstice, but Christmas Day is not observed in Islug marka. The worship of Catholic saints in Isluga did not simply replace previous religious observance, and M. E. Grebe Vicuña makes the important point that although, on some levels, Christ is identified with the sun and the Virgin with the moon, the saints are confined to *taypi pacha* (this world/time), whereas the sun and moon and other astral bodies also move through *arax pacha* (the upper world/time) and *manq"a pacha* (the inner world/time) (Grebe Vicuña 1983: 159).[2]

Thus, the communal observance of All Souls in the graveyard of Isluga marks the beginning of the rainy season, while Carnival is the riotous event with which it closes. It may seem paradoxical that a commemoration of the dead should begin the annual ritual cycle. However, a parallel is also to be found in the ritual calendar of the Aymara-speaking Laymi people of Bolivia. According to Olivia Harris, the calendar is organized round the two major festivals of the dead – All Souls and Carnival – which announce and conclude the rainy season, a period of agricultural fertility expressed metaphorically through the image of wild flowers.[3] As analysed by Harris, the celebration of All Souls serves to 'socialize' death and the graveyard: 'death is domesticated, it is made cyclical' (Harris 1983: 142). The last day of Carnival includes a rite of expulsion of demons, whose festival this is, and who are, according to Harris, in a sense, the spirits of the dead (ibid.: 143).

The celebration of All Souls in Isluga is a tearful and chaotic event, while Carnival is joyful, colourful and crazy: 'Carnival is crazy, crazy' proclaimed one of the participants. The Aymara word for this festivity is Anata, which also means 'game'. While both events are a time of ritual drunkenness, white flour and streamers are flung liberally at all people present for Carnival. During the five or six days spent at Isluga for this latter event, thanks are offered to the ground and to the Wirjin Tayka. People carry shawls on their backs, overflowing with flowering *quinua* stalks and different grasses, which they pull out and heap on to the ritual *misa* (literally meaning 'table') as they visit different houses. They sprinkle this vegetation with alcohol and greet the owners of the house. On the second day of Carnival (Tuesday), quinces are brought from Camiña, and the men pelt the church tower with them using their brightly woven and plaited slings (in Isluga the church is symbolically conceived as being female, *t'alla*, and the belfry, which stands a short distance apart, as male, *mallku*). In an act of bravado, and a display of their skill in using the sling, the young men pelt quinces against each other's backs. This is known as *mara ququ* in Aymara (Büttner and Condori Cruz 1984: 133).[4]

At the height of the rainy season is the summer solstice, the day dedicated to Saint Thomas, which intervenes between the two celebrations of the dead. The people of Isluga call themselves 'children of Santo Tomás', and Martínez reports a legend in which Señor Santo Tomás (Lord Saint Thomas) escaped from Camiña across a flat plain, named Islu. Camiña was his place of origin, but the people there ignored him. Instead, he settled in Isluga (Martínez 1975: 406). This claim to be descended from Saint Thomas, who in European Catholic traditions is said to have travelled to India to convert people to Christianity, is very interesting in an Andean context. In their association with the figure of Saint Thomas, Isluga people are inscribing themselves within a mythical tradition that they share with neighbouring Aymara groups. This deeply rooted tradition is embodied in the mountains and water sources of the area.

Thérèse Bouysse-Cassagne (1997: 162) has examined what she calls 'the construction of an Andean hagiography' by following the purported journeys of

Saints Thomas and Bartholomew in the writings of sixteenth- and seventeenth-century clerics and missionaries. Her perspective is global, from Mylapore in Asia to the various parts of Latin America (Mexico, Brazil, Patagonia as the 'Tierra de Santo Tomás', and the Andes). She develops her theme to argue that a Counter-Reformation hagiography emerged with a particular American character in which the figures of 'Thomas' or 'Bartholomew' incorporated pre-Hispanic Andean divinities. These figures she describes as a 'mixed product, resulting from a cultural compromise' that was the 'fruit of evangelization' (ibid.: 178). The materials with which she works are those of the Christian missionaries, whose objective was to recast Andean divinities in a framework that would convey Christian beliefs and values. Bouysse-Cassagne supposes that Andean peoples would have participated in the project of creating a hagiography with a geographical reference to Lake Titicaca, Río Desaguadero and the volcanoes of the western *cordillera* of the Andes. The result was the imposition of the lives of Thomas and Bartholomew on local natural formations (water sources and mountains) (ibid.: 196). Her final paragraph includes the claim that 'the indigenous myth that served the making of the Christian source is still partially valid' (ibid.: 196). This claim requires further comment, as some of the places in the north of Chile where Saint Thomas has been associated are not mentioned by Bouysse-Cassagne.

In 1986, a young man named Mario, of Caquena in the far north of Chile, told me about Lake Chungara, situated at an altitude of 4,500 m above sea level, near the frontier with Bolivia. He said that the lake was formerly the site of a village, which was visited by the Lord (*el Señor*), dressed as a poor man. No one recognized him; only a young woman offered him hospitality. He told her that he would punish the village, but she was to leave and not look back. The volcano erupted and the village was submerged under the lake. While the woman walked with her baby on her back and her spindle in her hand, she could not resist turning round to look. Immediately, she was turned into stone. Apparently, this stone exists near Lake Chungara, but Mario did not know of its whereabouts. The legend is analogous to one written down in 1613, an account of the wanderings of Tunupa, who appeared in human form as a tall, bearded, grey-haired man wearing a long white garment.[5] Santa Cruz Pachacuti Yamqui related that Tunupa travelled in the same set of clothing and often slept in the open air, until one day he arrived in the town of Yamquesupa, where the people rejected him. He cursed the town, and it was flooded, forming the lake, which the author called Yamquiçupacocha. Eventually, Tunupa arrived at Carabaya, where he was killed and thrown into Lake Titicaca, with his shawl serving as a raft. His body sailed to Chacamarca, where the earth parted, forming the River Desaguadero. His body went on to Lake Poopó (Santa Cruz Pachacuti 1968 [1613]: 284).

Tunupa, the name of a pre-Hispanic deity who shares characteristics with Saint Thomas, was apparently an Aymara deity related to the fire above (lightning) and the fire from inside the earth (volcanic lava). Figure 3.1 shows places that are traditionally connected with Tunupa/Saint Thomas.[6] There is an aspect of Tunupa's story that connects back to the oral traditions of Isluga people. The Jesuit priest Ludovico Bertonio listed in his Spanish–Aymara dictionary the following terms:

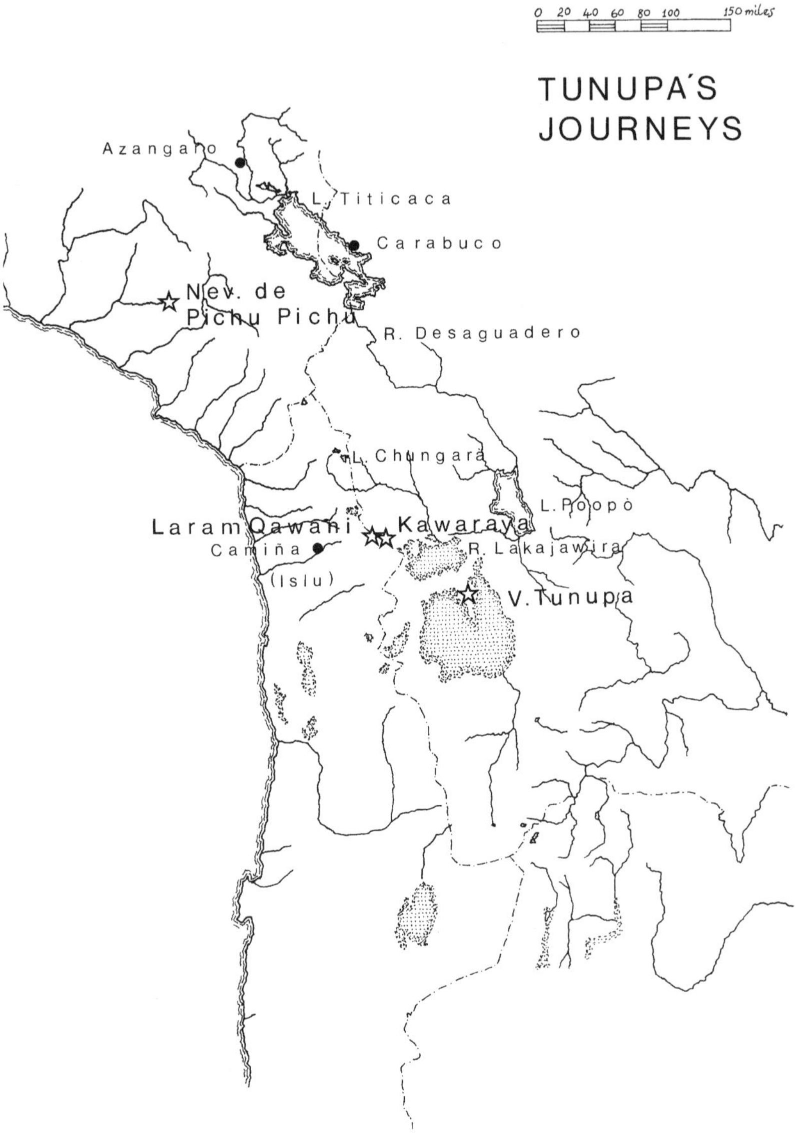

Figure 3.1 Places associated with Tunupa.

> *Quesi Chaulla.* Fish they call boga . . .
> *Quesintuu*: another kind of *boga*.
> *Quesintuu, Vmantuu*: These are two sisters with whom Tunupa sinned, as is told in stories of the Indians.
>
> (Bertonio 1984 [1612] Bk II: 291)

According to Gisbert (1980: 70, n.121), a myth collected in the second decade of this century among the Chipaya people refers to Ekako, another name for Tunupa in Bertonio (1984 Bk I: 192). Ekako had two wives: Chaulla and Umantuu, which are the names of the two Andean fresh water fish listed in Bertonio's dictionary. Gisbert mentions that young Chipaya girls tie bronze figures in the shape of *sirenas* (mermaids) in their hair (Gisbert 1980: 47). Chipaya territory is in Bolivia, in Qaranqas, which is immediately adjacent to Isluga, where the people have great respect for the ritual curers (*llatiri*) of Chipaya. Isluga people associate the Chipaya (to the east of Isluga) and the Chiapa (to the west of Isluga) with *chullpa*, that is, with ancient times (see Martínez 1976: 274, 279).

It seems, therefore, that Tunupa was a pre-Inkaic Aymara deity on to whom the image of the Inka deity Wiraqucha was later superimposed (see, for example, Betanzos 1968 [1551]: 11). Santa Cruz Pachacuti (1968 [1613]: 283) identified him with Santo Tomás, and Garcilaso (1966 [1609]) and Guaman Poma (1980 [1615]: 653–4) with San Bartolome. The author Santa Cruz Pachacuti came from a town immediately north of Lake Titicaca, and his identification is relevant here, where Tunupa/Tomás is associated with what Gisbert describes as the dorsal spine of this region: River Azángaro – Lake Titicaca – River Desaguadero – Lake Poopó – River Lakajawira – Lake Coipasa (Gisbert 1980: 47).[7]

The name of Tunupa is still remembered in the region but, in a complication not considered in Bouysse-Cassagne's 1997 article, in Carangas it has female associations. Torrico Prado retold a myth from Sabaya (visible from Isluga) in which the now extinct volcano called Tata Sabaya defies his older brothers, the giants Illampu, Illimani and Sajama, in his wish to marry the female mountain T'alla Tunupa. A volcanic battle ensues between Tata Sabaya and Sajama, following which there is a reconciliation, and the marriage takes place between Tata Sabaya and T'alla Tunupa (Torrico Prado 1971: 73–7). One of the volcanoes mentioned as taking part in the battle is Laram Qawani, known as Volcán Isluga on maps of the area, and below which is sited the village of Enquelga.

Lynn Sikkink (1998) has examined the human-like activities of the mountains as they are expressed in the myths circulated by the people of San Pedro de Condo in Uyuni, where the mountain Thunapa[8] stands. In these myths, Thunapa is almost always a female mountain, and Sikkink argues that the verbal accounts show an understanding of mountains and deities that is couched in 'the lived experience of the oral tradition (the feel and sound of it)' (Sikkink 1998: 40). It is this physical knowledge that gives character to the land forms that people frequent during their herding activities.

Within Isluga, people used to spend the greater part of the month of December in Islug marka. The festival of Santo Tomás was preceded by the festivals of Santa Bárbara, Concepción (the Virgin of the Immaculate Conception), Octava (the octave),[9] and Santísimo (Most Holy), each of which had ritual sponsors elected from each moiety to serve in the office of Alférez and Mayor Domo. Gisbert (1980: 27) states that Santa Barbara is associated with thunderbolts and storms. Therefore, this month of festivities opens and closes with the celebration of a divine person associated with lightning. However, it was said that during the festival of Saint Thomas, his effigy was processed round the town with that of Concepción. The hierarchical position of Saint Thomas in the sequence of saints' days was made clear in no uncertain terms, for it was said that Santo Tomás is 'He who commands all'. Yet the importance of Concepción cannot be denied, for her effigy was placed in the central and uppermost position in the niches on the wall behind the altar in the church of Islug marka.

In more recent years, the people of Isluga observe only the festival of Saint Thomas. The effigies of the saints Barbara, Concepción and Santísimo (the last, especially, was said to contain much gold) were stolen from the church and were taken to Peru, where they were sold, causing a great deal of anxiety and distress in Isluga. In December 1987, the ritual sponsors for the festival of Santo Tomás defaulted in their duties and there were no celebrations that year, although the festival has been resumed since then. Evangelical Protestants were blamed for disrupting this Catholic event and, indeed, they were alleged to be implicated in the theft of the saints. Recollections concerning the celebrations formerly held throughout December are becoming blurred: for example, many people were no longer sure of the exact dates of each festival. However, people who had served as Alférez or Mayor Domo remembered their fulfilment of these duties with pride and they could relate exactly how many llamas they slaughtered to provide meat for the ritual meals served to each moiety. Success in herding activities sustains the observance of the ritual calendar in Isluga. Conversely, a lack of success in herding is one of the contributory factors that encourage people to turn to Protestant cults.

As can be seen, the rainy season opens in Isluga with the celebration of All Souls, it reaches its climax with the festival of Santo Tomás, and it closes with the celebration of Anata, or Carnival. Other saints' days are observed during this period, but at the level of individual *estancias*. In Enquelga, Saint Andrew (30 November) and the Virgen de la Candelaria (Candlemas, the Purification of the Virgin, 2 February) were formerly celebrated. Saint Andrew is very much associated with rain and lightning. Monast (1972: 74, 80), who claims that there are no known images of him in Carangas, has dubbed him 'a rain devil'. In fact, there is no effigy of Saint Andrew in Enquelga. If the rains came late, a *wilancha* took place in the square in front of the church: that is, a llama was sacrificed and a ceremony was performed asking the saint, as 'Lord of the Rains', for rain. The dances took the form of lines of people making zig-zag movements across the square, in imitation of streaks of lightning. The participants also carried white flags, this being the colour of clouds, and it was also said to be the colour of llamas claimed by Saint Andrew.[10]

At the end of November, large clouds (*q'inaya*) accumulate over Isluga, coming westward from Oruro, bringing with them lightning and rain. The lightning is greatly feared and people avoid herding alone in the open and treeless landscape. Lightning, as previously mentioned, is an agent of death, which often strikes down grazing animals. In June 1987, Doña Natividad and Don Marcos told me that two of their white male llamas had been killed in this way the year previously. They wept over the loss of their animals, and she added that she had demanded of the clouds, 'Why did you kill my llamas?' He explained that one could not eat the meat of llamas killed by lightning, for this would cause boils to burst out on one's skin. Various people reiterated this. Isluga people were unanimous that one should not eat such meat, or even use the fleece; ideally, the corpses of animals killed in this manner should be burned or buried.

Candelaria (the Purification of the Virgin celebrated on 2 February) is one of the festivals associated with the Andean divinity known as the Pachamama, who is more often referred to as the Wirjin Tayka ('Virgin Mother') in Isluga. The name *Pachamama* means 'time/earth mother', and this independent divinity has her own self-sufficient and creative power to sustain life on this earth. Magical auto-genesis, or virgin procreation, is attributed to the Wirjin Tayka, who is without a male consort. Herders revere her as the benefactor of pasture, but she also has the power to withhold those benefits. The Christian Virgin Mary has, over the years, adopted some aspects of her pre-Inkaic counterpart, although Isluga people often distinguish between the two divinities (for example, the smoke of burning incense is offered to the Christian Virgin, while *coca* leaves are offered to the Andean Wirjin). Candelaria, the Virgin of Candlemas, is regarded as 'powerful' and may punish those who ignore her. In contrast, a symbol of the autogenerative powers of the Wirjin Tayka is to be found in the *ankañuku*, an edible tuber which grows entirely underground and which is ripe between Candelaria and Carnival. Another aspect of the Virgin Mary is celebrated on the feast of the Assumption (Asunta) on 15 August. This time of the year marks the period when potatoes and *quinua* should be planted (for further details see Martínez 1976: 281–2).

Corpus Christi is a movable Catholic festival that takes place between 20 May and 23 June. The Quechua Qoyllur Rit'i ceremonies with which Corpus Christi is observed near Cuzco have been interpreted by Randall (1982) as being of Inkaic origin, serving to honour the constellation of the Pleiades, which returns to the night sky after an absence of approximately forty-five days, between 18 April and 3 June (see also Urton [1988: chapter 6]). A resident of Enquelga mentioned that Corpus Christi was another name for the festival of Santísimo, in the month of December. However, he might have been thinking of another set of associations in which the sun represents God (Santísimo, the Most Holy Sacrament), as these two dates (Santísimo and Corpus) fall near the summer solstice in December and the winter solstice in June. The sun (*inti*) is also ritually called 'Qullan Awksa awatjistu' ('Our great Father who is herding us'), while the moon (*p"axsi*) is 'Taykas María awatjistu' ('Our Mother Mary who is herding us') (Martínez 1989: 132). Thus the sun herds by day and the moon by night, but the sun is also 'Qullan santísimo' and it is identified as the most holy sacrament that restores health (Grebe Vicuña 1983: 158).

Winter, the windy season, is also marked by various saints' days, which are celebrated by individual *estancias*. They provide further evidence for the complex beliefs concerning meteorological phenomena and their personification as saints who are remembered in the course of the herding year.

The patron of Enquelga is San Felipe (Saint Philip, 1 May). His effigy is combined on the same stand with that of Santiago Apóstol (Saint James), who is portrayed as a bearded horseman.[11] In the ethnographic and historical literature,

San Santiago is usually identified with Illapa, the Andean deity of lightning. This saint was the patron of the Spanish conquerors, and his identification with lightning follows the defeat of the Inkas in 1536 at the Inka fortress of Sacsahuaman outside Cuzco, when the saint was seen to appear in a great thunderbolt, as explained by Guaman Poma (1980 [1615]: 407).

In Isluga, San Andrés is blamed specifically for causing lightning during the first part of the rainy season. However, it should be pointed out that the associations between saints and meteorological phenomena are never simple correspondences, for Santa Bárbara and Santo Tomás are also implicated, as well as Santiago, in producing thunderbolts and lightning. A *novena* (this word is used to mean a 'shrine' in Isluga) is dedicated to San Santiago on top of a hill known as Q'urawani, above the *estancia* of Ch'api Qullu. The name of this hill means 'the place of slings'. Evidently, the noise produced by the whiplash of a sling is meant to imitate the crack of thunder. A ceremony used to be held on the summit to request rainfall in times of drought. It involved the sacrifice of a llama, a sheep and a cockerel. Among the necessary elements was a bucket of sea water, covered in cotton, like fleece.[12] Gisbert (1980: 29) discusses the iconographic replacement of Inka representations of Illapa as a man bearing a sling with which he breaks a jar of rainwater, by that of a Spanish soldier. Nevertheless, when speaking to the people of Enquelga, it seemed clear that the primary association that they made with San Santiago was that of extremely strong winds, since thunder and lightning occur less frequently in the windy season (*t"aya pacha*). The name T"aya Santiago indicates the saint's shamanic functions, related to the concept of animation by breathing. Whirlwinds beat the dry earth and create hollow depressions (*wayri*) that resemble the hollows made by llamas as wallowing places. Mothers warn their children against playing in such dangerous spots.

Also in the church in Enquelga are three small effigies of San Antonio (Saint Anthony of Padua, 13 June), one each for the three family names represented in the village.[13] In Isluga he is regarded as the patron of llamas, and since his festival occurs in winter he, too, is associated with strong winds, which whip up the earth 'like devils'.[14] San Antonio's status as a herder and his close association with the family names of Enquelga give him a local importance that contrasts with the wider significance of saints such as Tomás and Santiago in Isluga.

The cycle of ritual observance in Isluga demonstrates patterns of dispersion and congregation in unison with the herding cycle. It also takes into account patterns of meteorological phenomena that favour the reproduction of the herds or threaten their existence. Not all ritual observance is in connection with Christian saints (see the discussion of *wayñu* in Chapter 4). Figure 3.2 shows the annual ritual cycle of Isluga. Events which take place in Islug marka are marked outside the circle, while events observed in individual villages are entered inside the circle. However, no attempt has been made to include all the festivals of Isluga, and more events relating to Enquelga than to other *estancias* are included. Figure 3.3 presents the relationship between economic activities (and the exploitation of pasture as outlined in Chapter 2) and the annual cycle. Activities relating to the herding of camelids and sheep consist of:

- shearing (November to early December);
- the *wayñu*: that is, the ritual marking of the camelids, which occurs between January and Carnival, and the *wayñu* of sheep, which occurs immediately after Carnival;
- the *costeo*, the dispersion to the *precordillera* from May to August;
- the cleaning of the canals used for the irrigation of *bofedales* in September.

Activities dedicated to the growing of *quinua* and potatoes include:

- planting, usually in mid-August;
- the *aporca*: that is, the reforming of the drills so that the growing potatoes remain covered with earth, which is done two or three times in the wet season;
- the harvest (usually the last week in April and the first week in May); followed by
- the burning of the new fields in preparation for the subsequent planting.

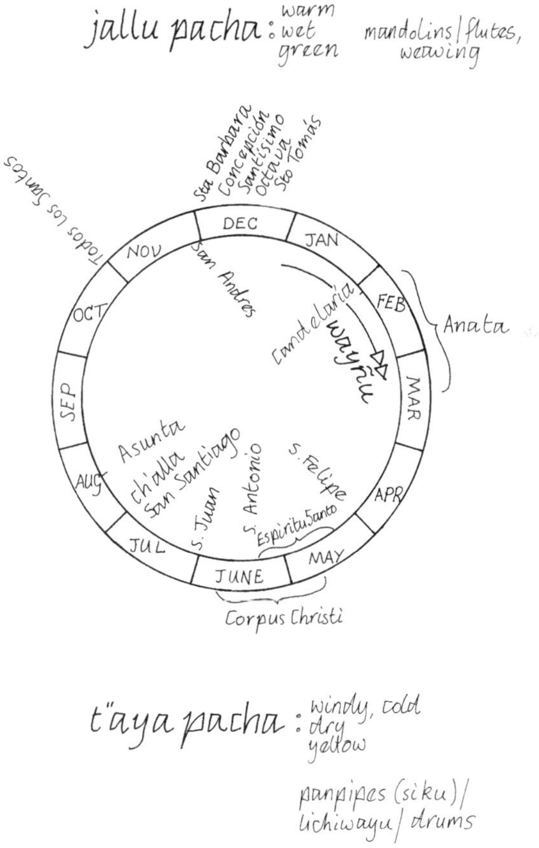

Figure 3.2 The annual ritual cycle in Isluga.

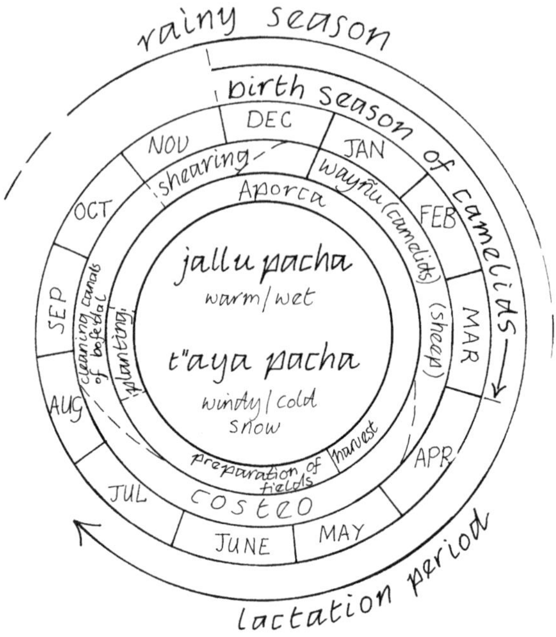

Figure 3.3 The cycle of herding and agricultural activities in Isluga.

Enquelga families celebrate a good harvest with a ceremony at the time of Espiritu Santo (Whitsun): a llama is slaughtered, a communal meal is served and songs are sung in Aymara to the potatoes and *quinua*. Such celebrations took place in mid-May 1986, but few families have had cause to celebrate their harvest yields in subsequent years.

The two seasons are related not only to cycles of saints' days and patterns of economic activity. Other associations may be noted. *Jallu pacha* is warm and wet, with green pasture, while *t"aya pacha* is windy, cold and dry, therefore the pasture is shrivelled and yellow, a sad colour in this context.[15] At times it is blanketed with snow. Weaving, as practised by the women using the Andean four-stake loom, is a ritual as well as an economic activity. It is associated more with the rainy season, as the women say they can devote more time to it then because pasture is plentiful and the llamas eat contentedly without wandering from the *bofedal*. Another distinction is to be found in the musical instruments played by the men. During the wet season, they play mandolins and flutes, but during the windy season they play *siku* (panpipes) in *sikuri* groups accompanied by a drummer, whose drum is formed by stretching llama hide over a wooden frame, and the *lichiwayu*, another type of wind instrument. Human breath blowing through these instruments parallels the wind produced by T"aya Santiago. In other highland communities, musical instruments may vary according to the season, as among the Laymi (Harris 1983: 144; Stobart 1996: 68). Gary Urton noted associations between the rainy season and drums and flutes, and the dry season and wooden trumpets, in the Quechua-speaking community of Misminay near Cuzco (Urton 1988: 30–1). However, the Misminay calendrical cycle is opposed to that of Isluga, for the herding patterns are dispersed during the wet season, and weaving is associated with the dry (ibid.: chapter 2).

The character of the land in Isluga

The people of Isluga have a very special relationship with the land within which they carry out the herding activities described in Chapter 2. For them, earth and configurations of the local geography constitute a truly animated landscape. Pachamama, the earth, is conceptually regarded as being of female gender, but specific places, such as hills and springs, may be gendered male or female. In a structural analysis of the Isluga landscape, Martínez (1976) distinguished four types of place classified into the following categories: *uywiri*-hill, *pukara*, *juturi* and *sireno*. *Uywiri*, meaning 'creator' or 'herder', is in this context the spirit of a hill considered to be the protector of a hamlet or family. Thus, each social group has a specific relationship with certain hills with which it is intimately involved. *Uywiri* may grant individual people llamas or good fortune or, conversely, they may 'eat' one's llamas. Places that are particularly beneficent are known as *aviador*, a word meaning 'provider', no longer used in modern Castilian Spanish (see Martínez 1976: 281). *Pukara* is a place where the whole village performs the ritual planting ceremonies in mid August.[16] *Juturi* is considered to be a deep hollow leading into the inner world (*manq"a pacha*); it is regarded as being a point of creation, a place where llamas may enter this world. The *juturi* have close associations with the *wayñu* celebrations discussed in Chapter 4. Finally, *sireno* is a wet place from where water splashes or makes a noise. Men take their musical instruments to these places for musical inspiration. The Spanish word *sireno* is the masculine form of 'siren'. However, one man claimed that he saw such a being, and he described it as a large animal, with the hind part in the form of a fish and with two breasts in front (ibid.: 288). This observation acquires great interest in the historical context of the relationship between Tunupa and Quesintuu and Umantuu discussed above.

Theoretically, these places may be further classified as male (*mallku*) or female (*t'alla*). In practice, one place may often consist of both *mallku* and *t'alla*, such as the double spring at Jalsuri mentioned in Chapter 2. This spring is listed by Martínez (1976: 302) as the *juturi* of Islug marka. However, not all of the residents of Enquelga agree with the definition offered by Martínez of *uywiri*, since some describe only the village's *juturi* as being *uywiri*.

Here I wish to discuss the immediate landscape around Enquelga, where people regularly pasture their animals. I consider three types of geographic feature: (1) the deep hollows known as *juturi*; (2) places connected with ancient time and known by the terms *chullpa/t"aqsu/qala*; and (3) the hills *mallku qullu/t'alla qullu*.

Juturi

Herders become familiar with the local pastures which their animals frequent through the herding routines discussed in the previous chapter. Families use fairly restricted circuits in areas of dry and moist pasture, depending on the location of their houses. The knowledge people have of their surrounding landscape is reflected in the density of toponyms used to designate what may seem, at first sight, featureless stretches of land. Enquelga is sited on dry, open and gently sloping ground between the marshy *bofedal* and the foot of the long lava flows which fan out from the bottom of Laram Qawani, the volcano (figure 3.4). These lava flows are clad in *waña* (dry) vegetation and each prominence is distinguished by its name, such as Sajima, Ch'api puljru or Añawani. Many such names refer to the predominant type of pasture: for example, the last-named place receives its toponym from the fact that a spiny shrub, eaten by llamas, called *añawa* grows there. Maransil juturi, considered to be *uywiri* and *aviador* (provider) of camelids, is a hollow in one of these slopes near Enquelga. The name is significant, for *maransila* is a medicinal plant used in childbirth against haemorrhage. The hollow is double, representing *mallku* and *t'alla*, and it forms a connection through dry land to the inner world. This is not the only *juturi* known in Enquelga, for there is also a *juturi* for sheep called Wila Qullu ('red hill'), located on a low triangular neck of land approximately half way between the village and the thermal spring. Another *juturi* called Taypi Qullchu is in a watery place, located in the *uq"u*. Thus,

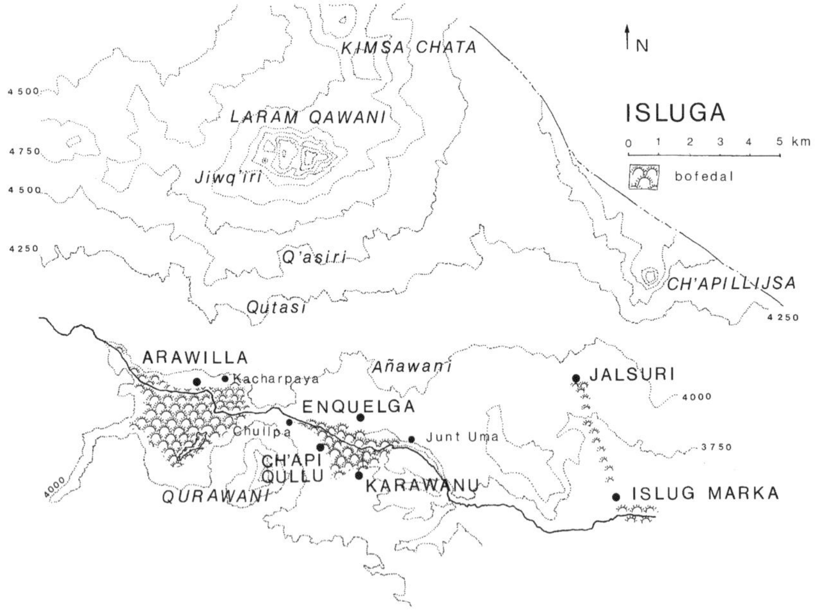

Figure 3.4 Pasture grounds in Enquelga, Arawilla and Islug marka.

some of the most important points of the landscape are inconspicuous and they do not constitute the most imposing of geographical features.

Chullpa/t"aqsu/qala

At the juncture between the gently sloping, sparsely vegetated ground and the rocky, steeper lava flows of Sajima and Añawani, are to be found *t"aqsu*, which are sometimes described as resembling small corrals, but which are said, emphatically, never to have served such a purpose. I do not know the exact locations of these enigmatic structures, since their position was usually indicated with a vague wave of the hand or a nod of the head. Urton (1988: 95) also discovered a general disinclination to point at anything sacred in the Andes when he attempted to identify different astronomical constellations. Nevertheless, the word *t"aqsuni* (meaning 'the place with *t"aqsu*') often occurs in conversations. These small enclosures are considered to date from ancient times and are called *chullpa*, a term also used to designate a type of archaeological funerary monument. They are associated with a previous world age, when there was no order. According to Don Ambrosio Condori: 'in ancient times, there were no *carabineros* (policemen)'. Moreover, the land was flat and the hills did not exist. The people fought each other; the daytime was lit only by the light of the moon, and where the village of Enquelga is sited was a seething mass of boiling red *quinua*. The people of those times – 'gentiles' – lived hidden in the *t'aqsu*. Their world age came to a catastrophic end when they were burned like fried lard by the sun. Such accounts of the mythical past clearly represent a version of the cataclysmic world reversal known to many Andean societies past and present, the concept of *pachakuti* (see, for example, Randall [1982]), known as *muntu t'ijrasiniw* ('the world turned upside down', or 'inside out') among the contemporary Laymi of Bolivia (Harris 1987: 96).

I am not aware of the use of the word *t"aqsu* in the ethnographic literature.[17] In Isluga, I have also heard the verb *t"aqsuña* used in contexts of turning out and replacing, in the sense of 'to turn something out and put it back again' or 'to take out and replace with something else'.

The *t"aqsu* are associated with the generation of human beings during a previous world age, while the *juturi* continue to be associated with the emergence of lineages of camelids and sheep in the present world age. The people of the previous generation also are identified with standing stones (*qala*), which are located round a dry area of slightly rising ground in the marshy ground below Enquelga, forming a ring round some llama corrals standing on this dry ground. These stones are not always silent, for they can make noises. Doña Natividad relished the sound when she told me: 'They can make a racket, boom, boom, boom. There is a "grandfather" near the llama corral. It can play the drum, boom, boom, boom' (personal communication, 31 May 1987). However, if you approached these stones it was said that you would see no one. In addition, these stones are understood to have the ability to cause one's teeth to hurt, or a woman's stomach to swell.

Mallku qullu/t'alla qullu

The places discussed thus far are frequented regularly in the course of herding activities. The higher slopes of the more imposing hills are less often visited, although the quality of these pastures may be very good if they are near permanent snow cover. However, separated herds of uncastrated male llamas (*tataqullu*, 'hill males') may be left to feed on these high-altitude pastures. The people of Enquelga use the slopes of a prominent, conical-shaped hill called Ch'apillijsa, on the slopes of Kawaraya, a huge permanently snow-clad chain of jagged peaks located in Bolivian territory. Herds using these high slopes would not be attended every day, but would be supervised on an irregular basis. Since they have ample opportunity to feed on good-quality pasture, they are said to produce much finer fleece.

Ch'apillijsa is regarded as being 'the herder' of flocks, and in some families it is the first-mentioned place when ritual libations are made and an animal is slaughtered.[18] Herders intone these place names and sprinkle the animal's blood on the ground in a long litany of names. The correct procedure when a llama or alpaca is slaughtered is to sprinkle at least some of the blood on the ground. The blood is offered to the Wirjin Tayka and to the *uywiri*, both *mallku* (male) and *t'alla* (female). It is splattered in different directions and the herders recite a litany of place names, listing the pasture grounds the animal used to frequent, Sajima – Añawani – Ch'apipuljru – Santapiña – usually in a whispered voice. Herders are very anxious to remember all the *uywiri*. They list the names of the great *uywiris*, and then add 'those remembered, those not remembered' in Aymara, lest any names be omitted (Martínez 1976: 275).

The character of the imposing Kawaraya Mallku is of a different nature, for it provokes feeling of unease among the people of Enquelga, from where its tall, snow-covered peak has the appearance of a large face. Some people showed signs of being perturbed that Kawaraya was constantly looking at them. 'Why is it looking at us and not at Bolivia?' one woman asked. In addition, the hill is reputed to appear in the form of a handsome young man, with an appetite for young women. It is said that a sister of a certain Celestino was herding llamas alone, on the lower footslopes of Kawaraya. An attractive young man followed her, silently. He seized her by the hand. She went home, but she fell sick and died. On occasions such as this, one should select a fine llama and sacrifice it ('you have to shed blood'). These powerful hills are perceived as demonic by nature, with certain anthropophagic qualities. They are supernatural, perhaps not strictly human, therefore not exactly cannibalistic. Martínez collected a few legends about the appetite of some hills for human beings (1976: 292–5). A place reputed to be anthropophagic is to be found near Enquelga, on the slopes of Laram Qawani itself. It is a valley, a huge scoop in barren volcanic rock at the very upper limits of vegetation. This is Q'asiri ('she who shouts'), a female supernatural who can shout in order to frighten people and who can also eat them. Like the *gentiles* of the *bofedal*, the hill is mute but not always silent; during the rainy season it is said to shout.

The outline of Laram Qawani encompasses a whole population of supernatural beings, both male and female, who possess awesome powers. The highest ridge, as seen from Enquelga, is called Manku Qapaq, who was the first Inka Emperor and who, according to Inka mythical traditions, emerged from a window-like cave with his three brothers and four sisters. A prominence to the left of this is called 'Kunturi peak', from the Aymara word for 'condor', which is *mallku*, therefore masculine. It has a celestial counterpart: a dark cloud constellation in the milky way, with its head formed by the four stars of the Southern Cross, is also called Kunturi, which in the previous world ate humans.

At this western end of the ridge is located the fumarole, from where hot steam wells out of green, yellow and blue rocks into the air. It is called *jiwq'iri* (from *jiwq'i*, 'steam'), and there is another smaller fumarole (*qallu jiwq'iri*) located lower down the barren slopes. These cavities in the hill, which emit sulphurous smoke, give rise to another name for the hill: *jiwq'ir qullu* ('volcano hill'). I have no evidence to support the assertion made by Martínez (1976: 315) that Laram Qawani is the secret name and that Jiwq'ir Qullu is the common name for this hill. In fact, the name Laram Qawani is known all over Carangas. The people of Enquelga use these names to refer to distinct aspects of the hill. Jiwq'iri refers to the fumaroles and Laram Qawani (the literal meaning of which is the 'place of the blue valley') to the conjunction of ridges and valleys: Q'asiri, Manku Qapaq, Kunturi peak, Qachun peak – this last would be female, judging from its name (*qachu*, 'female'). The volcanic aspect of the hill is greatly feared, and it is said that the two *jiwq'iri* become annoyed when approached. The hill is sometimes perceived as a devilish baker making bread, a similar image to that of 'cook' described by Monast for Wayllatiri, a volcano in Chilean territory to the north of the Salar de Surire (Monast 1972: 89). Water issuing from inside the hill is equally distrusted. Qutasi is such a place, below the *qallu jiwq'iri*; its water is regarded as 'evil'. The thermal spring (*junt' uma*) is understood to come from the volcano. Although regularly used by the people for bathing and washing clothes, it is said to cause illness. Its water is classified as 'female', but women especially are wary of bathing there, and they do not do so alone.

Some parts of Laram Qawani above the vegetation line are charged with colour: sky blue, red and yellow. In Enquelga, anything that is highly charged with colour is perceived as being dangerous, possessing ambiguous qualities. The bright colour indicates that minerals are to be found within, and, indeed, sulphur has been mined from Laram Qawani. Such hills may be reputed to swallow up humans. Cerro Culebra, which has a 'bald head', and which is located on the Bolivian side of the border, was said to have consumed ten Chipaya men who went to mine silver and gold: 'the hill ate them, they were shouting inside'. A similar story was said of Wayna Potosi, a red- and ochre-coloured hill nearly devoid of vegetation, located near the road to Chiapa (this is not to be confused with the larger, snow-covered Bolivian Wayna Potosi). 'Ten men' were sent to mine silver, tin and iron: 'they were eaten, the hill ate them' (Don Marcos, personal communication, 19 July 1987).

Thus far, two types of *mallku* and *t'alla qullu* (male and female hill) have been considered. First, there are the hills with pasture, used for herding camelids, such as Ch'apillijsa, which is used by the people of Enquelga. Q'urawani, with its shrine dedicated to San Santiago and used by the people of Ch'api Qullu, also belongs to this category. The names of these hills indicate an association with meteorological phenomena such as lightning accompanied by fertilizing rains.

Then there are the taller mountains that go beyond the upper limits of vegetation, such as Laram Qawani, Kawaraya Mallku and Cerro Culebra. These hills are associated with atmospheric phenomena, but some additionally display strong colours, indicating minerals within, and they may be actively volcanic, emitting steam and hot water. They are greatly feared and they are reputed to ingest both llamas and human beings. However, they may have some traits which link them to the saints discussed earlier in this chapter. Laram Qawani is said to appear as a devilish horseman, whose horse tramples boulders like dough at the time of the festival of the bearded horseman San Felipe.

Two other types of special place were discussed. The *chullpa*, *t"aqsu* and *qala* are ancient structures, or standing stones, that refer to the past and to the future. They are associated with the dead, who are said to lead a parallel existence alongside that of the living. However, Isluga people do make distinctions between the baptized dead – that is, their immediate ancestors – and the *gentiles* – the unbaptized dead of a previous world generation. The dead are 'fed' by the living; a woman makes token offerings of cooked food to the dead grandmother who bequeathed her the house in which she and her family are eating, by throwing small pieces into the fire. People also offer plates of food during the observation of All Souls. Thus the dead consume cooked food, unlike the powerful spirits who consume uncooked llamas and human beings. The nourishment of the *gentiles*, the grandparents of the previous world generation, has been described by Martínez (1976: 268); the living help ensure that the fertility of the soil is thereby enhanced.

Finally, *mallku juturi* and *t'alla juturi* are wet or dry places providing an exit from the inner chthonic world of darkness, from which domesticated animals are said to emerge. If the *juturi* is located in a wet place, the sign of its existence may be indicated by the mythical form of the *chullumpi* bird, discussed in greater detail below.

Thus it can be seen that the people of Isluga have a very special relationship with the landscape within which they carry out their herding activities. While walking over the land they are very conscious of the events that have been imprinted on it. These include events from recent times: for example, the footprints left by passing llamas – in which case the herders try to work out whether the impressions were left by their own or someone else's animals – and tracks left by birds such as the *avestruz* or partridge. All these imprints are laden with significance for the herders. They also are aware of older and ancient events, including places associated with their own ancestors and the *gentiles* of the previous world generation.

Herd composition and ownership of animals

Llamas and alpacas are herded separately in Isluga. Hybrid animals (*wayki*) remain with the mother, and a herd of llamas may appear to contain alpacas. These are llama–alpaca crosses whose phenotype is closer to that of the alpaca. It is very rare indeed for alpacas to be combined with a herd of llamas. I know of one case in Enquelga where a black alpaca was bought and made to join a herd of llamas, since the owners did not have any other alpacas. Its owner explained that this alpaca 'was used to walking with llamas'. The rest of the herd, approximately eighty head of llama, consisted of females, including mothers and their young, castrated males (*capones*), and the dominant male, which is called *sarija* or *relincho*. Young animals usually accompany their mother even after they have been weaned at the age of six or seven months. In another Enquelga herd, the mother of a young llama died, and this young animal was 'adopted' by another female described as an 'aunt' by the herder; the young llama kept close to the older female. If a young llama is reluctant to be weaned, the herders will tie an apron round the udder of the mother until the weaning has finally taken place.

In Isluga, herds of camelids are trained to act as a unit and to stay together. Some of the animals act as guides; they are called *adelanteros* or *tilantir*, and they are usually older females without young animals by their side. Animals which do not cooperate and which wander off to join other herds cause a great deal of time to be lost, as their owners have to search for them. Animals which do this regularly are slaughtered. Feral camelids are virtually unknown in the Andes. Moreover, llamas are trained to obey two commands: *kuti*, which means 'turn round!', and *piska*, which means 'keep going!' The herders also use a series of whistles or brandish hand-plaited ropes to encourage the animals to move in the desired direction.

Domesticated animals are inherited by individuals and cared for by households in Isluga. This conforms to the usual pattern of ownership in the Andes seen also in Ulla Ulla, Bolivia (Caro 1985: 32). Rights to possess animals as property are transmitted from parents to children through the reproduction of the herds themselves, and ownership is claimed for progeny in the uterine line. Animals, therefore, have known genealogies that reveal a network of social relations among herders. The owner of an animal may sell it, present it to another person, or keep it. The raw materials provided by that animal belong to its owner, who will use or dispose of them as he or she thinks fit.

Children participate in herding activities from a young age. One of the tasks in which they take a keen interest is watching over the newly born llamas and alpacas as the herds graze on the *bofedal*, lest the young animals fall into the water and drown. From a young age, they know the annual herding cycle and have a very good knowledge of the landscape with its pastures where their animals graze. Even very young children are able to recognize their family's herds and also those of other families. Parents make gifts of female llamas or alpacas to their children (to both girls and boys), in recognition of their help. These presents are offered on formal occasions, such as at the *wayñu* ceremony, or to mark a birthday celebration. When a young person marries, his or her animals will form separate herds along with the animals belonging to his/her spouse when the couple sets up a home together.

Both men and women herd animals in Isluga, although one author claims that men never herd sheep (Gavilán 1985: 36), while another (Grebe Vicuña 1984:

463) states that men delegate their responsibilities for herding to their wives and the women to their children. My fieldwork observations do not support these claims. In practice, both sexes of all ages are involved in herding the family's animals. Young boys are more likely to go on to the boarding school in Colchane, while many of their sisters actively choose to leave school at the age of 12, expressing the desire to herd and weave. It is this latter activity that enhances their prestige in the eyes of the community. Young people may migrate to the valleys or to the coastal cities and will not actively participate in the herding, but married men living in Enquelga are frequently seen on the *bofedal* looking after animals, although they will not be available to do such tasks if the heavy work of clearing fields to plant *quinua* has to be done. Much depends on the family concerned and the type and size of herds it possesses. Grebe Vicuña's analysis, made on the basis of fieldwork done mainly in Manq"asaya, presents men as traders, constantly absent in the pursuit of trading activities (1984: 462). However, it should be made clear that the percentage of families owning a truck is very much less than families without vehicles, even in Manq"asaya.[19] In addition, I have no evidence to support Grebe Vicuña's assertion (1984: 457) that herding is a low-status activity in Isluga.

The classification of camelids in Isluga

Camelids are either wild (*sallqa*) or domesticated (*uywa*). While *uywa* are cared for by human owners, wild animals are believed to belong to the spirits of the hills. The mountains Arintika and Pukintika to the north of the Salar de Surire are particularly feared as the demonic owners of vicuñas.[20] Table 3.1 shows how the camelids are classified in Isluga, and the range of hybrid crosses. Vicuña are rarely seen in some parts of Isluga, but they are found west of Mawk'i, on the slopes of Tata Jachura, and to the north they are abundant round the Salar de Surire. Guanaco do not occur in Isluga, but they are known to exist in the western valleys. Hybrid crosses between vicuña and domesticated camelids are rare, but a half-breed vicuña–alpaca cross was obtained by some families in Enquelga in the past. These animals are prized for their fine fleece, and also for their 'intelligence' (one herder spoke of the alpaca as being 'silly' in comparison). Such a cross is known merely as *wik'uña*, a term borrowed from the Quechua for 'vicuña'.

Table 3.1 The classification of camelids in Isluga.

	Camelid species	*Names of crosses between species*
Sallqa (wild)	*Wanaku* (guanaco) *Wari* (vicuña)	*Wik'uña* (vicuña–alpaca cross)
Uywa (cared for)	*Allpachu* (alpaca) *Qawra* (llama): – *t'awrani* (woolly llama) – *q'ara* (bare llama)	*Wayki* (alpaca–llama cross) *Waritu* (alpaca–llama cross) *Wakaya* (alpaca–llama cross)

In some areas of Peru, particularly in Quechua-speaking parts of southern Peru, two breeds (or, according to some authors, varieties) of alpaca are known: *suri*, which has long, wavy and lustrous locks of fleece, and *wakaya*, which is the more widespread form of alpaca found in the Andes. The *suri* breed is not known in Isluga, where this word has its Aymara meaning of *avestruz*, a small Andean ostrich which is sometimes to be seen in small groups around Isluga. The Quechua usage of the word *suri* may be a borrowing from the Aymara, and the long locks of the *suri* alpaca may be said to resemble the long white feathers of the ostrich. However, Aymara speakers do not use this word to refer to alpacas, and an alpaca herder from the community of Caquena in the far north of Chile who had heard of the *suri* breed, referred to it as *wari* ('vicuña') in the Aymara language. There is a great deal of linguistic confusion, for among Quechua-speaking herders, the word *wari* refers to an alpaca–llama hybrid (Flores Ochoa 1978: 1,007). Another ambiguous word is *wakayu*, which is one of the words for an alpaca–llama hybrid in Isluga, but among Quechua speakers it resembles the word *wakaya*, which refers to the widespread form of alpaca and which may be written as *wakayu*, according to Flores Ochoa (1978: 1,007). Unfortunately, many of the biological and veterinary publications have adopted an exclusively Quechua-based nomenclature, without recognizing that such words refer to different breeds or to different animals in Aymara (see, for example, Franklin 1982: 465).

In Isluga, hybrid crosses between alpacas and llamas are called *waritu*, if the phenotype is that of a llama, and *wayki*, if the phenotype is that of an alpaca. The word *wakayu* also may be used for a hybrid cross. These terms are listed in Table 3.1. There are two breeds (or varieties, according to some authors) of llama in Isluga: *t'awrani* (woolly) and *q'ara* (bare). The former may be recognized by the greater quantity of fleece round the face and neck, whereas the latter has less fleece, particularly below the throat.

The camelids are further classified according to four named age roles, of which the first three are sex neutral (that is, the same word applies to male and female). Old age in camelids, both llamas and alpacas, is designated by the sex-parallel terms of *achachi* ('grandfather') and *apachi* ('grandmother'). These age roles are listed in table 3.2. It should be pointed out that the Quechua term *ankuta*

Table 3.2 The age roles of camelids in Isluga.

	Llamas		*Alpacas*	
	Qachu ♀	*Urqu* ♂	*Qachu* ♀	*Urqu* ♂
Calf	*Jiska qallu* *Qallulla*	*Jiska qallu* *Qallulla*	*Jiska qallu* *Qallulla*	*Jiska qallu* *Qallulla*
6 months–2 years	*Ankuta*	*Ankuta*	*Ankuta*	*Ankuta*
Adult	*Qawra*	*Qawra*	*Allpachu*	*Allpachu*
Over 6 or 7 years	*Apachi*	*Achachi*	*Apachi*	*Achachi*

does not correspond to the Aymara usage of the same word. Flores Ochoa (1978: 1,009) lists *ankuta* in reference to adult llamas, alpacas and alpaca–llama hybrids. However, this age role in Isluga is simply designated by the word *qawra* which, if used without further qualification, implies an adult llama, and *allpachu* which, if similarly employed, implies an adult alpaca. The word *ankuta* in Isluga refers to a young animal as it approaches sexual maturity. Camelids, like human beings, are assigned age roles, and old age in both animals and human beings is designated by the same words, although it should be mentioned that it is more common to use the Spanish words *abuelo* and *abuela* to address human grandfathers and grandmothers. At first sight, it might seem that humans and camelids are accorded parallel treatment in linguistic terms. However, Hardman points out that vocabulary words are different for human and non-human animals, and as an example she lists the following:

> *jaqi*: 'people, person'
> *warmi*: 'woman, wife'
> *chacha*: 'man, husband'
> *wawa*: 'child, baby'
> *imilla*: 'girlchild'
> *yuqalla*: 'boychild'

and compares them with:

> *uywa*: 'domesticated animal' [i.e. 'cared-for animal']
> *qachu*: 'female'
> *urqu*: 'male'
>
> (Hardman 1988: 347–8)

She also distinguishes two sets of pronouns, one for humans and another for all else. The non-human pronouns may be used as adjectives, but not as pronouns for human beings. Hardman warns that to do so would be to assign animal status to human beings, which would be an insult in Aymara terms (ibid.: 347). She maintains that the Aymara language makes a clear demarcation between human and non-human animals, even though domesticated animals belong to lineages and are assigned age roles like their human counterparts.

In addition, there is a specialized range in gender-related vocabulary for camelids to account for castrated or sterile animals. These terms are presented in table 3.3. For comparative Quechua terminology, see Flores Ochoa (1978: 1,008). In Isluga, male camelids are castrated at the age of about two years, thus ensuring that the males are docile and can be herded alongside females and their young. Not all families keep separate herds of uncastrated males (*tataqullu*), because the number of adult members in a family may not be sufficient to supervise the different herds. At certain times of the year, for example, when the *quinua* is growing in unfenced fields, a family may be stretched to supervise all its animals adequately. At such times, the adults have to get up before dawn to ensure that llamas and alpacas left

Table 3.3 Gender-related classifications of camelids in Isluga.

Urqu (male)	*Qachu (female)*
Jañachu (sire)	
Sarija (dominant male)	*Tilantir usta* ('guide')
Tataqullu (males, beasts of burden)	*Mamaqullu* (females herded separately)
Urqu capon (castrated male)	*Qachu machorron* (sterile female)

out on the hillside do not feed on the growing *quinua*. Disobedient animals which feed on neighbours' crops are chased out by their owners, who throw stones at them in punishment. The entry of animals into corrals where potatoes and *quinua* are growing is a source of great tension between families, and an aggrieved party may demand a financial compensation for lost crops. Since everyone in Enquelga can recognize which animals belong to whom, the culprits for causing such damage are immediately identifiable, even if their owner is not present at the time of what the human beings consider to be the wrongdoing of the animals.

The naming of camelids

From the foregoing, it can be seen that llamas, alpacas and hybrid animals are identified according to age, gender and variety. They may be further identified according to colour, and this is discussed further in Chapter 5. Here, it is my intention to discuss the naming of camelids according to the disposition of patches of colour on the body of the animal.[21] This nomenclature is discussed in terms of the configuration of light and dark patches, but it should be made clear that 'light' and 'dark' may correspond to several different colours (for example, 'light' may be white, beige or light grey; 'dark' may be black, brown or dark grey). The same principle for naming may be applied to both llamas and alpacas. However, llamas in Isluga show more patchy colour combinations than alpacas, which are more frequently of one single colour than llamas. Additionally, llamas display a fuller range of colour than alpacas in Isluga. For these reasons, the drawings in figures 3.5 to 3.9 are of llamas. The drawings presented here do not claim to be an exhaustive presentation of all the possible names for camelids. Undoubtedly, more exist. These drawings are based on the animals I observed while accompanying herders in Enquelga, and many of the depictions represent an actual animal.

A pure white camelid is called *timpla or timpla janq"u* (figure 3.5i), whereas an entirely black animal, without a single spot of another colour, is called *llunka* (figure 3.5h). A camelid which is entirely brown is known as *ch'umpi kuypa*. Such animals are rare. White llamas often have splotches of colour on the face, but they may simply be called *janq"u* ('white'). However, the majority of camelids are characterized by patches of colour, the patterning of which is identified by name.

One of the main types of patterning is termed *allqa*. This Aymara word means 'black and white', and this single term implies the coexistence of two contrasting colours. It is applied not only to llamas, for the adult plumage of an Andean bird of prey which displays large patches of black and white is called *allqa allqa* or *allqamiri* in Isluga (elsewhere in the *altiplano* it is called *allqamari*). When a llama displays large patches of light and dark (not necessarily black and white) over the surface of its body, but not in equal proportions of each colour, then it receives the name of *allqa* (figure 3.5c). This term may be further qualified. If the llama's face is dark, the top of its head down to the withers is light, and the rest of its body is dark, the llama is known as *sawsi allqa* (figure 3.5a). I also heard this name in a human context. During a *wayñu* celebration, a woman dressed in the traditional dark dress and wearing a light-coloured shawl round the top of her head and over her shoulders was called *sawsi allqa* by the wife of one of her uncles. If the dark patch covers the hindquarters of the animal, it receives the name of *kayti allqa* (figure 3.5b). Quechua-speaking herders in the Department of Cuzco, Peru, also use the category of *allqa* (Flores Ochoa 1978: 1,010).

A llama with a dark head and neck, and with the rest of its body of a light colour, is called *chiwana* (figure 3.5g). If the dark colour extends beyond the withers along the back over the spine, it is called *paru*. The top half of its body is dark and the lower half light (figure 3.5d). The converse distribution of light above and dark below is known as *linkwaru* (figure 3.5e). So far, the terms discussed have dealt with large patches of light and dark, but in the case of *mulluchi* (figure 3.5f), the dark colour is greatly reduced in comparison with the light; this word is applied to an animal with a dark face and a light body.

Figure 3.6 shows some possible configurations created by smaller patches of colour. With the exception of one example, all the rest show patches of a dark tone against a light background. Figure 3.6d shows a scatter of light patches against a dark background. Such an animal is called *jawas p"aqalli*, or 'bean flower'. In this term, the Aymara word *jawasa* is borrowed from the Spanish *haba* ('bean', *Vicia faba*). The converse, that is, dark spots on a light background, is known as *t'axllu* in Isluga (Aymara speakers elsewhere say *ch'axllu*). The Quechua equivalent for this term is *muru* (Flores Ochoa 1978: 1,013). The animal represented in figure 3.6e is called *puku t'axllu*, perhaps after a small duck known as *puku puku* in Aymara. If small spots are combined with a large area of plain, dark colour, the animal may be known as *t'axllu allqa* (figure 3.6g). These three examples are all of small patches of colour. However, larger patches of colour occur frequently on llamas. One of the most common types is *wankalli*, in which a fairly large patch of colour covers the back of the animal where the load would be placed if the animal was a beast of burden (figure 3.6b). The word *wankalli* is also the name of part of Laram Qawani, a gently rounded ridge near the summit which is covered with snow for a greater part of the year. Amongst the Quechua-speaking herders of the Department of Cuzco, Peru, the word *chullumpi* is applied to camelids that are called *wankalli* in Isluga. *Chullumpi* is an Andean duck, but a different aspect of its appearance is applied to camelids in Isluga, as represented in figure 3.9d (see p. 77). If the *wankalli* patch of dark colour extends down to the upper forelimb and if it makes the animal look as though it were wearing a vest, then the name becomes *saliwan wankalli* (figure 3.6c). However, if the dark patch is larger and extends as far as the rump or hindquarters, the llama is called *allqa*.

Figure 3.5 Camelid nomenclature in Isluga: (a) *sawsi allqa*; (b) *kayti allqa*; (c) *allqa*; (d) *paru*; (e) *linkwaru*; (f) *mulluchi*; (g) *chiwana*; (h) *llunka*; (i) *timpla*.

Figure 3.6 Camelid nomenclature in Isluga: (a) *pallarisa*; (b) *wankalli*; (c) *saliwan wankalli*; (d) *jawas p"aqalli*; (e) *puku t'axllu*; (f) *chimputa*; (g) *t'axllu allqa*; (h) *llanamati*.

An animal with two patches on its back receives the name of *pallarisa* (figure 3.6a). This word incorporates the Aymara *paya* ('two'), but I am not aware of the full meaning of this term. Medium-sized splotches of colour would cause an animal to be called *tyutti*, but this is the name of a bird, and the configuration is included with the bird names of figure 3.9. A narrow patch over the withers gives rise to the name *chimputa*, since this is where *chimpu* wool is tied to the backs of female camelids during the *wayñu*. The animal shown in figure 3.6f combines, in addition, the characteristics of figure 3.5g, therefore, its full name is *chiwana chimputa*. Similarly, figure 3.5h shows an animal that might have been named *wallata* (compare figure 3.9a), but Doña Soria called it *llanamati*, the meaning of which I am unaware.

Often the names are descriptive. For example, *ch'illa ch'umpi* in figure 3.7a receives its name because it is a black llama apart from the area where its intestines (*ch'illa*) are located; this area is brown (*ch'umpi*). *Wanaku* (figure 3.7b) is named after the guanaco because its colouring resembles that of the wild animal; in particular, the underparts of the body are white and its face is grey. Another name deriving from the animal's anatomy is shown in figure 3.7f. In this case, there is a ring of dark colour round the base of the tail, giving rise to the name of *chinqi* ('vagina'). A llama with a white disc shape on its chest may be called *qullqini* (figure 3.7c), while a dark animal with white ears may receive the name *qullq jinchu* ('silver ears') (figure 3.7e). A white crescent moon shape on the brow of a llama gives rise to *p"axsini*, a name derived from the Aymara word for the moon (*p"axsi*). The full name of the llama shown in figure 3.7d was *charaju p"axsini*, because of the rear white leg. The colouring of legs is discussed further below. A dark-coloured llama with light hindquarters may be called *sarqa* (figure 3.7g), while a dark-coloured llama with a long white patch beneath the throat may be called *limachiru* or simply *chiru* (figure 3.7i). Finally in this group, is *q"ach'uma*, as represented in figure 3.7h. This is the name of a dark-coloured animal with a white area round the mouth. Its name probably derives from the grass known as *q"ach'u* which grows in *bofedales* and high on the hill slopes near the snow line, as though the animal's mouth were stained from eating this moist pasture.

The next category considered deals with the animal's legs. A dark-coloured llama with a light-coloured rear leg is known as *charaju* (often pronounced *charachu* in Isluga) (figure 3.8b). If the foreleg (or legs) of a dark animal are light coloured, the llama may receive one of three names: *parjanq"u* (from *ampar*, 'hand' and *janq"u*, 'white'), *ch"uch"ullu* (the Aymara word for the legs of an animal), or *panadera* (a Spanish word for 'baker', which evokes the image of a person with flour-covered arms) (figure 3.8c). A dark-coloured llama with the lower parts of all four limbs in a contrasting light colour may be called *wutasani*, meaning 'with boots', an Aymaraized version of the Spanish word *bota* ('boot') (figure 3.8d). If the light-coloured limbs are 'crossed' – for example, if the front right leg is pale coloured and also the rear left leg, as in figure 3.8a – the llama will be named *kalsaru*. Quechua-speaking herders also employ the word *kalsa*. Flores Ochoa (1978: 1,012) derives this word from the Spanish *calza*, which refers to a male garment, the short, wide trousers formerly worn by the men in Isluga, which are

Figure 3.7 Camelid nomenclature in Isluga: (a) *ch'ila ch'umpi*; (b) *wanaku*; (c) *qullqini*; (d) *p"axsini*; (e) *qullq jinchu*; (f) *chinqi*; (g) *sarqa*; (h) *q"ach'uma*; (i) *limachiru*.

Figure 3.8 Camelid nomenclature in Isluga: (a) *kalsaru*; (b) *charaju*; (c) *parjanq"u* (or *ch"uch"ullu* or *panadera*); (d) *wutasani*; (e) *wallksani*; (f) *trinsan allqa*; (g) *wistallani*; (h) *k'ili*; (i) *trumpiñu*.

still worn in some parts of the Andes. However, the Aymara *kalsaru* might derive from the Spanish *calzado*, which means footwear or leg-covering of any kind.

There are further analogies derived from accoutrements more usually associated with human beings. Figure 3.8e shows a dark animal which looks as though it is wearing a necklace (*wallka* in Aymara), hence the name *wallksani*, 'with a necklace', is applied to this llama. *Trinsan allqa* in figure 3.8f was a llama with a dark line down the back of the head and neck which resembled a plait of hair (*trenza* in Spanish). The white forequarters and the dark hind parts show that this llama may be further classified as *allqa*. If a light-coloured llama has a more or less rectangular patch on its flank, it receives the name of *wistallani*, as though it were carrying a *wistalla*, which is a small woven bag used by men and women in Isluga for carrying *coca* leaves (figure 3.8g). A light-coloured animal with a black mouth (this is the reverse of the configuration portrayed in figure 3.7h) is perceived as though it were playing a trumpet, and so it is called *trumpiñu* (figure 3.8i). The remaining figure in this group is 3.8h, which represents a dark grey llama with a white triangular shape beneath the throat, with the point stretching down over the chest. This particular llama was called *k'ili*, an Aymara word for the spinal column, but it is also applied to herringbone-like configurations on textiles, and especially to a type of woven belt worn by men, women and children.

Figure 3.9, the last category to be discussed here, shows llamas with avian characteristics. A duck has previously been mentioned with regard to figure 3.6e, *puku t'axllu*. One of the most commonly used terms is that of *wallata*, an Andean goose, which has a white body and black wing tips. At least one pair of male and female *wallata* is a characteristic sight in the *bofedales* of the Andean highlands. Llamas with a dark-coloured rump and tail are named *wallata*; the full name of the calf depicted in figure 3.9a was *irpa wallata*. Smaller birds associated with watery places are depicted in figure 3.9b, c, d and h. The *such'urqa* has a white chest (figure 3.9b), and the *chullumpi* a spotted face (figure 3.9d). The *chullumpi* is an Andean duck, very much associated with *bofedales*. In the Quechua classification analyzed by Flores Ochoa (1978: fig. 6), *chullumpi* camelids are characterized by a patch on the back, the equivalent of the Aymara *wankalli*. However, some Quechua terms correspond with the Aymara usage, such as the previously mentioned *wallata*. The *ajuya*, a black bird (*tagua gigante*) with a white face which lives on high-altitude lakes, is similarly used by both Aymara and Quechua speakers (Flores Ochoa 1978: 1,011) to describe llamas or alpacas with a dark body and a light face (Figure 3.9c). This is the configuration that reverses the proportion of light to dark known as *mulluchi* (figure 3.5f). Yet another bird to frequent the *bofedales* is the *tyutti*, which has dark-coloured bars on its wings. Its name is given to llamas with medium-sized splotches of colour (figure 3.9h).

Birds of the air lend their names to camelids, too. A dark llama or alpaca with a white ring round its neck resembles a condor, therefore it is called *kunturi* (figure 3.9e). Eagles are also seen in the Andes; the llama in figure 3.9f is called *q"usi ajila*, since the dark patch is a beige colour (*q"usi*) and this particular colour configuration resembles that of an eagle (*ajila*). The llama in figure 3.9i is called *liqi* on account of the patch of dark colour on the back of the head; it is thus called after

Figure 3.9 Camelid nomenclature in Isluga: (a) *irpa wallata*; (b) *such'urqa*; (c) *ajuya*; (d) *chullumpi*; (e) *kunturi*; (f) *ajila*; (g) *qillwa*; (h) *tyutti*; (i) *liqi*.

a bird of prey. Figure 3.9g shows a llama called *qillwa*, which is the name of an Andean gull with a jet black head. Quechua speakers also relate this configuration of dark head and upper neck and light body to the *qillwa* bird (*qellwayto* in Flores Ochoa [1978: fig. 3, no. 5]). I collected three more bird names which are applied to camelids but which are not drawn in figure 3.9. *Siljiru* ('linnet') is a brown animal with a white chest, and *ch"uqamiri* ('widgeon') is a medium-brown animal with brown feet. Finally, *parina* refers to a particular beige hue, which is considered to be the equivalent of the pink of the Andean flamingo, the *parina*. The llama known by this name had recently died, so I did not see it; it was described as having a 'red' back, with white fleece underneath and *q"usi* ('beige') on top.

In addition to the names listed above, which Flores Ochoa regards as the proper names of the camelids (Flores Ochoa 1978: 1,014), the camelids may possess other features that enable them to be identified. Llamas with short ears are described as *muru*, this being the word for a 'cut' (the verb *muruña* is 'to cut' [Büttner and Condori Cruz 1984: 141]). Those with intermediate-sized ears are called *q'atura*. A short animal may be named *chuqullu* ('dwarf'). Camelids should have two digits on each foot, but an inherited trait may be polydactilism in some or all four feet; such camelids have three or four toes on some or all of their feet. These are called *t"ala* in Aymara. Another inherited characteristic is that of clear or blue eyes. This is called *laram nayra* ('blue eyes') or *q"ispi* ('crystal') in Aymara.

Herders in Isluga never count their llamas or alpacas to see if all are present, although they may count their sheep, which are not easily distinguished from each other. No matter how large the herd, and even if for some reason a herder is also caring for a relative's herd in a combined group which contains twice as many animals as usual, the herder checks that all are present by observation. Any one animal may be pinpointed by combining the relevant terms and categories as presented above. Usually, identification is straightforward. One of the Enquelga herds contained two female llamas with the marking depicted in figure 3.8g. They were identified as *chinqi wistallani* and *qillwa wistallani*, since they also displayed other prominent characteristics. In the same herd were two llamas called *q"ach'uma* (figure 3.7h). These animals were distinguished by their colour, since one was black and the other dark grey; hence their full names were *char q"ach'uma* and *uqi q"ach'uma*, respectively. The minimum number of categories that are sufficient to enable identification will be combined to name an individual animal in a herd. Herein lies the key to the herder's ability to recognize and remember so many animals, for the herder remembers each animal as an individual, a member of a particular lineage, possessing characteristics and even behavioural traits or 'personality' not shared by other animals. The possible permutations from combining the names listed above are greatly in excess of the number of individual animals in the herd.

Camelids, fleece and water

Terms derived from bird nomenclature are evidently important for naming camelids in Isluga. To recapitulate, the avian and camelid names that are shared

Table 3.4 Terms derived from bird nomenclature for naming camelids in Isluga.

Camelid name	*Bird name*	*Spanish*	*English*
Allqa	*Allqa-allqa; allqamiri*	*Gallinazo*	Falcon
K'ili	*K'ili-k'ili; mamani*	*Halcón*	Hawk
Kunturi	*Kunturi*	*Cóndor*	Condor
Ajila	*Ajila*	*Aguila*	Eagle
Liqi	*Liqi-liqi*	*Ave centinela*	?
Siljiru	*Siljiru*	*Jilguero*	Linnet
Puku t'axllu	**Puku-puku*	*Gallito de los Andes*	?
Wallata	**Wallata*	*Ganso andino*	Goose
Such'urqa	**Such'urqa*	?	?
Ajuya	**Ajuya*	*Tagua gigante*	Coot
Chullumpi	**Chullumpi*	*Anade andino*	Duck or grebe
Ch'uqamiri	**Ch'uqamiri*	*Gallareta*	Widgeon
Parina	**Parina*	*Parina; flamenco*	Flamingo
Qillwa	**Qillwa*	*Gaviota andina*	Gull
Tyutti	**Tyutti*	?	?

Note: * Indicates a bird associated with water.

are listed in table 3.4. Of course, there are other birds to be observed in Isluga. Those frequenting dry land include the *kullu-kullu* (*tórtola*, turtle dove); *q'acha*, a small orange bird with black head and wings, whose singing in August and September announces that it is time to sow *quinua* and seed potatoes; *kiwla* (partridge); and *suri* (*avestruz*, the Andean ostrich); and those frequenting the *bofedales* include the *llak-llak* (a duck), *lap'ara* (an ibis) and *unkayllu*. I did not hear these names applied to camelids, but what is noteworthy is that many of the bird names also used for camelids are birds of watery places. These names are indicated by an asterisk in table 3.4.[22]

Some of the birds listed above may be regarded as good or bad auguries. The eagle, for example, is regarded as the bearer of good fortune, and some families in Isluga possess a stuffed eagle which has paper money attached to its talons. The *qillwa*, on the other hand, when observed flying over the village of Enquelga was considered to be a bad portent, but good if flying over the *potrero*. However, I have no evidence to suggest that llamas who bear the name of *ajila* or *qillwa* are regarded in any way as the bearers of good or bad luck. The same applies to camelids called *chullumpi*. A mythical *chullumpi* exists, with awesome powers, associated with a hill called Tumawi near Mawk'i. I heard mention of it only once, in a scared whisper, for it is said to have the ability to transform itself into llamas and also to be able to consume humans. The *chullumpi*, as previously mentioned, is a sign that may indicate the presence of a *juturi*. Martínez recorded a dialogue with an Isluga man who identified a *juturi* in the middle of wet pasture from the presence of a *chullumpi* bird:

> 'A chullumpi doesn't fly. . . . it comes out on dark days . . . like this, when the moon is full. In that period it comes out, on the Day of the Godfather, Godmother.'
>
> 'And does water come out of it?'
>
> 'Yes, it is a hollow like this, like a spring . . . a hollow which gets lost inside! So, from there a chullumpi comes out . . . then you see lots of llamas, a small herd of llamas . . . there it goes for the hill. . . . An animal like this, like . . . like a 'ghost'? Like that. Whitish are the animals . . . you see them clearly. Then you're going to see llamas . . . and all the animals disappear. So you don't see a single animal there. That's how you know that that is a *juturi*'.
>
> (Martínez 1976: 284)

In the sequel to this account of the *juturi* offered by Martínez, the relatives of the story-teller attempted to hold on to the llamas. In the struggle to retain the llamas they seized handfuls of fleece to hold them back. The following sunrise, there were to be seen only scattered *chullumpi* feathers; the llamas that they had seized were, after all, *chullumpis*. The one llama they succeeded in tying down overnight turned out to be a *chullumpi*, and it was the progenitor of a whole lineage of blackish-brown llamas, which the protagonist of the story possessed in abundance (Martínez 1976: 284). A similar myth occurs in llama mating songs Arnold and Yapita recorded in Qaqachaka, Bolivia. In the songs, a woman herder says that she has given birth to a baby llama, but on looking backwards, she observes that in reality it is a *chullumpi* or another kind of bird, the same colour as the llama fleece (Arnold and Yapita 1998: 336–44). The courtship rituals of chullumpi birds are accorded special significance in Qaqachaka in the context of ceremonies during which herd animals are mated or crossed (*qarw jarq"ayaña*) (Arnold and Yapita 1998: 358–62).

This examination of the herding way of life practised in Isluga demonstrates how each new generation of llamas and alpacas is co-opted into the systems of relationships established between people. Herders recognize their animals as individuals, and by caring for them they bring llamas, alpacas and sheep into the category of 'domesticated animal' used by the archaeologists and anthropologists whose writings were discussed in Chapter 2. However, the Aymara term *uywa*, meaning 'cared-for' animals, is extended by Aymara speakers to encompass different contexts. Gary Urton was somewhat perplexed to find that in the Tarabuco-Candelaria area of Bolivia, weavers used the term *uywa* for animal forms that he designated under his categories of 'non-domesticated' (condors, foxes, vizcachas, pumas, monkeys and ducks) and 'domesticated' (horses, dogs, cats and chickens) (Urton with Nina Llanos 1997: 130). It seems that in the process of producing visual images of animals in a textile medium, the weavers combined groups of animals that can be distinguished according to other criteria. In Isluga, vicuñas, guanacos and rodents such as vizcachas are the *uywa* of the spirits of the hills, while llamas, alpacas, donkeys, cats, dogs and chickens are the *uywa* belonging to human owners.

There is a strong sense of season and location that is associated with the process of caring for animals. Isluga herders assume responsibility for their herd animals in the context of a landscape that they perceive as possessing its own specific character. Human beings interact with non-human animal species in geographical configurations of Isluga terrain that constitute a truly animated landscape with which the people enter into a relationship (Dransart 1996: 35). Isluga people establish particularly intense relationships with areas that are intimately involved with their daily activities, places where they go to herd animals and gather firewood. The Wirjin Tayka is ever present, and the *uywiris* are credited with a voracious appetite. Religious observance must be rendered to them in the course of the herding year lest they 'eat' one's llamas. Similarly, the cult of saints in Isluga provides another aspect of the calendar in which herding practices are conducted. Like the Wirjin Tayka and the *uywiri*, the Christian saints are at times beneficent, and at other times they are vengeful to herds and their herders.

Tim Ingold has argued that hunter-gatherers perceive the environment and its resources as being imbued with 'personal powers', and that 'if they are to survive and prosper', they must 'maintain relationships with other human persons' (Ingold 1994: 9). The people of Isluga also consider their land to be nourishing terrain that incorporates land forms perceived to have 'personal powers'. These entities are clearly gendered, although their 'human' or non-human status is more ambiguous. They, in turn, have to be nourished by human beings. This act of nourishment is featured in the elaborate *wayñu* ceremony discussed in the next chapter.

4 Flowers of the herds

The *wayñu* ceremony in Isluga

This chapter examines the *wayñu*, or marking ceremony, which reveals two sets of relationships. First, I consider the relationship between human beings and their domesticated animals, on the one hand, and the land and its pasture, on the other. An important focus is the transformative process in which pasture and water are converted into camelid fleece. The ceremony symbolizes the transformation of pasture into fleece, and this aspect is closely related to the material practices of spinning yarn, plaiting ropes and weaving cloth, which will be discussed further in Chapter 5.

Second, I address the relationship between a married couple with their children and their domesticated animals with their progeny. My understanding of the relationships between human beings and herd animals in Isluga is that the former consider their own (human) lineages to be, in many respects, parallel to those of herd animals.[1] The *wayñu* ceremony is a rite of passage for herd animals. It is performed by human families to regenerate procreative vitality among their animals, which nevertheless have shorter generations than humans.

To a large extent, people understand their territory through their own close association with their llamas, alpacas and sheep, animals which are closely associated in everyday life, and through ritual libations, with its water sources. The practice of the *wayñu* ceremony enables the human owners of animals to comprehend the landscape that they inhabit.

The wayñu in Isluga

The culmination of the herding cycle in Isluga is the *wayñu*, a celebration held by the nuclear family (mother, father and children) in honour of their animals and of the Pachamama and the *uywiri*. In the Aymara language, *wayñu* refers to traditional song or dance; elsewhere, Aymara speakers may call this ceremony *k'illpa* or *uywa k'illpaña* ('the livestock-marking ritual'). In the Quechua language, the equivalent ceremony is referred to as the *señalakuy*, and in Spanish, the *floreo* ('flowering') or the *marca* ('marking') of the domesticated animals.[2]

My concern here is to evoke the Isluga *wayñu* in the context of its environment: that is, in the context of the categorization and evaluation of the landscape, meteorological phenomena and fauna (camelids and birds) made by the people of

Isluga, as discussed in the previous chapter. These categories are culturally coded and colour how an inhabitant of Isluga perceives the insertion of both human and animal society into the nourishing terrain. The marking ceremony is a vehicle of symbolic thought; it enables the presentation of images and sounds which are, in effect, constellations of significant signs displaying varying degrees of abstraction. This symbolic thought pervades all other areas of life and all aspects of herding technology. Peter Gose regards rituals as an outgrowth of the 'profane' activites of everyday life that 'draw us further into the cultures from which they spring' (Gose 1994: 4). He cautions, however, that rituals themselves need to be interpreted.

In Isluga the flowering or marking has two forms, the *wayñu* and the *ch'allta*, the latter being an abbreviated version of the former, which does not include an all-night vigil during which mournful songs are sung. However, both forms of ceremony involve a considerable investment in financial terms, and also in the time required to prepare the essential material items. Isluga families generally celebrate two such ceremonies: one for their camelids, which takes place between New Year's Day and Carnival, and another for their sheep, if they possess any, which usually takes place immediately after Carnival.

The ceremony takes the form of a symbolic investiture of the animals, and the ritual 'dressing' of the animals is accompanied by songs and mandolin music which are designed to ensure the fertility of the herds. Each family has its own date for the *wayñu* or *ch'allta*, but it always takes place on a Thursday or Saturday. These days of the week are known as Día Compadre and Día Comadre; the former refers to the day of the *mallku* ('lord'), and the latter to the day of the *t'alla* ('lady'). As a result, several ceremonies may take place at the same time in Enquelga. Although these are organized by individual families, all other members of Enquelga are freely invited to attend. Families may choose to observe the *wayñu* one year and the *ch'allta* the next. Some families hold a celebration every other year. Those families with a small rectangular *kancha* corral situated among the large corrals (in which potatoes or *quinua* are grown) surrounding the village of Enquelga itself, conduct the *wayñu* when these large corrals are not being used for agricultural purposes. In the intervening years, when the corrals surrounding Enquelga are planted with *quinua* and potatoes, families owning a *churu* corral on the rising ground surrounded by the *qala* standing stones in the *bofedal*, hold their *wayñu* ceremonies there. The word *churu* is used to designate the dry-stone llama corrals on this rising dry ground. In appearance they are identical to the *kancha*, but they are never referred to as such. Individual families have some flexibility in deciding how often and when to observe the *wayñu* and *ch'allta* ceremonies. The health and well-being of their animals is a decisive factor, for if they suffer ill health or meet with misfortune, a family may well decide to change its usual practice in the observation of these ceremonies. The following account is based on the preparations for and celebration of ten *wayñu* or *ch'allta* ceremonies in February 1987 and one *ch'allta* on 1 February 1997.

Preparations for the ceremony begin with the making of *chicha*, a maize beer, and of the *sarsillu*, *chimpu* and *wistalla*, the items with which the animals will be 'dressed'. *Sarsillu* are the earpieces, tassels of coloured wool, in which red predominates. Four different types are made for different categories of camelid: *tilantir sarsillu, tama, wantilurita* and *sombreros*. *Chimpu* is unspun llama or alpaca fleece dyed red, pink or orange; sometimes yellow is also used. It will be tied to the backs of the llamas and alpacas. The third item, *wistalla*, is a neck piece which consists of a series of coloured yarns in tones of pink, green and orange hanging from a thick cord which will be tied round the neck of the camelids (the word *wistalla* also is used for a small bag used by men and women for holding *coca* leaves). It is important that all these colours be strong and clear; any faded colours are rejected as being useless for this ceremony. The responsibility for making the *chicha*, the *sarsillu* and the other items lies with the women of the household, but the men often help make some of the items, especially the *sarsillu* called *wantilurita* and *sombreros* (which will be worn by the male animals) and also the *wistalla*. Neighbours may lend a hand with these preparations.

The *wayñu* begins after nightfall as the family gathers in a specially cleared house where a ritual *misa* ('table', which is described in further detail below) has been laid out. The vigil begins with the burning of incense. As the participants chew *coca* and drink *chicha* and cane alcohol, the singing of dirge-like *wayñu* songs begins. During the darkness of night, the father and mother of the household may go to visit a *juturi* or pastures which the herds are accustomed to use near other houses the family may own, if they are not too distant. The woman wafts incense over the pastures to prevent the *uywiri* from 'eating' her llamas. Offerings to the *juturi* may also include burnt *parina* (flamingo) feathers, made in the hope that the camelids will behave like *parina* birds, which have the habit of clustering in close, cohesive flocks. The feathers are burned so that the llamas will not wander, but will stay close together.

Early in the morning, the llamas and alpacas are released from the large corral where they have spent the night. As they leave, their owners sprinkle the ground with libations of *chicha* and pure cane alcohol. They are herded across the *uq"u* and onto *waña* pasture by a young couple (*tawaqu* and *wayna*, a young woman and a young man of marriageable age) who will later bring them back to the *kancha*. This will take place sometime after about midday, when the whole family congregates outside the *kancha*.

At first sight, the *kancha* (or the *churu*) seems like a simple rectangle casually built of drystone walling (*pirka*), which is often in a state of disrepair. However, it is designed to suit the needs of the *wayñu* ceremony, and it is furnished with some significant features (figure 4.1). Most *kancha* in Enquelga are constructed so that the entrance faces east, towards the rising sun, like most of the houses in the village. Some, however, have an entrance facing north, looking towards the mighty Kawaraya Mallku. Many are furnished with a small, more or less square stone which is placed in the approximate centre of the interior space of the *kancha*. They all have a large, flat stone placed outside, to the left of the entrance as one looks in, on which is laid the *misa*, or ritual table. Some corrals are provided with a niche (*t'uju*) in the middle of the rear wall, low down on the ground, directly opposite the entrance. This is the place where the dishes containing substances

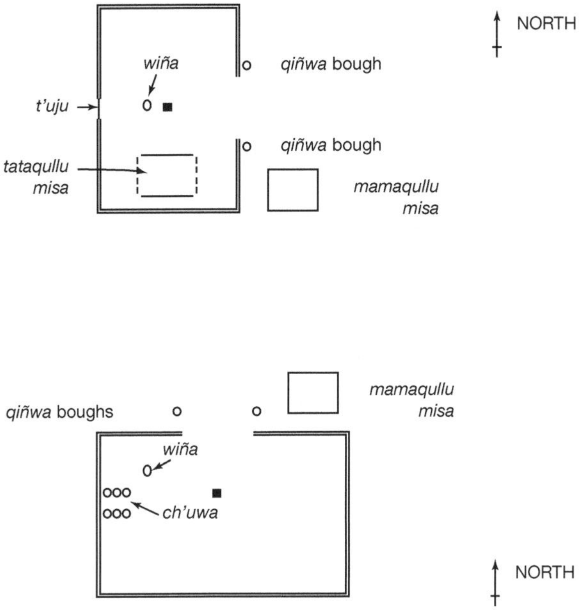

Key

Wiña Hollow for containing the sacrificial blood of a llama.
Qiñwa A tree which grows at high altitudes in the Andes.
T'uju A niche in the corral wall at ground level.
Mamaqullu misa The ritual table which is laid out on a large flat stone for the part of the ceremony when the female animals are symbolically dressed.
Tataqullu misa The ritual table which is laid out on the ground inside the corral for the part of the ceremony when the male animals are ritually dressed.
Ch'uwa A series of dishes which are filled with the so-called *ch'uwa* libations.

Figure 4.1 The design of the *kancha* (corral for camelids) in Isluga.

used in libations are placed in a row. There was no niche in a *kancha* that faced towards Kawaraya. When it was used for the ceremony, the dishes containing the *ch'uwa* libations were placed in a double row at the western end of the interior space. Consequently, this row of dishes did not face the entrance. During the marking ceremony, a hollow is dug in the ground between the central stone and the *t'uju* or row of *ch'uwa* dishes. This is called *wiña*, and will be filled with the sacrificial blood of the llama. For the purposes of the ceremony, branches of *qiñwa* tree are placed on both sides of the entrance, which is closed when the animals are inside by means of a rope held taught across the walling, over which a blanket is draped. *Qiñwa* is a high-altitude tree which grows in sheltered places, including in the valley of Q'asiri on Laram Qawani; its sturdy wood is used to roof the houses of Enquelga.

The focus of the elaborate *wayñu* ceremony turns to the interior space of the *kancha* or *churu*, where more songs are sung, libations are made and the symbolic dressing of the animals takes place, to ensure that lineages of camelids multiply. The *qiñwa* tree, besides providing a genealogical metaphor, is also a symbol of the source of life and of longevity. As a strong tree which is resistant to the cold, it was probably the symbol of Cuzco in the fifteenth century, when the city formed the centre ('the navel') of the Inka Empire (Sherbondy 1988: 112).[3]

After the llamas (and alpacas, if the family has any) have been driven into the *kancha*, the mother of the household enters bearing a tin or shell containing glowing embers of incense. She moves in an anticlockwise direction round the central stone and wafts the fragrant smoke towards each of the four corners, before placing the embers on the central stone. Then she and her husband kneel at the entrance of the *kancha*, looking out to Kawaraya Mallku, and they make libations of *chicha*, alcohol and *coca* leaves to the spirits of the hills, both male and female, from whom they ask permission to continue with the ceremony.

The sacrifice

Before the marking of the animals begins, a llama over which libations are made (plate 4.1) is sacrificed and its blood is offered to the Wirjin Tayka and to the *urwiri*, both *mallku* and *t'alla*. The father of the household in the meantime grinds twelve white and twelve black or grey maize kernels into flour, which he mixes with water in separate cups. He then places this *ch'uwa* liquid in front of the niche.[4] He digs the *wiña* and a relative or neighbour cuts the jugular of the sacrificial llama, to whose back *chimpu* has been tied, and over which maize flour, *chicha* and alcohol have been sprinkled. Droplets of blood of the dying llama are sprinkled over the ground of the *kancha* by the father, then the rest is allowed to fill the *wiña*, to which *coca* leaves are added. The father then spreads the black and white *ch'uwa* liquid over the ground, and all persons present lie down, head to the ground, in an act of obeisance to the *uywiri*. This act of obeisance usually is performed along the rear wall of the *kancha*.

The participants leave the *kancha* to gather round the ritual table outside, the men sitting behind with their backs to the wall, on a low row of stones. The women sit round about on the ground. Two young men butcher the llama inside the *kancha* (in the north-eastern corner), and some of the women cook the intestines with onions, a dish which is served with toasted maize. The intestines are always eaten first when a llama or alpaca is slaughtered, as I was told, 'for the Wirjin, for the *uywiri*'. Small pieces of intestines and heart are given to them, as an offering, before eating commences. This is also normal practice when an animal is slaughtered at other times of the year. During the marking ceremony, tiny pieces of heart, liver, meat and hide are also placed on a flat stone, and then pieces of llama fat and herbs stored in a small woven cloth are added to the mixture. These symbolic offerings are sprinkled with *coca* leaves, sugar and alcohol, and the stone is placed on the ritual table. The animal's fleece is placed in front of the *misa*, and the forelegs on top.

Plate 4.1 Husband and wife making libations of *chicha* (fermented maize drink) over a llama which will be sacrificed during their *wayñu* ceremony, Enquelga.

The ritual investiture of the herd animals

Following the sacrifice, the marking of the females and of the males takes place. During this time, the camelids receive the ritual name of *p"aqalli* ('flower'). The marking may be completed in one day, but often that of the males is done the following morning. In any case, the order is always the same: first the *tilantir* ('guides'), then the other females, next the *sarija* or *jañachu*, who are known as the 'fathers' of the herd, and finally the castrated males. All the human participants are given a length of *chimpu*, which they kiss, and then they form a long line and dance into the *kancha* in an anticlockwise direction round the central stone. This is the initial direction taken on ritual occasions for dances and processions, an observation also made by Rivière (1983: 49) in the neighbouring Bolivian community of Sabaya. The dancers turn round and move in a clockwise direction, before dancing on the spot to finish the dance. Four *tilantir* llamas are selected, and they are lined up to face west. (In the case of the *kancha* which faced north, the lines of animals were made to face north-west, looking towards the row of *ch'uwa* plates and *wiña*.) Libations are sprinkled over their backs, songs are sung, accompanied by the mandolin, then all present tie *chimpu* to the withers of the animals, while some women insert *tilantir sarsillu* by piercing the ears of the llamas with a needle. The father of the household ties a *wistalla* round the animals' necks. Next, four females who are not guide animals are selected, and the process is repeated, but *tama sarsillu* are inserted in their ears (*tama* means 'herd'). If the family owns alpacas, four females are selected and libations made, but there is a different song for the alpacas. In all these cases, the animals are female, and the *chimpu* is tied in a restricted area over the withers of the animal (*chimpu* means 'mark' or 'sign'). Then all the other female animals are 'dressed', as long as they are over one year old. In addition, animals born in the previous year's birth season have an owner's mark cut into one or both ears. Each member of the human household possesses his or her mark which is cut into the ears of his or her animals. The females are released from the *kancha* and they are herded towards the hill.

The human participants then leave the corral and congregate round the ritual table, which has, all this while, served as the *mamaqullu misa* (the ritual table for the female llamas), and which has been raised off the ground as it is spread over the large flat stone. Now the woven cloth and its contents are transferred to the inside of the *kancha*, where it is spread on the ground. Its next duty is to serve as the *tataqullu misa*, and all the male animals will be marked. The *sarija* and *jañachu* are lined up, in the same manner as described previously, libations are made over their backs and a new song is sung. They are dressed with *chimpu*, which is tied down the back of the neck along the spine to the tail, a *wistalla* and a double set of *sarsillu*, the *wantilurita* ('flags') and pompom-like *sombreros* ('hats'). Finally, all the remaining males are dressed and they, too, are released and allowed to join the rest of the herd.

The people dance round the central stone and, touching it with the right hand as they leave the corral in a line, they dance round the *misa* stone and most of them return to the village. A small group of men make a symbolic journey to the *likiña*. This Quechua word means 'valley', and the men role-play a herder and a caravan of llamas which will transport four miniature sacks of goods, representing:

1 tonelada azúcar (a ton of sugar);
1 tonelada maíz (a ton of maize);
1 tonelada mote (a ton of moté maize);
1 tonelada arroz (a ton of rice).

These goods are brought back from the 'valley' to the house where the *wayñu* is being observed. The leading men carry bells as though they were the leading animals of the caravan, and one of them carries two bottles of water. They also collect branches from shrubs which they carry on their backs. As they enter the village, the herder attempts to keep his 'llamas' in order, but they escape to chase young women as they endeavour to indulge in sexual buffoonery by mimicking the copulation of llamas. The miniature sacks are ceremonially handed over to the household, and there is much applause. The *wayñu* finishes with a communal meal to which all are invited. This is known by the Spanish terms *comilona* ('feast') or *boda* ('wedding'). A woven shawl (*awayu*) is laid on the floor of the house and four plates are placed on top, each containing soup and a quarter of the sacrificed llama's head. The roasted breast of the animal is laid between the dishes, and round the edge of the *awayu* are laid out joints of roast llama meat and a large circle of *p"iri* (cooked maize flour). Before the assembled people begin to eat, the father of the household carefully selects a small piece of each type of food which he places on a spoon. He leaves the house and offers this food to the Wirjin Tayka. The meal is followed by dancing in foursomes round a bonfire started off with the branches brought back by the llama caravan.

The transformation of pasture and water into fibre

Many different levels of interpretation of the complex *wayñu* ceremony are possible. First, the llamas are 'flowers' in ritual language – they are the *p"aqalli* of the herders. Don Marcos explained some of the meanings of the term by indicating the *muru p"aqalli* flower design woven by women in their shawls and blankets, and by referring to 'llamas, both male and female' (personal communication 25 February 1987). During the *ch'allta* of his llamas and alpacas on 28 February 1987, Don Ambrosio gestured expansively and commented: 'My *p"aqalli* are my bank'. Elsewhere in the Andes, the earpieces are called *tika*, which means 'flower' in Quechua (Tomoeda 1985: 290). The flowering ceremony is performed to help ensure that the lineages of camelids flourish and multiply, an imagery that is also conveyed through the gnarled and deeply symbolic *qiñwa* trees. Both herd animals and trees represent a stock, whether livestock or rootstock.

In Aymara culture the notion of flowering is laden with resonance; it has to do with both memory and inspiration.[5] The Aymara verb *p"anchayaña* can be used for the opening of a bud and the opening of the heart, as one is inspired to create something. Denise Arnold observes that in Qaqachaka, Bolivia, women say 'heart, make me flower'[6] when they seek inspiration in their weaving (Arnold 1997: 108). There is a gendered aspect to this notion of flowering because women, in particular, act as sources of memory and inspiration, as well as having the capacity to bring forth lineages. Similarly, female llamas and alpacas bear animal lineages. Herd animals, too, are encouraged to flower, just as their ears flower with blood when their ears are notched.[7] During the butchery of a sacrificed llama, the manner in which the blood drips from the heart is examined as an augury of good fortune. With the intestines, the heart is one of the first internal organs to be eaten by the human participants at the *wayñu* ceremony.

The flowering of the herd animals is enhanced by their ritual investiture with significant items (the ear and neckpieces and the dyed fleece). Nevertheless, these items will not have effect without the singing of certain songs. Mournful sounds are the essential complement of the strong colours (red, pink, orange, and the tonal gradations of green, orange and red). It should be remembered that Isluga people regard brightly coloured hills which are devoid of vegetation as ambiguous by nature, possessing dangerous qualities but also containing minerals. Anything which is glittery or highly charged with colour is held in awe: for example, rainbows, which have the power to enter the stomach of a woman, causing severe stomach pains, or to make people ill. Traditionally, women weavers incorporated very restricted narrow bands of strong, dyed colour in their textiles, which contrasted with wide expanses of natural colour. Nowadays, younger weavers are introducing more colour in their weavings, and they are extending the range of tones in a specific combination of colours called a *k'isa.*[8] Verónica Cereceda regards the Isluga *k'isa* as the 'mirror' of a rainbow, but it is not an exact copy (Cereceda 1987: 215). In my analysis, the tonal gradations of colour are the visual equivalent of music, hence the importance of *k'isa*, which are used in the neckpieces and two of the earpieces (the *tilantir sarsillu* and *wantilurita*) worn by the female guides and the males, respectively. In textiles, *k'isa* are incorporated into certain woven items, and weavers say that it is important that the colours should 'sing out'. They are therefore the visual counterpart to the making of music by men. In some parts of the Andes, the Aymara word *k'isa* means 'wrinkled fruit' or, in other words, fruit which is sweet and over-ripe, indicating that, on another level of contrasts, the almost over-bearing sweetness of the colours is opposite to the sombre and plaintive character of the sounds of the music. As the ceremony progresses, the owners begin to cry for their animals; the *wayñu* is a sorrowful event.

Colourful meteorological phenomena like the rainbow are perceived to be awesome, but also propitious because of their association with beneficial rains. A distrust for strong colour is paralleled by the Bororo in lowland South America; Lévi-Strauss (1970: 321) mentions their avoidance of colour and pattern in their clothing and pottery. Similarly, older people in Enquelga show a marked reluctance for too much polychromaticism in everyday contexts, and their preference for natural colours is in marked contrast to the textiles produced elsewhere in the Andes. However, the use of chromatic colour arrangements in the *wistalla*, which are tied round the llamas' necks, and in the *tilantir* and *wantilurita sarsillu*, takes place in a context designed to promote the fertility of the animals, and in a ceremony which pays homage to the spirits of the hills.

It is important that, with the abundance of strong colour during the ceremony, human beings should not be silent when confronted with the superiority of the *mallku* and *t'alla*. Perhaps herein lies the significance of ritual drunkenness during the marking ceremony. This is the one time of the year when women allow themselves to become drunk; during other ritual events they take alcoholic drink but they maintain their sobriety and act as intermediaries between their drunken menfolk, who may become aggressive towards each other in such a state. During the *wayñu*, the mother of the household is expected to imbibe an excess of alcohol: 'You have to drink until you fall down', Doña Luisa instructed me. Alcohol gives courage to call on the fiercest of the *uywiri* with enthusiasm, as related to Martínez by an Isluga man who visited a powerful *juturi* alone, under the darkness of night: 'With a little drink you . . . have more head . . . you call on the *juturi* with more enthusiasm, *Jutur Mallku! . . . Jutur T'alla!*' (Martínez 1976: 286–7). The need for lower-status humans to speak respectfully to higher-status *uywiris* also helps to explain why songs are sung all night and for a greater part of the time spent in the corral. There is one moment of respectful silence when obeisance is made to these spirits, when the human participants lie down, head to the ground, in an act of humility. This is followed by the offering of black maize *ch'uwa* to the *t'alla* ('lady' *uywiri*), and white to the *mallku* ('lord' *uywiri*).[9] Respect for the *uywiri* is also shown at the very beginning of the ceremony before the all-night vigil, and again at the entrance to the corral, when the married couple who are celebrating the *wayñu* kneel and ask the spirits for permission to proceed with the ceremonies.

During the daytime activities at the site of the *kancha*, people periodically return to the flat stone on which is laid the ritual table (plate 4.2). The layout of ritual items is an important aspect of the ceremony, and not only in Isluga.[10] There are three main objects which form the basis of the *misa* in Isluga: at least two plaited ropes and an *inkuña*, a small rectangular cloth containing *coca* leaves and sugar. Ropes are spun and plaited from camelid fleece by men. For daily use, they are made from naturally coloured fleeces, but for the ritual table they may include one or two dyed red or orange strands. The *inkuña*, on the other hand, is one of the typical products woven by women. When the ritual table is laid out, the *inkuña* is placed between the two ropes. These items are arranged on a carrying cloth (*awayu*), which is also the product of women's work. This cloth wraps the items of the ritual table when they are not in use, and for the greater part of the year the bundle of potent items (*q'ipi*) is stored in the dark corner of a house. Both women's and men's success in herding is reflected in these highly charged items that are the product of their own labour – spinning and weaving by women, and spinning and plaiting by men.

For the purposes of the *wayñu*, many more items are piled on the ritual table. When the *awayu* is spread over the large, flat stone outside the *kancha*, different

Plate 4.2 Laying out the items (*chimpu*, *wistalla* and *sarsillu*) with which the llamas will be ritually dressed during the *wayñu* ceremony. In the background is the *mamaqullu misa*, the ritual table laid on a large flat stone for the ritual dressing of the female animals. Note the stuffed wild cat, spiritual guardian of the camelids.

types of *waña* (dry) and *uq"u* (wet) pasture are placed in front. During the ceremony, the lower legs of the slaughtered llama are piled on top of the ropes, and the whole ritual table is regularly sprinkled with libations of *chicha*, alcohol and *coca* leaves. Other items included in the *misa* are a bell on a rope (formerly, these were used by the leading animals of a llama caravan) and mummified wild cats and *chullumpi* birds. A family may own one or two such mummified felines, stored inside the *q'ipi* bundle. They are placed on each side of the ritual table. *Titi* is the Aymara word for a wild cat, but during the *wayñu*, it is never referred to as such. Instead, it is called *awatiri* ('the herder'), for Isluga people regard this wild animal as being the supernatural herder of the llama and alpaca herds. *Chimpu* is tied round the neck and inserted into the eye sockets of the mummified *awatiri*. Similarly preserved *chullumpi* are also wrapped in lengths of *chimpu*, recalling the ropes with which the mythical bird was constrained in the narrative reported by Martínez in Chapter 3.

One family had suffered a great deal of bad luck and had lost many llamas during the year previous to its *wayñu* in 1987. As the mother of this family spread out the *awayu* and its contents on the *misa* stone, she picked up a rope and symbolically 'whipped' each of the *awatiri* while she berated them in Aymara for being 'lazy' and for not protecting the llamas. The *wayñu* is a time of heightened emotions, when herders attempt to put right all that is held to be wrong. Similarly, another woman vehemently berated one of her *tilantir* llamas and castigated it by kicking the animal for constantly wandering.

In contrast, Don Apolinario and Doña Luisa's family had had more success with their herding. I attended their *wayñu* on Saturday, 21 February 1987 and their *ch'allta* on Saturday, 1 February 1997. By 1997, their herd of alpacas produced two animals with a very light brown-coloured fleece (plate 4.3). Up until that year, people had been insistent that although alpacas sported the full colour range in other parts of the Andes, in Isluga alpaca herds do not contain animals with very light brown fleece. In contrast, it is commonly found in herds of llamas within Isluga. Various people in Isluga have told me that Sajama in Bolivia is an area where alpaca herds contain the full colour range in their fleece. The fortunate family with the two light brown alpacas is convinced of the effectiveness of the marking ceremony. During the *ch'allta* in February 1997 Don Apolinario's alpaca was referred to as his 'vicuña alpaca'. It was decorated profusely with dyed *chimpu* fleece down the length of its spine and also on its flanks. The herd, or *tama*, was seen as the generative source of this unusual vicuña-coloured alpaca, but the family also recognized the beneficence of their *uywiri* following their *wayñu* in 1996, to which they paid tribute during their subsequent *ch'allta* in 1997.

A theme that is made evident through the ritual of the marking ceremony is the symbolic transformation of pasture into fleece. Tomoeda (1985) discusses the use of *coca* leaves and *waylla* grass as symbols of the increase and abundance of wool among the herders of llamas and alpacas in the Apurímac region of southern Peru (in Isluga *waylla* is the name of a grass found alongside certain rivers). Tomoeda argues that, figuratively, wool derives from vegetable material (ibid.: 296). Thus, pasture is transformed into fleece, which is spun into yarn, which is

Plate 4.3 Light brown-coloured alpaca belonging to Don Apolinario Castro during the *ch'allta* ceremony on 1 February 1997.

then converted into clothing. In the Isluga marking ceremony, different types of pasture are given a prominent place in front of the ritual table, and the fleecy skin of the sacrificed llama is placed alongside the plants after the animal has been sacrificed. *Coca* leaves are also chewed and offered to the Wirjin Tayka by the human participants. Some families possess a special bag for holding *coca* leaves, normally stored inside the *q'ipi* bundle, and it is worn round the neck of the father of the family during the marking in the *kancha*. This bag, which is called *k'illpaña* in Isluga, is made of camelid skin with white, lustrous fleece. It resembles the *pukuchu* described by Flores Ochoa (1977: 215), a bag made from the skin of alpacas which die at a very young age, hence the beautiful sheen of the fleece. Every time the rows of llamas or alpacas representing the different categories of animals (*tilantir* guides, female llamas, female alpacas, llama and alpaca sires, castrated male llamas and alpacas) are selected and lined up to receive the songs and libations, animals with particularly long and lustrous fleece are chosen. Such animals are called *saxsali*, the significance of which is discussed further in the next chapter. The colour of the individual animals thus selected is important, too: they are also selected for their 'good tone'.

Thus the marking ceremony may be seen as the celebration of two different kinds of relationship: that between the *uywiri* (*mallku* and *t'alla*) on the one hand, and human beings and herd animals on the other. Human beings also celebrate the relationship between the married couple and their children, with that of their llamas and alpacas. The Wirjin Tayka and the *uywiri* constitute an animated landscape with regenerative powers and creative potency seen, for example, in the *juturi*. The design of the *kancha* and the use of the ritual table and of *sarsillu*, *chimpu* and *wistalla* help foster the fertility of the animals. All these items are made potent in ensuring increased fertility because of the magical situation created by the performing of the marking ceremonies.

The *wayñu* as a rite of passage

Coloured tassels and multihued ropes have been described by Browman (1974: 192) as 'property markers': that is, as identification marks. However, they do not serve to identify individual animals in a herd. By referring to their tassels, herders may distinguish herd animals, owned by human beings, from wild animals, which belong to the spirits of the hills and are not marked in such a fashion. Animals cared for by human owners are not allowed to go feral in Isluga. However, during the Isluga marking ceremony, human owners cut property marks in the ears of their animals, as described above. Paradoxically, they do not use these marks to identify their animals because they recognize them as individuals. They never check the property mark cut into the animal's ear.

Initially, I was puzzled as to why herders should bother to cut an owner's mark in the ears of llamas and alpacas since they are apparently redundant. Gómez Parra reported that during the marking ceremony in the community of Toconce, in the Upper Loa area of Chile, the male herder cut out small pieces of the animals' ears and kept them in a small woven bag (Gómez Parra 1975: 349). An earlier account given by Boman of the marking ceremony in Susques, north-west Argentina, also shows that the herder placed the cut pieces of ear in a *ch'uspa* (a small woven bag) which contained *coca* leaves. Later, the pieces of ear were offered to the Pachamama, and they were placed in a stone altar called *kiuri* (Boman 1908: 494–5). During the ceremony observed by Spahni in the community of Santiago de Río Grande, in the Chilean Atacama, marks were cut into the ears of llamas, goats, sheep and donkeys (in that order). The ear fragments were placed in three small woven bags, the first made from vicuña fleece, one of which held the pieces of llama ear, and the other two made from llama fleece, which held the pieces of sheep and goat ear, and the other the pieces of donkey ear. The following day, there was another ritual inside the corral where the marking had taken place, in which the herders formed three small 'enclosures' using pieces of yarn, one of sheep wool, and two of llama hair. Inside the space enclosed by the piece of sheep's yarn they placed the fragments of llama ears, and inside the spaces enclosed by the llama yarn they put the fragments of sheep and goat ears, and the fragments of donkey ears (Spahni 1962: 32). The precise and elaborate nature of these practices seem to contradict Nachtigall's unsupported assertion that the cutting of earmarks, and hence the name of *señalada* or *marca* for the ceremony, derives from the Spanish habit of cutting marks in the ears of sheep which, in post-Hispanic times, was extended to llamas (Nachtigall 1966: 195). He undermined his own argument by commenting that camelids are distinguished by their colouring and do not require supplementary means of identification.

In the Andean marking ceremony, the cutting of notches in the ears of the herd animals is not the only form of marking involved. The term *chimpu* also means 'mark', and the tying of dyed fleece to the backs of the animals is another form of marking, albeit less permanent than making cuts in the ears of llamas and alpacas. If pieces of *chimpu* cling to bushes or fall to the ground in the course of the year, herders say that they are sprinkled (*p"awaña*) on the Wirjin Tayka. There is, in any case, a profusion of synonyms listed in Aymara dictionaries for the Spanish *marcar* or *señalar* that may be used for the marking ceremony: *ch'ikullaña, chimpt'aña, p'ikiña, p'ukucha* (Büttner and Condori Cruz 1984; Ayala Loayza 1988).

It is true that the use of earmarking is a very widespread phenomenon worldwide. In reference to the herding of reindeer, Tim Ingold suggests that social relations between households may be given formal expression through the use of a socially accepted code of earmarks or brands, which serve to demonstrate which households own which reindeer. He maintains that the existence of a close bond of attachment between a herd animal and its owner renders the need for marking superfluous, while the absence of such close bonding renders it necessary (Ingold 1980: 114). The two ethnographic examples he gives are the Tungus and the Chukchi. The former do not use any property mark and they recognize their reindeer as named individuals. He contrasts them with the Chukchi, who never count their reindeer and who are not able to identify their animals by their looks (ibid.: 114). Earmarking provides evidence of ownership, he comments, but it in no way guarantees the good behaviour of the animals. For Ingold the earmarking of reindeer contained in round-up fences is symptomatic of a ranching economy, and it characterizes the breakdown of pastoralism (ibid.: 122).

Why should Andean herders, then, cut earmarks in the ears of their animals, at a time of great ceremony? As I indicated in Chapter 3, Isluga herders recognize their camelids as individuals, and their animals learn to obey two verbal commands. They train their herds to use pasture grounds, and they do not contain them in round-up fences. From scattered references in the literature dating from the colonial period, and from surviving pre-Hispanic artefacts, it seems that the Andean marking ceremony is of some antiquity. Ecclesiastical investigations into the so-called 'religious error' of Andean peoples in the Archdiocese of Lima (the Extirpation of Idolatry) uncovered religious observances dedicated to cults that included rituals for enhancing the well-being of herd animals. Kenneth Mills examined documentation about Juan Chapa and his consort María Ticlla in the province of Yauyos (Mills 1997: 65–6). In his evidence, Juan Chapa maintained that he cut the tops of the ears of the young llamas. Then he cooked the pieces in a fire, to which he added *coca* leaves and coloured maize. He offered the mixture to the mountain peak called Pata Caca. Both María Ticlla and Juan Chapa performed this ritual twice a year, despite the punishment meted out on them by the ecclesiastical authorities.

In pre-Hispanic times, the *napa* was an elaborately dressed white llama with golden earrings which preceded the Inka rulers (Sarmiento 1942 [1572]: 40). The Huarochirí oral narratives collected in Quechua at the end of the sixteenth or the beginning of the seventeenth centuries tell of the decoration of llamas with bells and earrings for the deity Pariacaca (Taylor 1987: 383). There are also iconographic representations of camelids with pierced ears. Rydén (1936: fig. 117a) discusses a Candelaria-style modelled and painted vessel from the Department of La Candelaria in northern Argentina. The zoomorphic form represents a pair of copulating camelids, and of interest is the tassel drawn by means of incised lines hanging beneath the right ear of the male camelid. However, the ears do not have any notches cut into them, and the vessel may have been intended for a camelid mating ceremony, rather than a marking. Classic Tiwanaku 4 pottery vessels with modelled representations of camelid heads often show a rope encircling the head, passing through the base of one ear. Even more remote from the present, and far afield geographically, some Moche ceramics from the north coast of Peru depict modelled camelids with patterns of cut marks round the ears (Donnan 1978). The link with present-day practice seems tenuous.

Nevertheless, I consider that making the ears of domesticated animals bleed was, and is, an indigenous ritual practice, a rite of passage for the llamas and alpacas at the time of their initiation into sexual maturity. As such, it would have paralleled the ear-piercing ceremonies and the ritual investiture of the breechcloth for noble Inka boys, which formed part of the Qapaq Raymi celebrations at the time of the winter solstice. This ceremony was known as the *warachikuq*, but Christian evangelization has eradicated the Andean practice of making boys' ears bleed in such ceremonial rites of passage. However, such practices for camelids have continued under the guise of European-style earmarking. A point of interest in the Isluga marking ceremonies I observed is a gender division in the tasks performed: men cut the ears of the llamas and alpacas, while women use a needle to pierce and stitch the earpieces in place.[11] Both actions produce blood. The Isluga blood-letting ceremonies are a celebration of the social bonding which is maintained by the herders with their animals, and which is transmitted from one generation of animals to the next. They do not indicate a breakdown of such relations, rather a celebration of the social identity of the domesticated animals, especially those born during the previous birth season, whose ears are made to bleed for the first time. The parallel activities conducted by women and men in the ritual investiture of female and male herd animals with gender-specific forms of 'dress' is a particularly salient aspect of the marking ceremony.

The celebration of human and animal lineages is an important aspect of the marking ceremony. It is celebrated by a married couple who respectfully ask *Mallku* and *T'alla* for permission to proceed with the event. In the interior of the *kancha*, rites are performed to ensure that the lineages of camelids multiply. But the sponsors of these ceremonies are a couple (*chachawarmi*) who themselves are progenitors of a lineage. Human and animal procreation would seem to parallel one another, and fertility, whether of human beings or of animals, is dependent on luck (*suerte*).[12] One Isluga woman, who had suffered a series of miscarriages over five years, explained: 'It's *suerte*. The llamas multiply with the *marca*. Like women, there are some families with many, and some with few.' In this respect, luck may be a limiting factor controlling human existence or the multiplication of one's animals, and the *uywiri* are considered to have power to grant or withhold *suerte*.

The focus is on animal fertility during the marking ceremony, but the contrast between human and animal lineages constitutes the cultural matrix in which the events unfold. When I attended the *wayñu* and *ch'allta* ceremonies during February 1987, the impression I formed was that the events centred on activities intended to promote the lineages of herd animals. Since then, Isluga people have inserted a ritual intended to remember the souls of human beings into the marking ceremony. While the llama was being sacrificed and its internal organs cooked for human consumption in the *ch'allta* I attended in 1997, the hosts took an enamel cup filled with alcohol and set it alight with a match. Don Apolinario, Doña Luisa, their eldest son Alex, and his new bride Vilma took it in turns to burn *coca* leaves in memory of their ancestors. They, and other human participants present, then recited in low voices a Christian prayer for the souls of the dead, at the end of which they crossed themselves. The ritual of burning *coca* leaves (a *coca kintu*) is one that used to be restricted to other occasions, such as the observance of All Souls Day.

With this emphasis on the importance of lineage, the meanings of two significant terms can be explained in a greater historical depth. The name of the divinity on to whom the cult of Santo Tomás was grafted, Tunupa, derives from the root *tunu*, meaning 'ancestor, main root', or 'stock' of a tree (*cepa* in Spanish) (Ayala Loayza 1988: 178). It is Santo Tomás who is remembered as the progenitor of Isluga people in present times. The word *wiña*, the hollow which is filled with red blood and green *coca* leaves in the *kancha*, is another term that shares a linguistic root with concepts of lineage. *Wiñaya* is the Aymara and *wiñay* the Quechua word for 'always, eternal, for ever'. This word appears in similar contexts in both languages in early seventeenth-century dictionaries:

> *Viñay*. A century which is one generation from father to son, grandson, great-grandson etc or the half of a lifetime, to middle age.
> *Viñamaci*. Those of the same age, or those who are brought up together . . .
> *Viñay*. The generations and descent.
> (González Holguín 1952 [1608]: 352)

> *Viñaya*; Always.
> *Viñaya viñayapa*. Saecula saeculorum.
> *Viñaya*; Placed after proper nouns, it means of the same age.
> (Bertonio 1984 [1612] Bk II: 388)

Bertonio's translation of the Aymara *wiñaya wiñayapa* into the Latin 'generations of generations' is interesting, since *saeculum* means 'the period of one human generation' – that is, about thirty-three and one-third years – but the plural *saecula* means 'successive generations [of people or of animals]' (Simpson 1973: 529). Therefore, the filling of the *wiña* with blood may be regarded as a means of enhancing vitality, which is regenerated in camelids through the ritual performance of the *wayñu* ceremony. The procreation of new life is important, especially because the lifespan of camelids is much shorter than that of human beings. Llamas and alpacas are said to be able to live for up to twenty years, but many are slaughtered at the age of about seven. Hence herders have to make sure that there are sufficient adult llamas in the herd to guide the younger members and to ensure the maintenance of good discipline.

Richard Tapper (1988) has argued that animals provide an ideal model for human beings to conceptualize their own humanity. In his argument, the herds are replicas of society, yet they are matrilineal and uxorilocal (the female is the stable core of the herd), whereas human beings in pastoralist societies organize themselves in patrilineal, virilocal terms (the men are the stable core of the community). He states that human beings and animals are identified on one level but differentiated on another. He regards what he sees as the prevalence among pastoralists as 'patriliny and patriarchy' as the necessary way for human society to organize itself so as to provide 'an otherwise absent distinction of humanity from animality' (ibid.: 55).[13]

This generalization does not really apply to Isluga, nor to the 'carnivorous pastoralism' defined by Ingold as the type of reindeer herding practised in arctic and subarctic Eurasia. Ingold lists a series of traits, including 'diverging devolution of property' and 'bilateral systems of kinship reckoning' as characteristic of carnivorous pastoralism, in contrast with 'milch pastoralism' in East Africa, which he associates with 'unilineal devolution of property' and 'agnatic systems of kinship reckoning' (Ingold 1980: 25). It is these latter traits that have contributed to traditional views of pastoral societies as being both patrilineal and patriarchal. Despite the fact that there may be a tendency towards virilocal residence after marriage in Isluga, this is not always the case, and kinship reckoning is cognatic. Parents devolve property, especially animate property in the form of animals, to both female and male children. Hence, pastoralism in the Andes seems to have something in common with carnivorous pastoralism of subarctic Eurasia. Where it differs from Ingold's notion of carnivorous pastoralism is in the relationship between herding leaders and their assistants (ibid.: 165–9). Assistantship is not a common practice in Isluga.

According to León Campbell, throughout the colonial era in all of Latin America, the male head of the family received reinforcement from both Church and State in determining the status of his offspring (Campbell 1985: 164). The present-day system in Isluga may be skewed towards patriliny, but Isluga society is not entirely characterized as 'patrilineal and patriarchic'. In the past, the Inka system of kinship reckoning interpreted by Lounsbury suggests a system of bilateral filiation, which was matrilineal and patrilineal at the same time, because men considered themselves to be descended in the agnatic line and women in the uterine line (Lounsbury 1978: 1,000).[14] In her detailed study of *ayllu* Qaqachaka in Bolivia, Denise Arnold has demonstrated the complexities that are 'generated by *matrilineal* descent-constructs and descent-ordered rules' in matrilineal forms of social, political and kinship organization in a contemporary patrilineal setting (Arnold 1988: 3, original emphasis). Commenting that anthropologists have frequently assigned unequivocal political control to men in the Andes, she explores the contexts in which women's political authority has important symbolic functions that may be greater than that of their husbands.[15] In any case, the traditional annual elections to a one-year term of office in Isluga, as in other parts of the Andes, conferred prestige on the office bearers without necessarily concentrating control in the hands of a few leaders. The dispersed, pastoral way of life in Isluga runs counter to the consolidation of power by a central authority. Yet, within the household, the father's role as patriarch – traditionally thought to be characteristic of pastoral societies – cannot be taken for granted. Arnold investigates the matrilineal and patrilineal systems of descent that occur within Qaqachaka society in order to demonstrate that representations of kinship are 'located not just in the kinship domain, but in the rituals, practices and metaphors of everyday life, and in both reproductive and productive spheres of activity' (ibid.: 38). My appraisal of the marking ceremony in Isluga likewise emphasizes the processes of gendering that occur among lineages of both human and herd animal participants.

My rejection of the universality of Tapper's argument concerning pastoralist societies should not be taken to imply that Isluga people do not differentiate between humanity and animality. The mythical origins of human beings and animals are associated with different features of the landscape (herd animals with *juturi* and human beings with *t"aqsu*). In addition, the Aymara language itself makes clear distinctions between human and animal.

The taming of camelids is not something which happened once and for all several millennia ago. It is a process by which herded animals are brought into a particular form of animal life. Moreover, it is a process that must be constantly renewed. During the *wayñu* ceremony in Isluga, the coming into sexual maturity of the young animals is recognized. This ceremony focuses on a perceived need to promote the regenerative, procreative vitality among camelids and sheep. It is undoubtedly an elaborate undertaking, although it is but one event in the process involving the maintenance of the social bonds of taming. It is an important part of the bonding between human beings and herded animals, the process of taming experienced by all camelids and sheep in Isluga. Llamas and alpacas are trained to behave in certain ways, and their human owners have certain expectations of them.

As Chapter 3 made clear, these social relations do not take place in a meaningless environment. To a great extent, people understand their landscape through the herding of their camelids. Llamas and alpacas are closely associated with water, wet places and, metaphorically, with birds. In fact, bird feathers are seen as the counterpart of locks of fleece. However, llamas and alpacas receive the ritual name of 'flower' in the *wayñu* ceremony, thus symbolizing their potential for procreative fertility.

Such is the complexity of the marking ceremony as observed today in Isluga, it seems to have responded to Christian imagery introduced to the Andes in the eighteenth century, as well as incorporating pre-Hispanic elements. Eighteenth-century paintings of the Virgin Mary as the Divine Shepherdess depicted Mary seated and surrounded by a flock of sheep, feeding her sheep with roses (Carcelén 1992: 469; Dransart 1998: 138–9). In this iconography, the sheep stand for

Christians, and the roses for the spiritual nourishment of the rosary. In the Isluga *wayñu*, herd animals are encouraged 'to flower' and, when they are covered with red and pink *chimpu*, they might be the visual counterparts of sheep with roses in their mouths. However, in my understanding of the relationship between human herder and herd animals in Isluga, the latter cannot 'stand for' human beings, given the parallel character of human and animal lineages. Isluga people offer spiritual nourishment to their herd animals *and* to human beings. Apparently similar semiotic signs produced between cultures, such as the imagery of flowering, can distract from the different social and moral messages conveyed by the outwardly similar visual codes (Klein 1995: 263; Dransart 2002).

On another level, in the Isluga *wayñu*, water and pasture are metaphorically and literally transformed into fleece. I will now turn my attention to the fleece itself, how it is perceived in Isluga and how it is transformed into yarn and fabric. Later on, in Chapter 6, I will consider some pre-Hispanic evidence that helps shed light on how people in societies of the past perceived this raw material.

5 The transformation of fleece into yarn

Chapters 3 and 4 considered the incorporation of herd animals into the cultural life of Isluga. This process of incorporation makes possible the production of fleece, which plays such an important part in the material culture of the community. Isluga people mediate relations between themselves and their environment through the organizational, ideological and technological aspects of their culture. This is clearly demonstrated by the methods used by the people of Isluga in identifying their llamas and alpacas outlined in Chapter 3. Such classification systems are not only indicative of the social relationships perceived by the people between themselves and the animals (camelids and birds) of their environment, but they are also a good example of what Van Kessel calls Aymara technology (1980: 323; 1988).

Pierre Lemonnier has written that technical systems involve 'materials, sequences of action, "tools" (including the human body), and a particular knowledge' (1986: 154). He describes the latter element as comprising knowing how to do things, manual skills, procedures, and 'a set of cultural representations of "reality"' that influence technical behaviour (ibid.: 154–5). To rephrase Lemonnier, it might be said that classifications of camelids and birds in Isluga form part of the classification of fleece as a material for technical activity.

The analysis of the marking ceremony highlighted the transformation of pasture and water into fleece. For Andean herders, the increase in abundance of camelid fleece is an important concern. Animals with long, good-quality hair are known as *saxsali*, and such animals receive particular attention during the marking ceremony in the *kancha*. Raw fleece is subsequently converted into yarn, which in turn is converted into the clothing and blankets with which people are wrapped. Thus the symbolism of a *saxsali* animal is a powerful means of expressing the theme of abundance. It is the purpose of the present chapter to examine the properties and characteristics of camelid fleece, first using methods devised in Western technology for the study of animal fibres, therefore using 'outsider' terminology, and second using the 'insider' classifications employed by the people of Isluga.

Table 5.1 presents a description of the coat of each of the South American camelids. All camelids are similar in that there is variation within the fleece over the body: the body, neck and upper legs are covered with the longest fibres, which consist of a mixture of finer and coarser hairs. In all four animals, the fibres are

Table 5.1 General description of the fleece of the South American camelids.

Guanaco (*Lama guanicoe*)

Gilmore (1950: 449) describes guanaco as being rich tawny brown or faded brown dorsally and laterally; white ventrally and medially (including neck); the head is smoky grey-brown, with whitish orbital rings, chin, lips and ear edges. He states that albinism occurs, but melanism is absent. The fleece consists of fine wool fibres with a length of 3 to 45 mm and a diameter of 15 to 16μ (microns) over the body, neck and chest, and a varying amount of coarser overhair which measures from 100 to 140 mm in length and 25 to 27μ in diameter (Duga 1985: 130). These wool and hair fibres are coarser over the belly – Duga gives measurements of 18 to 19μ for the former and for the latter over this part of the body (ibid.: 130). Franklin (1982: 469) states that 95 per cent of the fleece has a diameter of 16.5μ. Over head and legs, the fibres are very short, and the axillae and groin are almost bare. The overhair fibres often appear to form a ridge down the back of the neck, and winter specimens from Patagonia and Navarino Island have a profuse covering of rich brown overhairs up to 15 cm in length which nearly hides the wool underfibres. Gilmore (1950: 449) suggests that the fleece is probably renewed in a complete annual moult during the summer. Juvenile animals are prized for their pelts, the fleece of which consists of fine woolly fibres, which do not display the dimorphism observed in the coats of adults. The colour of guanaco fleece has been said to vary according to the particular terrain of its habitat through mimetic adaptation (Guerrero 1986: 13).

Vicuña (*Lama vicugna*)

Vicuña are a rich cinnamon brown colour dorsally and laterally and on the neck and head; they are white ventrally and medially and on the chin, under the eyes, on the flanks and the tip of the tail; there is a small circle of black, stiff hairs around the eyes (Gilmore 1950: 451). The northern subspecies (often called 'Peruvian') is characterised by a large bib of long white hairs which cover the chest region (Franklin 1982: fig. 4). This bib consists of fibres some 10 to 30 cm in length, and Gilmore (1950: 451) states that it is longer in both sexes during winter. The larger and somewhat lighter-coloured southern subspecies (known as Atacaman or Argentinian) lacks this conspicuous white tuft. Vicuña fleece consists of short, curly underwool fibres, 30 to 40 mm long, on the body and neck, interspersed with hairy fibres. Franklin (1982: 474) gives an average of 11 to 14μ diameter for the fleece. Legs and face are covered with short, thick fibres, which gradually decrease in length from the body fleece (Gilmore states that this is in contrast with llamas and alpacas, where there is an abrupt change in length); the axillae and groin are almost bare (Gilmore 1950: 451). A single annual moult of the entire coat takes place in summer.

Llama (*Lama glama*)

Llamas are uniform or multi-coloured white, brown, grey or black, but they may also display colouring characteristic of guanacos. Piebald animals are more common than is the case with alpacas. The fleece is heavy and long on the body, shorter over the neck, and very short on the lower legs and head. Unlike the wild camelids, there is an abrupt change in length between the head or leg fibres and the body fibres (Gilmore 1950: 435). Axillae and groin are almost bare. The coat consists of finer wool fibres interspersed with coarser hairy fibres, which are easily distinguished. These hairy fibres tend to be longer and form a ridge down the back of the neck (especially in younger animals or in the 'bare' variety of llama). Gilmore says the fleece attains a length of 100 to 200 mm if sheared annually, and 500 mm if not sheared; the fleece grows continually and sheds like the mane and tail of a horse, although some individuals shed more during summertime (ibid.: 435). The fleece of juveniles of up to four to six months is characteristically fine and does not display the dimorphism of fine and hairy fibres seen in adult coats. Two breeds or varieties of llama have been detected: 'woolly' (*t'awrani* in Aymara and *ch'aku* in Quechua) and 'bare' (*q'ara* in both Aymara and Quechua). The former is recognised by the fuller fleece covering its neck and on its forehead. The latter has much shorter fleece round its face and down the front of the neck.

Alpaca (*Lama pacos*)
Alpacas are white, brown, vicuña-coloured, grey or black. They may also be multi-coloured, but they tend to be so less often than llamas. The fleece is long and heavy over the body, neck and top of the head, chest, belly and upper legs. There is an abrupt change to short fleece over the face and lower legs, and the axillae and groin are almost bare. According to Gilmore (1950: 442), the fleece grows to a length of 400 to 500 mm, but if it is left to grow over a period of many years it may reach the ground, up to a length of 750 mm. There is no annual moult and the fibres grow continuously, shedding like the hair of a horse's mane or tail. Alpaca fleece is somewhat greasier than llama (but not as much as sheep's wool), and it lacks the obvious dimorphism between wool underfibres and hair overfibres observed in llama fleeces. However, the fibres covering the belly are coarser than those over the rest of the body. Two breeds or varieties of alpaca are reported in the literature, named after the Quechua terms *suri* and *wakaya*. *Suri* animals are only known in parts of southern Peru. They have a greater density of fibres per sq. mm (140 to 170), whereas the *wakaya* have a density of 90 to 100 fibres per sq. mm (Gilmore 1950: 442). *Suri* animals have a coat of long, wavy fibres, which are said to cover the body in much the same manner as those of Lincoln sheep, but *wakaya* animals have a shorter, curlier coat, said to resemble that of Corriedale sheep (Novoa and Wheeler 1984: 126). *Wakaya* animals tend to have a heavier live body weight than *suri* animals at the same stage of development and, correspondingly, their fleece yields are slightly higher (Avila Felipe 1979: 2; Blanco Velásquez 1980: 1–2; contra Gilmore 1950).

---

coarsest where they cover the belly. In contrast, the axillae and groin are almost bare. Very short hairs short cover the face and lower legs. Variability in fibre diameter in the body fleece permits a range of products to be made using different parts of the fleece, as described in further detail below. However, with the domestication of llamas and alpacas, important differences have arisen between these two animals that distinguish them from vicuña and guanaco. Perhaps the greater range in colour variation displayed by llamas and alpacas is the most outwardly striking feature, but no less an important difference are the changes in staple structure occurring in the domesticated animals. Ryder (1978: 8) defines staple as 'variation between fibres [in the same fleece]'. It should be noted also that in the wild camelids the transition from the longer staples covering the neck and body to those covering the face and lower legs is gradual, whereas the corresponding transition in domesticated camelids is abrupt.

South American camelids possess a fleece that is highly spinnable. Its generally fine diameter is an important characteristic in this respect, but no less important is the degree of crimp (waviness) evident in alpaca *wakaya* fleece and in the woolly fibres of vicuña, guanaco and llama. Camelid fibres also have the physical properties of durability, elasticity and resilience, and herders in Isluga are unanimous that camelid fleece is warmer than that of sheep wool. Ryder (ibid.: 13) points out that animal fibres are composed of the substance belonging to the same group of proteins that forms hair, wool, horns and hooves. This substance is keratin, and he claims that this is one of the most fascinating products of the animal kingdom. The following section of this chapter discusses how the people of Isluga perceive the fleece of their llamas and alpacas.

The classification of fleece in Isluga

The people of Isluga have a rich vocabulary for describing different aspects and qualities of fleece. Generally speaking, wool (which in Isluga includes sheep and camelid) is called *t'awra* and fleece is known as *jawi*, but this second term also means 'ointment'. Thus, to prepare fleeces is *jawichaña*, and to anoint with fat *lik'imp jawintaña*. Clarified llama fat has medicinal uses, in particular to cure skin conditions, but it is also an important ritual offering to the ritual table during the *wayñu*. For this purpose, small pieces of llama fat (*lik'i*) are kept in an *inkuña*, a small cloth woven from camelid fibre yarns, along with small pieces of fragrant plants from the valleys. Fleece, like fat, is soft and it can be teased out to make it spread further. According to Martínez (1980: 88, 100), the word for 'fleece' (*jawi*) and the word for 'river' (*jawira*) share the same root, and he claims a semantic connection between the two. He describes fleece as being made up of a continuum of discrete units, as it consists of a mass of individual filament-like fibres. On the other hand, a river is characterized by continuous movement, but Martínez refers to its celestial counterpart, the milky way, which is also called *jawira* by the people of Isluga; it consists of a mass of granular units, or stars. The two words may be homonyms, but Zuidema and Urton (1976: 67) also explore the significance of *jawi* with the double meaning of 'fleece' and 'wet' or 'water', as listed in the seventeenth-century dictionary of Bertonio, in the context of a myth concerning a celestial llama called Yacana.[1] They relate the Quechua myth, recorded in Quechua at the end of the sixteenth or beginning of the seventeenth century, to a black constellation in the milky way, which is considered by Quechua speakers today to be a llama suckling her calf. This being the case, the celestial river has a diaphanous character, a quality shared by the *ch'uwa* liquid of ground maize flour suspended in water which is offered to the *uywiri*, both *mallku* and *t'alla*, during the Isluga *wayñu* ceremony. Liquid libations are made over the llamas and alpacas during this ceremony, as previously described, in the constantly reiterated symbolic association between camelids and water, but grains of sugar and maize flour are also sprinkled over the backs of the animals, tiny granular units nestling in the fleece. There is here a whole realm of tactile qualities which acquire symbolic meaning in certain contexts, and which are designed to enhance the desired characteristics of camelid fleece.

While *llamp'u* is an adjective usually employed to describe softness in substances such as very finely ground maize or *quinua* flour (as a noun it is used as another word for llama fat), the adjective most often used to describe particularly soft camelid fleece is *quña*.

Length is another highly desirable characteristic in fleece, thereby increasing fibre yields if not necessarily of itself making the fleece more spinnable. Long fleece is called *jach'a t'awra*. *Saxsali*, of course, is the word applied to prized animals with abundant and good-quality fibres.

Spinners identify llama and alpaca fleece on the basis of the presence or absence of guard hairs. *T'awra uywini* is wool containing guard fibres and, according to this definition, it refers to llama fleece. Alpaca fleece (*allpach t'awra*), on the other hand, is identified by the absence of hairy or guard fibres, even though it may be quite coarse in the belly and chest regions of the animal. Fleece grown by hybrid crosses is often simply classified as llama or alpaca, depending on its appearance. In practice, much of the fleece handled by the people of Isluga has been shorn from their own animals, which they know as individuals and which they may well remember long after the animal's death. Doña Natividad was reminded of one of her llamas that had drowned in deep water near the thermal spring several years previously when she looked at the blanket spun and woven from its fleece. This blanket kept her warm at night, but she still mourned the tragic loss of her animal.

Fleece in good condition and which is considered to have a particularly soft handle may provoke the exclamation: 'how beautiful that wool is!'[2] but fleece may also possess less desirable qualities. Very coarse fleece is described as *lank"ut t'awra* ('coarse wool'), while *t'akara*, meaning 'rough' (also implying 'tangled'), and *ch'urku* ('hard'), are adjectives which may be applied.

In any case, the staples of both fine and coarse fleeces have weathered, matted tips known as *p"ichu*, a term also used for areas of matted fleece: for example, from under the belly. *P"ichu* in its widest sense is wool that cannot be spun, because in order to spin successfully, the fibres should lie more or less parallel to each other. However, such wool is not entirely unusable since the people of Isluga exchange their unwanted *p"ichu* for groceries supplied by Bolivian traders – the rate of exchange is weight for weight, for example, a kilo of *p"ichu* is exchanged for a kilo of rice. Eventually, *p"ichu* is used for stuffing mattresses. It is interesting to note that this word also refers to a wad of *coca* leaves chewed in the mouth. Just as in the case of *p"ichu* fleece, where the individual fibres are converted into a homogeneous wad, the individual *coca* leaves are mashed up together in the mouth and broken down into a uniform whole.

The people of Isluga also classify camelid fleece according to colour. I have already referred to *k'isa* (Chapter 4): that is, a gradation of colour arranged in a narrow sequence from light to dark, or *vice versa*. Some of the textiles in Isluga display *k'isa* of brightly dyed colours; however, this concept is also applied to natural colours. Women in the neighbouring territory of Sabaya, in Bolivia, use such arrangements of natural colours in their woven bags (*wayaqa*). Thus the colours of camelid fleece may be organized in an extensive *k'isa* from white to black, passing through the various tones of brown, as follows:

Janqu: white
Janqu q"usi: light fawn
Q"usi: fawn
Wari q"usi: light brown (like vicuña)
Paqu: vicuña colour
Ch'umpi: brown

Ch'ar ch'umpi: dark brown
Ch'ara: black

I also heard some less frequently used terms referring to brown fleece colour: *liriu q"usi*, which describes a very light brown tending towards pink; and *chupika*, for a resonant reddish-brown which often is seen in alpacas in Isluga. Both these colours are, of course, undyed. Whereas *k'ura* is the Aymara term for undyed colours, dyed colours are known as *p"ana*. However, when asked to describe the colour *liriu q"usi* in further detail, Doña Soria Mamani Challapa resorted to defining it as a light *p"ana* colour. In this case, an undyed natural colour is perceived as taking on a quality shared with dyed colours.

Tones of grey may also be arranged in a *k'isa*, from light to dark:

Janqu: white
Janqu uqi: very light grey
Chilka or chuchi: light grey, *azucarina* in spanish
Uqi: grey
Ch'ar uqi: dark grey
Ch'ara: black

According to one Enquelga woman, *chuchi* is a Bolivian word for the same colour known in Isluga as *chilka*, which is the name of a plant in the *precordillera*. In dictionaries, *chuchi* is listed as 'brown colour' (in camelid fleece) and also as 'badly dyed, discoloured' (Büttner and Condori Cruz 1984: 31; Ayala Loayza 1988: 82). Camelid fleece may also appear to be half grey, half brown; *ch'uñup"uti* is a dull colour, not particularly prized. It resembles the colour of cooked *ch'uñu* potatoes, which are freeze-dried in the Andes to preserve them, and this process turns them blackish-buff in colour. More favoured grey colours are *kupala*, a light colour named after *kupala* resin used for medicinal and ritual purposes; *q'illuwari*, a slightly yellowish grey colour, which like *chilka* is sometimes translated into the Spanish *azucarina*; and *chinchilla*, a dark grey like the fur colour of the chinchilla. Other variations employed for black are *ch'iyara* and *ch'ar aceituna*, from the Spanish for 'olive'. In Isluga there is a greater variety of grey colours in alpacas than in llamas, which tend to be a darker grey or *ch'uñup"uti* coloured. In all, Doña Soria recognizes eighteen different natural colours of fleece.

It can be seen that there is an extensive colour terminology, which may be called upon to describe the colour of a given fleece. However, colour perception varies from person to person and there may be disagreement between people in deciding on what colour to call a particular fleece. Much may depend on the intended use of yarn spun from such fleece, since the natural colours of llamas and alpacas have very specific roles in the traditional garments of Isluga. For example, the vicuña colours of *wari q"usi* and *paqu* are traditionally used for men's ponchos, but *janqu*, *ch'umpi* and *ch'ara* are used in bags called *wayaqa*. The same brown yarn may be called *paqu* if it is woven into a poncho, but *ch'umpi* if it is used in a *wayaqa*. Similarly, Doña Soria referred to Don Apolinario's 'vicuña alpaca' as being *q"usi* coloured, perhaps because her own herd lacks fawn coloured alpacas, and she has to use narrow stripes of llama yarn to complete the gradations of natural colours in an otherwise all-alpaca textile.

Shearing

Both camelids and sheep are shorn during November and early December, at the beginning of the rainy season when the temperature starts to become warmer. However, shearing is delayed until later if cold winds persist and the rain-bearing clouds are slow to arrive, as most herders are unwilling to expose newly shorn animals to cold night-time temperatures. It is normal practice to shear camelids every other year, but if an individual animal has a slow rate of fleece growth it is shorn at three-year intervals. Herders are reluctant to allow alpacas to remain unshorn for longer periods, as the fibres become twisted and tangled and the fleece is lost on shrubs and bushes. In the far north of Chile, in herding communities near Lake Chungará, camelid herds are larger in size and families may own many more alpacas. These families may not shear live llamas at all, and usually wait until the llama is slaughtered before using its fleece. One herder in that area said that llamas should be shorn for the first time at the age of five years and alpacas at the age of four. A variety of opinions was expressed on this matter, and in Enquelga the consensus seemed to indicate that the first shearing should be at the age of two years. This is in contrast to the unanimity reported by Bustinza Choque and Sapana Valdivia (1985) in their examination of the quality of alpaca fibre produced by peasant communities. They state that alpacas are shorn 'almost obligatorily' at the age of one year, reasoning that such a practice allows rapid fibre growth in subsequent years and, additionally, that external parasites are released from the bodies of the shorn animals, reducing fleece loss (Bustinza and Sapana 1985: 101).

In general, few llamas and alpacas are kept beyond the age of seven or eight years. Thus, an animal may only be shorn two or three times during its lifespan. Indeed, llamas may be shorn only once, or not at all, depending on the family's requirement for fleece and whether the family owns an adequate number of alpacas. This is in contrast to the practice currently employed by Research Stations and the Empresas Asociativas in Peru, where animals are shorn on an annual basis. Since fibre becomes gradually coarser after being shorn, as the animal grows older the fleece grown by animals owned in the peasant sector might well be finer than that produced by the large estates, even though the estates have access to larger and better-quality pasture lands. The report by Bustinza and Sapana (ibid.: 108) shows that alpaca fibres measured from Peruvian communities, and one Bolivian community, produced excellent results, the majority of the fibre diameters falling between 25 and 30μ.

At least two or three people are required to shear an alpaca or llama. A plaited rope is tied round the legs of the animal partially to immobilize it, and one person holds down the head by grasping the ears. The shearer proceeds to cut the fleece from the rear flank, working in a forward direction, pulling a handful of fleece with the left hand and cutting with the right. The cutting instrument used in many peasant communities is the lid of a tin can, one part of which is folded over so as not to cut the user's hand, and the outer arc or opposite edge is sharpened on a convenient stone. If another shearer participates, he or she will begin by cutting the fleece from the neck, working towards the other person. The shearers cut the fleece from one flank of the animal, which is then turned over to cut the other side. The fleece is removed leaving about 50 mm in length of unshorn fleece over all the body, in order to protect the llama or alpaca from the cold and rain. The cut tends to be irregular since it is not easy to shear the handfuls of fleece all at the same level with a tin lid. As a consequence, the fleece will have longer and shorter portions. The resulting difference in fibre length does not cause problems later on for the spinner. However, a more regular cut can be obtained with the use of shears, which improves the presentation of the fleece, and a few families in Enquelga have now acquired shears. Once shorn, the fleece is gathered together on top of a carrying cloth (*awayu*). Each half is twisted from one end to the other and then back on itself, and the ends are tied together. Thus a complete fleece is composed of two parts, which are stored with others in large sacks. A fleece is the personal property of the owner of the animal that grew it; the owner may use it, sell it for money or exchange it for other goods.

To remove the fleece after the animal has been slaughtered, the carcass is first butchered. Alpacas and llamas are slaughtered (with the eyes covered by a handkerchief) by making a slit through the ventral side of the neck, which cuts through the throat and severs the spinal cord. Some or all of the blood is sprinkled on the ground round about as an offering to the Wirjin Tayka, and the herders recite a litany of place names, listing the pasture grounds that the animal used to frequent. The animal is then rolled on its back and, first of all, it is skinned, starting with an incision from the sternum backwards to the anus. Other cuts are made through the hide from the sternum upwards towards the throat, from the sternum along the inside of the front legs, and also from the ventral cut along the inside of the rear legs. Cuts are made through the articulation of the joint between the carpals and the metacarpal (in the case of the forelegs) or metatarsal (in the case of the rear legs), then the lower bone is bent forward to break the joint and remove the lower part of all four limbs. Skinning proceeds by pulling the hide from the flesh underneath by prising it away from the flesh with the blade of a knife, then tugging. Where the hide is attached to muscles, the skin is stretched tightly and hit with a clenched fist. A helper pulls the proximal end of what remains of the leg in the opposite direction and the hide comes away cleanly and easily. There is no need to scrape the inside of camelid hides at a later stage. First one side is stripped, then the animal is turned over and the other side is skinned. The butchery then takes place, the carcass being cut into pieces as it lies in its own skin.[3] Afterwards, the meat is stored safely away from marauding cats and dogs. The hide is doubled over, with the fleece outside, and it is left to heat in the sun, being turned periodically (plate 5.1). The hide is complete at this point, and the neck is not cut off as reported by Miller (1979: 47). If the thick hide of llama skin is to be used to make sandals or lassos it will be removed at a later stage.

Plate 5.1 Laying out the hide of a sacrificed llama, Enquelga.

Plate 5.2 The late Doña Gabriela Mamani Challapa shearing the fleece from the hide shown in plate 5.1, using a piece of broken bottle glass.

One or two days later, the fleece is removed from the hide. The fibres can easily be pulled free with the hand if the fleece has warmed up enough in the heat of the sun. If, on the other hand, the hide becomes overheated, pieces of skin will adhere to the roots of the fibres and these will have to be cut off with scissors before spinning can commence. If the fibre does not come free easily the hide will be left in the sun for a further period, but if it is only firmly attached in restricted areas the shearer will cut the fibres close to the level of the skin using a piece of broken and discarded bottle glass (plate 5.2). Fleece removal is done in a manner similar to shearing, one side at a time, working from the rear flank forwards. It is usually possible to detect whether a fleece has been shorn from a live animal or whether it has been pulled from a hide. The cut ends of the fleece are cleaner in shorn fleeces, while the whitish roots of the fibres are visible in pulled fleeces. In addition, the fibres are longer, since they are removed at the level of, or from below, the surface of the skin.

Spinning

Before fleece can be spun it must be prepared adequately. The spinner takes one or two large handfuls of fleece and begins to tease it out, holding the fleece under the right arm with the cut ends of the fibres towards the body and the tips of the fibres uppermost. She or he proceeds, quickly pulling out all the matted *p"ichu* tips, and also the *uywi*, or hairy fibres, if it is a llama fleece. The aim is to clean (*p"ichuña*) the fleece and tease it out (*tisaña*) without causing the fleece to disintegrate into small pieces (plate 5.3). This process also allows burrs and bits of vegetation trapped in the fleece to drop out. Once the cleaning has been completed, the fleece occupies much more space than previously, and it is usually easier for the spinner to sit down and let the fleece drape over his or her lap. She or he will turn the fleece over to ensure that no more *p"ichu* tips remain, and then will pull out the fleece into a long roving, wrapping it round the fingers of the left hand to form a wound ball (plate 5.4).[4] The prepared fleece may be stored in this form, with the end tucked in to prevent it from becoming unrolled. If stored for very long periods the fleece is said to become difficult to spin, and it may be said to become 'nervous' in such a state.

More often, the spinner proceeds to spin (*qapuña*) the fleece into yarn as soon as the roving is ready. To do this, the roving is wound round the wrist of the left hand in a movement away from the spinner's body. A starting length of fleece is pulled out by the fingers of the right hand, the spinner winds it round the spindle shaft (*tisi*), often low down near the whorl (*p"iraru*) but sometimes near the top, and then she gives the spindle a sharp twist with the right hand. The spinner may spit on the shaft to moisten it so that the end of the roving will adhere. Often the spindle will be made to spin balancing on the ground or on some other surface to ensure that it will not fall. Using both hands, the spinner draws out the roving until a few inches have been spun. Then a simple hitch (*muq'ara*) is made at the top of the shaft, forming the knot over the top of the right thumb.

Plate 5.3 Teasing out and removing the hairy fibres (*uywi*) in the fleece shown in plate 5.1.

Plate 5.4 Preparing a roving.

Spinning proper may now begin.[5] Ideally, wool should be spun in warm conditions, in the heat of the sun or in the warmth of a fire, when it is less liable to break. The spinner twists the top of the shaft between the thumb and the first two fingers, and immediately releases the spindle, allowing it to drop. Using both hands, she draws out the roving, and the twist caused by the rotating spindle passes into these drawn-out fibres, producing a continuous strand of thread (plate 5.5). When the spindle reaches the ground the newly spun yarn is wound figure-of-eight fashion round the fingers of the left hand. Most women in Isluga wind the yarn round the index and the middle fingers, but in Enquelga I observed some men winding the yarn round the index and second finger. The spinner then lifts the spindle in the right hand and winds the yarn round the shaft (plate 5.6), before reforming the half hitch and repeating the operation again. Spinning in this manner enables the spinner to perform other activities, such as herding animals, at the same time. However, yarn may be produced more rapidly if the spinner sits down and lets the bottom end of the shaft rotate in a ceramic dish or in the bowl of an enamel ladle, the handle of which has broken. In this case, the newly spun yarn is wound on to the top of the shaft, and more yarn will be spun without forming another half hitch. The yarn is allowed to accumulate at the top of the shaft until the spinner winds it figure-of-eight fashion on to her left hand and then transfers it to the neatly wound cone of yarn which rests on top of the whorl. She then reforms the half hitch and continues to spin until the spindle is full. At this stage, there is a tendency for the yarn to break due to the weight of the cone of yarn. Spinning may be continued for a short while by removing the whorl to reduce the weight of the laden spindle.

When two spindles have been filled in this manner, the spinner winds the two yarns together – to double yarn is *paytaña* in Aymara. The spindles are held in place between the big and second toes of each foot, and the spinner winds off one ball with the two yarns lying side by side, using the right hand to wind round two or three fingers of the left hand to start off the ball. More often than not, the two spindles do not contain exactly the same amount of yarn. The spinner may spin some more to make up the difference, but frequently she or he simply doubles back the longer end until it meets the shorter one. This means that the end of the plied yarn will consist of a turning loop, and the join will be invisible. Having wound the two single yarns into a firm ball (*muruq'u*), the spinner proceeds to ply (*k'antiña*) the yarn, using the same spindle as before and starting in a similar manner but reversing the direction in which the yarn is wound round the spindle shaft. To ply, the spinner gives the top of the shaft a sharp twist, using the palms of both hands. The spindle is then dropped, and the spinner uses both hands to feed through the doubled yarn from the ball, which is often suspended from the clothing at the level of the left shoulder by means of a safety pin to prevent it from rolling away. Although some people spin and walk at the same time, this is more difficult than walking and plying, especially if it is windy. Consequently, many spinners build up very large balls of doubled yarn which they ply when herding animals in more distant pastures.

The spindles used in Isluga consist of a tapering wooden shaft inserted through a flat wooden whorl, turned on a lathe. These can be bought in the Bolivian market

Plate 5.5 Drawing out and spinning the fibre using a drop spindle.

Plate 5.6 Winding a length of spun yarn on to the spindle.

held once a fortnight on the border (Pisiga, Bolivia), but spinners have their own collection, acquired over the years. Spindles (*qapu*) may also be improvised: I saw one young girl spinning with one made from a wooden shaft inserted through half a plastic yo-yo. If a spindle is bought, the purchaser will first check that it spins true by twisting it on the upturned palm of the left hand. A warped whorl will not spin correctly, but it may be corrected by soaking the whorl and leaving it to dry for two days, buried in the ashes in a corner of the hearth. Afterwards it is placed under a flat stone in the sun. If a spindle does not spin well it may be called 'lazy' (*jayra*), but a spindle that spins well is said 'to dance'. However, women do not spin during Carnival, at a time when they have to dance. Similarly, Lynn Meisch was told in Tarabuco, Bolivia, that it is 'forbidden' to spin during a festival at which *chicha* is drunk, since it would make one's head whirl (Meisch 1986a: 28). She also reports that Tarabuco spinners prefer larger spindles for plying (ibid.: 27). Obviously, such spindles hold more yarn, but more often than not in Isluga spinners use what is at hand. In the past, people spun finer yarns using spindles with smaller whorls.

Yarn may be spun in a clockwise or in an anticlockwise direction, as seen by the spinner looking down on the rotating spindle. The resulting yarns are different, being designated Z or S, respectively, according to the /, or Z slant, produced with a clockwise rotation of the spindle, or the \, or S slant, produced with an anticlockwise rotation, when the yarn is viewed in a vertical position. A single yarn tends to be unstable and it will twist back on itself when not held under tension, particularly if it is tightly twisted. Most of the yarns produced in Isluga are spun with a Z twist, and then two such yarns are doubled and plied in an anticlockwise direction producing an S ply (the spinner uses the palms of both hands to ply, pushing the right away from the body and pulling the left towards the body, thus creating an anticlockwise rotation). Such a yarn is described using technical notation as Z2S. In Isluga, this type of yarn has no special name, it is merely referred to as *ch'anka* ('yarn'). However, other types of yarn are produced, mostly for very specific reasons, as described in further detail below.

Ch'anka

This is standard two-ply yarn, made as just described above. However, it can be spun in a variety of diameters, and finer yarn takes much longer to spin than coarse yarn. The tightness of twist and time invested in both spinning and plying depends on the intended use of the yarn, and to some extent on the individual spinner. In theory, since camelid fibres have a long staple when compared with, say, sheep's wool, the degree of twist may not be as great. Some people spin more loosely than others, and women say that men produce yarn that is both finer and more tightly spun than their own. However, loosely spun yarn is soft, and this is a desirable characteristic. Very tightly spun yarn is described as *q'aralla*, and it feels harsh. Yarn is plied loosely if it is to be used as weft (*qipa ch'anka*) or tightly if it is to be used as warp (*saya ch'anka*). The weft should be soft and pliable, and in many textiles, such as *awayu* (carrying cloths), and bags, it should be fine in diameter.

Since loosely plied yarns tend to have a fuzzy character, the weaver often takes a needle and rubs it along the weft yarn which she holds tight by tensioning it round her foot. This action removes much of the fluff and makes the yarn slightly less substantial. Warp yarn, on the other hand, should be very firmly plied and have a smooth outline. This is made easier if the spinner first soaks the ball of doubled yarn and plies it while still wet. There is a practical reason for this, as fuzzy yarns catch against each other during the weaving process and the friction causes the yarns to break easily. In addition, women find the appearance of firmly woven, plain-weave, warp-faced cloth (that is, weaving in which the warp threads are so closely spaced that the weft is hidden) aesthetically pleasing if the warp yarns are smooth and the structure of this type of weave is plainly visible. They describe such cloth as *pelado* ('bare'). Fuzzy warp yarns obscure the structure of the fabric. Blankets are woven with much coarser yarns, and *ikiña* is the general term for a blanket, many of which are woven from brightly dyed sheep's wool yarns. Another type is called *chusi*, and it is woven from thick, but regular in diameter, naturally coloured alpaca yarns. If the yarns to make a blanket are very thick, irregular and, incidentally, very rapidly spun, the blanket is known as *jap"u*, which also means 'soft' and, by extension, may refer to a dry, soft *bofedal* (Mamani 1985: 20).

Ch'añu

Yarn plied using two strands of different colour is called *ch'añu*. For most purposes, spinners try to ply strands of the same colour. However, the weft for a type of bag used for storing food (*wayaqa*) should ideally have a *ch'añu* weft made from one white strand and one black or brown strand plied together. Weavers explain that such a weft is necessary because the weft in a *wayaqa* has to pass through warp threads organized in stripes of white, brown and black in a design that characterizes this type of textile. Even though the weft is scarcely visible in a warp-faced fabric, it is very important that the colour contrasts should be very clear and well defined. Weavers say that an all-white weft diminishes the blackness of the black stripes, and an all black weft diminishes the whiteness of the white stripes.

Ch'iqa ch'anka

This is yarn spun in an anticlockwise direction and plied in a clockwise direction. The notation for such a yarn, if two strands are used, is S2Z, and so it is spun and plied in directions that oppose those used in the construction of a normal yarn, which is Z spun and S plied. One spinner said she would spin *ch'iqa* yarn using her left hand to operate the spindle, but another spinner described the process 'as spinning as though you were plying and plying as though you were spinning'; she used the spindle in her right hand to produce *ch'iqa* yarn. Nowadays, many weavers do not spin this type of yarn for their textiles. Nevertheless, some women continue to use it in some very specific roles. For example, it is used as warp in the side selvedges of an *awayu* (shawl or carrying cloth); in one, two or three narrow strips inside the coloured stripe that forms the hem of an *urk"u* (women's dress); and in a type of border known as *k'anipa* applied to the edges of a *wayaqa* after the fabric has been woven and the sides seamed (figure 5.1). When *ch'iqa* yarn is used in an *awayu* or *urk"u*, the appearance produced is one of herringbone stripes (plate 5.7). The use of *ch'iqa* yarn has a practical purpose in helping to keep the corners of a textile from curling up, since the opposite spin direction counteracts the tendency

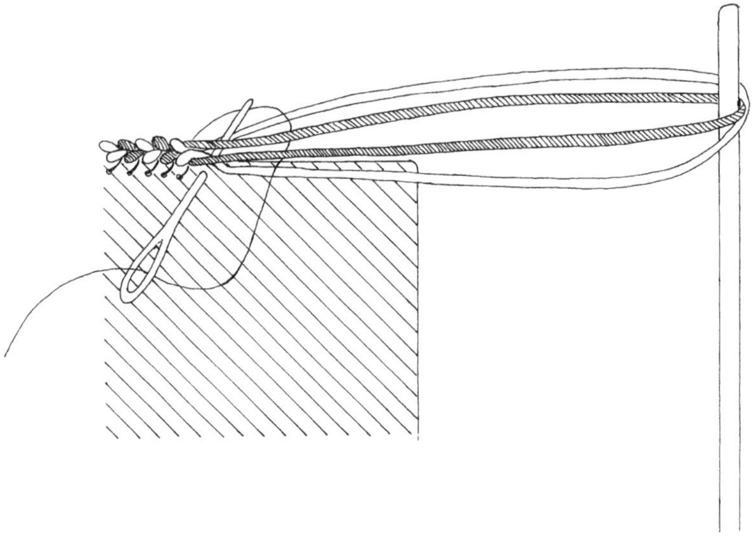

Figure 5.1 Adding the border known as *k'anipa* to the outside edges of a woven bag (*wayaqa*).

Plate 5.7 Yarn spun to the left (*ch'iqa*) in a woman's dress (*urk"u*).

for the corners to twist back on themselves in a warp-faced textile with yarns firmly plied in only one direction. When used in the edging of a *wayaqa*, a pair of white warp threads is crossed over a pair of black or coloured warp threads, each crossing being held in place by a weft yarn which is stitched in place through the fabric of the bag. The result is a vertical design sometimes also called *k'ili*, which means 'herringbone' or 'spinal column' in Aymara. Weavers use *ch'iqa* yarn in borders in the belief that such yarn is more durable and protective than yarn spun in the normal direction.

Ch'iqa ch'anka literally means 'left yarn': that is, yarn spun to the left. Among the Quechua speakers of Tarabuco in Bolivia, S-spun yarn and S-spun and Z-plied yarn is called *lluq'i* (Quechua for 'left'). Meisch found that most people were reluctant to discuss anything but the practical applications of such yarn, but she observed men and women wearing black and white *lluq'i* yarn plied together and tied round the wrist or ankle, and she concludes that this type of yarn in Tarabuco serves to protect both the person (the wearer) and the textile itself (Meisch 1986a: 27; 1986b: 250). A belief in the magical properties of *lluq'i* yarn has also been reported for the Cuzco area by Goodell (1968: 7).

Pichuña

The construction of *pichuña* involves a step further than normal yarn, as it is spun, plied and then re-plied. A length of plied yarn is doubled back on itself and then it is re-plied. The spinner twists the spindle with a clockwise rotation: that is, in the same direction as the initial spin. Such a yarn is technically described as Z2S2Z. *Pichuña* is made from natural-colour camelid fleece, often brown or white, and it is sewn to one of the top corners of a sack or bag for storing food, so that the sack may be closed by gathering together the edges of the top opening and winding the *pichuña* cord several times tightly round the top of the sack.

Other yarns used in Isluga have the same Z2S2Z construction, and they are named in accordance with their function. For example, *jinchu ch'anka* is spun, plied and re-plied using white fleece and it is used to suspend the *sarsillu* (ear tassels) from the ears (*jinchu* means 'ear') of the llamas and alpacas during the *wayñu* ceremony. Another yarn using this construction is called *anku* ('nerve' or 'tendon'). This yarn is used in the making of the *wistalla* that is tied round the necks of llamas and alpacas during the marking ceremony. Unlike *pichuña* and *jinchu ch'anka*, *anku* is made from dyed yarn, often spun from sheep's wool. The most commonly used colours are tones of red, orange and green; purple and blue are very rarely chosen. *Anku* is cut into short lengths and these are arranged and suspended from a length of *mismiña* (see below) by means of a larkshead knot. Women using the indigenous-style horizontal loom also use two types of yarn with this same construction. *Ch'ukurkata* is used to lash the warp selvedge to the breast and far beams of the loom, while *pulu* (sometimes also called *supulu*) is made simply from a double length of weft re-plied back on itself to form the first weft, which is beaten into the turns of the warp against the breast beam. While both these yarns are undyed, the lashing cord is much more tightly re-plied than the first weft.

Paychulla

Strictly speaking, this is not a true yarn, as it consists of two or three long lengths of drawn-out, unspun fleece which are then wound into a ball. It consists of *chimpu*: that is, teased-out and dyed camelid fleece. The colours used are red, pink, orange and sometimes yellow. The dyed fleece is left out to dry and then it is pulled into a long roving. Two different coloured rovings are placed side by side and wound round the fingers of the left hand to form what is called a *paychulla* (from *paya*, meaning 'two' and *chulla*, 'one of a pair'). In this form, it is easy to roll off lengths of *chimpu*, which are broken off and given to the participants of a *wayñu* ceremony to tie on to the backs of the llamas and alpacas.

Mismiña

This is a technique for spinning which does not involve the use of a drop spindle. Yarn produced in this manner also receives the same name. The fleece is prepared in the manner described above, but the spinner uses a straight stick, which he or she holds horizontally. The end of the roving, which is wrapped round the left wrist, is pulled out and twisted round the middle of the stick. The spinner then rotates the stick away from the body while drawing out the fleece, and the movement of the stick introduces a twist into the fibres. As it is spun, the yarn is wound round the stick. The spinner then takes two such yarns (or doubles back one yarn on itself) and winds them into a firm ball for plying, which is done in the same manner but with the stick turned towards the body (plate 5.8). The resulting yarn may be described as S2Z. However, it is never referred to as *ch'iqa* but always as *mismiña*. Both men and women produce this type of yarn. The women often use this technique to improvise a back-strap when using a small back strap loom for weaving belts or small bags. However, *mismiña* is usually associated with men's work. Men make very regular, coarse yarn with a high degree of twist in this manner to make into plaited ropes and slings.

Juñi

Strictly speaking, *juñi* is the Aymara word for 'hank (of yarn)'. Nowadays, women buy commercially spun, synthetically dyed yarn in the Bolivian market, or from Santiago, and this is often referred to as *juñi* because it often comes in the form of a hank. This yarn is designed as a yarn for knitting garments, and it is unsuitable for weaving without modification. First, the spinner winds such commercially produced yarn into a firm ball, which she dampens in water. Then she overplies the yarn: that is, she plies the already plied yarn using a spindle, which she twists by using the palms of both hands as described above. It is important not to carry this process too far, as the yarn kinks back on itself when not held under tension. This also applies to handspun and plied yarns, and the kinks in such yarns are known as *ch'ila*, which is also the word for the small intestines of an animal.

Plate 5.8 Making *mismiña* at Paniri, near the Vega de Turi in northern Chile. In this picture, the woman is spinning a hobble for a donkey.

Spinning is a condition of life in Isluga; women, men and children are all actively involved in producing yarn. Doña Juana put it very succinctly when she observed, 'spinning is our work'. There is apparently a pronounced pragmatism towards the task in the minds of the practitioners. However, spinning processes are also profoundly implicated in systems of cultural representation in Isluga, and they are connected with potent visual imagery. The dance movements people make in the *kancha* during the marking ceremony mimic the rotations involved in spinning and plying yarn. When I mentioned this observation to Doña Soria, she agreed that the movements inside the *kancha* are, indeed, like spinning *ch'iqa* yarn (personal communication 19 July 1995). Since the action of plying unites paired strands of yarn, it is perhaps significant that a human pair – mother and father – accompanied by their children host each marking ceremony.

It is probable that spinning has long been regarded as a meaning-rich activity in the Andes. Spinning is a kinetic activity, combining movements in time and space. Mary Frame suggested that the spiralling movements of strands in a plied yarn were perhaps viewed by ancient Andean peoples as being analogous to the apparently spiral pathway of the sun in its daily and annual cycle, as seen from a fixed viewpoint on earth (Frame 1986: 56).

Dyeing

Natural dyes are no longer used in Isluga. Weavers buy powdered aniline dyes by the ounce in the Bolivian market. Yarn that is to be dyed must first be wound into a hank, and it is wound figure-of-eight fashion round the spinner's left hand and foot, or round the outstretched hands of a helper. Winding the yarn in this manner introduces a cross (*q'ata*) into the hanks and prevents the yarn from tangling when it is washed. The yarn is washed in hanks, usually in cold water to which commercially produced detergent has been added, although some people use a naturally occurring white powder (*chuka*) instead of detergent. Dye powder is added to boiling water (often empty fruit tins are used for this purpose), and then the hanks are pushed into the water and left to boil on the hearth for about quarter of an hour. A naturally occurring nitrate called *millu* is used as a mordant, and it is added at this stage. Then the cans are lifted from the heat and the hanks are left to soak overnight. The following morning they are hung out to dry. The take-up of the dye may be made more efficient by leaving the hanks to soak overnight in water to which lemon juice and rind has been added, or in stale urine (in which case, boy's urine is preferred), prior to dyeing the yarn.

It is easier to dye yarn made from sheep's wool than from camelid fibre, since the cuticle of the fibres of the former is thinner and therefore it is more easily penetrated by the dye. However, many women complain that the aniline dyes sold now are not as effective as those sold in the past. Nowadays, weavers prefer to buy commercial synthetic yarns, as they find the uniformity and fastness of the colours attractive. Younger weavers, in particular, are increasing the use of colour, by means of these commercially produced yarns, in their textiles. This is in contrast to middle-aged and older women, who are reluctant to talk about colour. Indeed, they may even not admit to dyeing camelid fibre. Gavilán quoted a statement made to her by an Enquelga woman: 'Sheep's wool is good for dyeing, but that of llama does not take well, it only gets dirty' (Gavilán 1985: 38). In actual fact, camelid fleece is successfully dyed in the making of *chimpu* for the marking ceremony, although Gavilán's respondent was reluctant to admit this to her.

Unfortunately, the knowledge of dyeing using natural sources as practised a century ago has not been passed down by the generations of weavers. Doña Natividad acknowledged that alpaca is easier to dye than llama. Her mother, from Karawanu, had been an expert dyer and dyed her llama yarn a bright orange-red colour, using aniline dye. As a teenager, Doña Natividad wove this yarn into the borders of an *awayu*, using grey alpaca for the main part of the shawl. This dyed llama yarn retained a good hue for some fifty years. Her mother also knew how to use natural dyes, including a source obtained from a cactus flower in the *precordillera*, but she herself no longer remembered the recipes. Traditionally, dyed yarns (usually of sheep wool) were used only in certain textiles, and they were always employed in small quantities.

Wherever dyed colours are used, they should be strong and true. Faded colours are considered to be very ugly and useless. When preparing the items for her family's *wayñu* ceremony, a woman found three *paychulla* left over from a previous ceremony. As they had faded, she immediately ordered one of her sons to dispose of them by hiding them in a drystone wall, instructing him to make sure they were well hidden.[6] The items used in the marking ceremony must bear clearly differentiated colours. Equally, when making textiles used in everyday life, if the weaver makes colour choices that are regarded as lacking in contrast they are described as *qaqa*. This word used as a noun means 'lead', and as an adjective, 'discoloured, faded'; it corresponds to the Isluga Spanish adjective *pavoso*, which presumably corresponds to the modern Castilian *apagoso* ('muted'). These aesthetic judgements apply equally to dyed colour and to natural colour, as indicated above with the discussion of bi-coloured *ch'añu* weft which is designed to maintain the colour contrasts of the overlying black and white warp.

Weaving and plaiting

In Isluga, women weave items using the Andean loom, which is either erected horizontally on four stakes (plate 5.9), or, if smaller items are to be woven, the loom is tensioned with the aid of a backstrap. The loom is the same in both cases; it is only the means used for maintaining tension on the warp that differs. Men,

Plate 5.9 Doña Soria Mamani Challapa (outside her house in Jalsuri) weaving one half of a carrying shawl for her daughter.

on the other hand, make a range of plaited items (slings and ropes). Previously they also wove *bayeta*, a plain-weave or twill cloth woven on a type of treadle loom, which was originally a European introduction. This latter activity is no longer practised but, in the late 1980s, men developed an interest for knitting jumpers and caps, another technique that was introduced by the Europeans to the Andes. There is clearly a sexual division of labour in the production of cloth and plaited goods in Isluga, although both sexes have a thorough understanding of the techniques used by the opposite sex. Weaving and plaiting are enormously important activities, and the garments produced by the women are especially responsible for maintaining ethnic integrity, as they bear the visual signs by which Isluga society is most readily identified.[7] Textile production is also a time-consuming activity, the various processes of which are perceived as significant in terms of the practitioner's understanding of his or her own experience.

Children learn to spin and, when they have mastered this, to weave or plait, at an early age. They learn by looking at their elders. In the case of weaving, the training of a daughter is mainly the task of her mother. Most girls start weaving small items by the age of five or six. As the mother warps up a ribbon or small bag using her daughter's yarn, the girl notices that her mother picks up a stone and lightly taps the balls of handspun yarn with a stone. The explanation given is, 'so that the yarn won't run out', and the daughter gradually learns that the process of weaving is imbued with a sense of bringing into being cloth which is not regarded strictly as inanimate.

The side selvedges, formed by the turns of the weft, are known as the 'mouth' (*laka*) and, with every weft pass, the weaver lifts the heddles or moves the shed rod in order to open the mouth, enabling her to enter the weft. The mouth is closed as she changes the shed, and then it is opened again. This process means that life is breathed into the fabric and that the animate quality is an inherent property of the cloth. This interpretation is strengthened with the naming of the central stripe of the *wayaqa* bag as *chuyma* (the 'heart'), in a metaphorical sense, since this word literally means 'lungs'.[8] The *wayaqa* bag is closed by gathering together the edges of the top opening, and winding *pichuña* cord several times round the top of the bag. The space enclosed inside the bag is evidently not regarded as inert, as it would seem to have a womb-like quality. *Pichuña* has, as its root, the word *pichu*, which can refer to the last child to be born: that is, the birth that closes the womb (Bertonio 1984 [1612], Bk II: 264; Büttner and Condori Cruz 1984: 154). Moreover, the broad vertical stripes woven into the fabric of the bag are known as 'mothers' (*tayka*), and the narrow stripes which insert themselves in between, as 'baby animals' (*qallu*). In this sense, the bag is imbued with procreative powers, thus echoing the powers of autogenesis (virgin procreation) attributed to the Wirjin Tayka, the earth mother. Infant llamas and alpacas are referred to as *qallulla* or *jiska qallu* (see Chapter 3), but the baby animals of this type of bag are not necessarily camelids, since *qallu* can be the offspring of any animal. Cassandra Torrico (MS) says that among the Macha people of northern Potosí in Bolivia, sacks are thought to be symbolic toads. Cereceda (1978) analyzed the Isluga woven bag. However, Cereceda adopted an approach based on a nature/culture opposition, whereas I prefer an approach that emphasizes the dynamic, generative symbolism of both the weaving process and the finished item.[9] I first suggested an interpretation of the active principle of the opening and closing 'mouth' during the weaving process in Dransart (1988: 44), and developed it further in Dransart (1995: 228, 238). Since then, Sophie Desrosiers has argued that we should be studying precisely such symbolic meanings in the movements of threads and in the means for realizing such movements (Desrosiers 1997: 325).

The character of the yarns and that of the fabrics produced in Isluga are closely interrelated. Spinning is gendered, in that men are responsible for producing the yarn they require for plaiting items and for knitting jumpers, while women are responsible for producing the yarn they require for weaving garments, bags and blankets. As indicated above, spinners have a pragmatic approach to producing yarn, which is spun for specific purposes. The spinner has a blueprint in mind as soon as he or she picks up fleece and a spindle. A common question on meeting a spinner is, '*What* are you spinning?', implying that the spinner has already conceived of the finished item.

The Aymara names of the different types of yarn listed above also indicate that yarn is woven for specific purposes. Indeed, some of these yarns are only used as part of the elaborate material culture of the marking ceremony. There is a tendency for the Aymara terms for these yarns to be named according to their function.

Nearly all the textiles woven in Isluga are warp-faced, resulting in the need for firmly plied, smooth warp yarns as described above.[10] Women's textiles are characterized by large areas of plain weave in natural colours, offset by narrow bands of *k'isa* (tonal gradations of colour, executed in plain weave), flanking narrow stripes that bear repeat motifs, executed in two-colour complementary-warp weave techniques (plate 5.10). Weaving is a public activity, and there is much discussion, and even rivalry, between weavers regarding their work. In the 1980s older women were critical about the amount of dyed colour used by younger women. They pointed out another difference between their own weaving and that of younger women in the traditional women's dress known as the *urk"u* (plate 5.9). This garment has been worn in some Andean areas for many, many generations dating back to distant pre-Inka times. During this timespan it has undergone modification and, indeed, it is no longer worn in many parts of the Andes. Since the late 1980s it has been gradually undergoing replacement by Western-style clothing, which is now worn by many women in Enquelga (this process began much earlier in the villages of Manq"asaya). When asked why they no longer use the *urk"u* on a daily basis, reserving its use for certain festivals, women reply that it is very heavy compared to imported clothing made from synthetic fibres. However, it is also the case that these traditional garments, as woven by middle-aged and younger women, have a very high warp density compared with those worn by old ladies. The garments of the younger women are heavier and more rigid compared with the light and flexible *urk"u* of their grandmothers. In the past, women used very little dyed colour in their garments. In their younger days, older women used handspun and hand-dyed sheep wool for the narrow bands of colour in their textiles. The more recently introduced synthetic yarns tend to be finer in

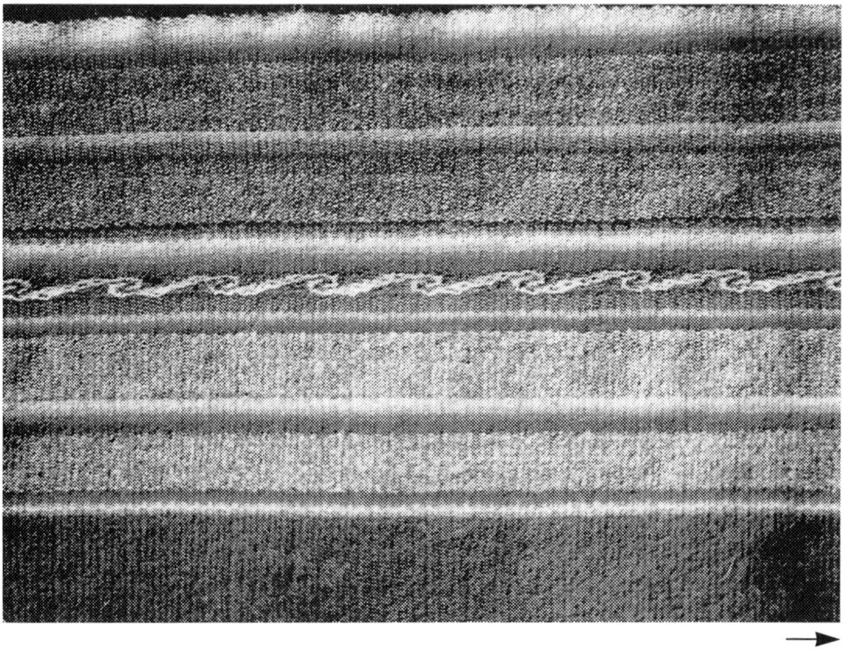

Plate 5.10 Detail of the coloured band of a carrying shawl (*awayu*) woven by Doña Luisa Castro Castro of Enquelga in the 1960s, showing narrow plain woven *k'isa* bands flanking a band of complementary warp-weave pattern (*salta*). The arrow indicates the direction of the warp.

diameter, and they are limp because they are loosely plied (since they are designed by the manufacturers as knitting yarns). Consequently, they have to be overplied, giving a high degree of twist. Natural-fibre yarns combined in the same textile have become more tightly plied and densely spaced to balance the highly twisted synthetic ones. The difference between finer machine-spun yarns and coarser handspun ones creates technical problems during weaving. As the weaving advances, the weft may no longer lie at right angles to the warp, and this has to be corrected by inserting partial wefts. A turn in the weft in order to straighten the fell is called *q'iqi*, and it sometimes has to be used in textiles woven from handspun yarns because of the difference in diameter of the warp, but the problem is aggravated when synthetic yarns are used.

As noted above, natural camelid colours are used in a culturally specified manner. Changes in traditional practice may be noted. Vicuña-coloured fleece is reserved for men's ponchos, which are plain and undecorated; more recently, the use of light grey has been permitted and narrow bands of patterned design may be included. The women's *urk"u* was always black or a dark brown, but younger women may use a medium brown if they cannot obtain dark-coloured fleece in sufficient quantity. The large, central, undecorated field of an *awayu* (carrying cloth or shawl) is referred to as the 'body' of the textile; it used to be a fairly dark colour, brown or dark grey, but now it is always a very light colour, such as light grey, light fawn, or even white. A small cloth for carrying *coca* leaves or money (*inkuña*) should traditionally have grey in the central undecorated field ('body'). Black, white and brown are kept for weaving the characteristic stripes in a bag for storing food (*wayaqa*), and also for sacks (*kustala*). However, younger women now include light or medium-grey stripes in the design of these items. The ropes plaited by the men combine white and natural-coloured strands (black or brown) if the rope is to be used for everyday purposes such as herding, or collecting firewood, but the white may be combined with orange or red-dyed strands if the rope is to be used for ritual purposes.

This chapter has demonstrated how the people of Isluga respond to the camelid fleece that they have at their disposal, a raw material that they perceive as possessing certain inherent qualities and possibilities. Fibre grown by animals is shorn and undergoes a series of transformations, which turns it into clothing and blankets to wrap human bodies; it is also used to make bags, sacks and cloths for containing foods and other items. These garments and items, the products of human hands, which work and transform the raw material, seem to be part of a living system. This sense of organic unity between maker and product is identified by Taussig (1980: 36–7) as a kind of fetishism which occurs in the economies of precapitalist societies: the products appear to be imbued with life or animated because they seem to embody the social milieu from which they emerge. Garments bear the visual signs which identify the wearer as a member of a particular ethnic group (that is, of Isluga), as well as expressing gender relationships. In addition, clothing one's body is a particularly human activity. A lazy girl who is reluctant to spin may be chided as *jayra* ('lazy') and be likened to the *jararanq"a* ('lizard'), which is naked and does not need to clothe itself. The intimate association between clothing and the human body is an important one; in fact, clothing may be said to stand in a 'metonymic relationship to the self' (Schneider and Weiner 1986: 178; Dransart 1992a: 145–6). Such an association may be carried to an extreme in the Andes.[11] In Isluga there is an added dimension, since the fleece grown by one's own animals provides the necessary raw material for much of this clothing.

Spinning, weaving and plaiting as practised by the people of Isluga link human activity with what they perceive as the cosmological order. Spinning, no less than the apparently more complicated tasks of plaiting slings or weaving complementary warp-weave designs, is a conscious activity.[12] Indeed, the separate components of a textile, as discussed above, are seen to constitute a microcosm of aesthetic and tactile values. To the people of Isluga, fleece is perceived as a composite entity, consisting of individual fibres – a continuum of small discrete units. It has been compared to the Milky Way, which consists of a mass of stars, and to *ch'uwa* liquid, granules of maize flour suspended in water, which is sprinkled over llamas during the *wayñu* ceremony. When fleece is spun into yarn and woven into cloth the warp must not be fuzzy or blur into an amorphous whole. Instead, it should have the appearance of discrete components which are united in the one fabric. In other words, the structure of the weave must be clearly visible. Equally, colour contrasts must be clearly defined, for certain colour combinations are considered inappropriate and 'ugly' if they do not generate sufficient clarity of contrast. This is the case even for the woven image of the rainbow (the *k'isa*), which is woven as a series of colour stripes separated by short intervals. The colours in a modern *k'isa* consist of little individual steps. Thus homogeneity, or a blurring of boundaries, is undesirable. It is as ugly as a wad of mashed-up *coca* leaves in the mouth or the colour of lead. Instead, a unity of discrete components is considered to be desirable. Beauty resides in the products of human labour, which demonstrate the clarity of such combined contrasts.

6 Historical perspectives on herding technology

A discussion on Isluga herding technology is juxtaposed in this chapter with material collated from documentary sources dating from the sixteenth and seventeenth centuries and with material culture from pre-Hispanic times. Chapter 5 demonstrated how the classification of fleece in Isluga as a material for technical activity results in the particular types of yarn that are spun in the community, and how different natural fleece colours are used for very specific purposes. A study of the manner in which fleece is exploited may reveal technical activities that reflect specific herding practices. An important issue explored in the present chapter is whether or not selective or restrictive breeding practices were undertaken in the Andes. This helps to set contemporary practices in Isluga in a historical context.

Animal breeding and artificial selection

The classificatory terminologies for camelid fleece discussed in the previous chapter deal with culturally determined responses to the fleece itself which, in turn, favour certain techniques for converting fibre into yarn, and yarn into fabric. Therefore, spinners and weavers exploit the characteristics of fleece grown by llamas, alpacas and sheep, and the products of their labour reflect the types of fleece available in Isluga. The fleece also provokes an aesthetic response when it is spun, and then woven and plaited in what is perceived to be the 'correct' manner. Changes, of course, have taken place in the appearance of 'traditional' textiles. I have noted that younger women favour the use of larger areas of dyed colour in their garments, and that there is a tendency for handspun yarn to become more firmly plied and more densely spaced. Both these changes have occurred in parallel with the introduction and increasing availability of commercially produced synthetic yarn. Nevertheless, it remains the case that garments and plaited items are found to be aesthetically pleasing if the maker has chosen colours which are distinct and do not blur into each other (this applies to both natural and dyed colours), and if the outwardly visible yarns are smoothly plied and do not obscure the structure of the fabric. Camelid fibre is still the most important raw material for producing woven and plaited goods in Isluga, and its qualities continue to influence yarn and fabric technology.

Given the importance of fleece as raw material, it is worth asking whether animal breeding through artificial selection takes place to promote an increase of animals with particularly prized qualities of fleece. Herd protection and selective exploitation practised by the herders ought to have an effect on the genetic make-up of the animals. To what extent can these practices be said to affect fibre quality in Isluga? This complicated and controversial question cannot be examined without taking a longer-term view that also considers past practices. There is some evidence available for Inka herding practices, but in other periods evidence is lacking and it must be inferred from material culture. However, this is a central concern for the problems under examination in this book, and the present discussion aims to form a bridge between the ethnographic evidence from Isluga and the archaeological evidence from Tulan, which will be presented in the following two chapters.

Isluga herders do not select white or any other colour in animals for reproductive purposes in an attempt to increase a certain colour especially relative to others. Selection on the basis of colour is practised by the large highland estates in Peru (the Empresas Asociativas). However, as stated above, the full range of camelid fleece colours is not present in Isluga. Herds of llamas tend to be more frequently piebald, but the range of grey is more restricted than that present in alpacas. With the exception of the two *q"usi-* or *vicuña-*coloured alpacas mentioned in Chapter 4, herds of alpacas, on the other hand, do not display light fawn colour. Moreover, herds of llamas or alpacas owned by individual families often contain a concentration of animals with a similar fleece colour, or with similar combinations of colour. Two white llamas killed by lightning, mentioned in Chapter 3, were members of a herd that was characterized by a high proportion of white animals. Herds of domesticated camelids, like wild vicuña social groups, consist of largely endogamous breeding units, but there are exceptions to the general state of affairs.

In the 1980s and 1990s, Isluga herders tended to castrate most of the males in the family's herds. An uncastrated male (*jañachu*) can serve about ten to twenty females, and most herds have only one or two sires. Isluga herders therefore do select which male animals will be allowed to breed, and they make an artificial selection through the males. Such interference will not reduce the capacity of the females to reproduce, and these castrated males may be culled while the females are allowed to reproduce and enhance the growth rate of the herd. However, people in Enquelga say that in the past there were more *tataqullu* herds: that is, uncastrated adult male llamas, which were herded in more distant pastures to keep them away from the females. These males were prized for their fine fibre, which was attributed to the good quality of pasture available below the snow melt line. It is not clear how often these herds of males came into contact with female llamas, but they were certainly brought together for the family's *wayñu* ceremony, which coincides with the rutting season. This practice would have been in contrast with the more recent trend, which prevents the majority of males from siring offspring.

Isluga herders do not oversee the mating of llamas and alpacas, in contrast to reports from other Aymara-speaking areas. In Chinchillapi, southern Peru, herders undertake a ceremony in January and February, called by Palacios Ríos (1981: 221) the *walquecha* ('sowing of the animals'), which presumably derives from the Aymara *wallq'i* ('pregnancy'). The herders introduce males into a corral, where the males are made to serve the females, which have been made partially immobile by binding the hind legs with ropes. The herders then make a series of ritual offerings and libations. Palacios Ríos explains:

> Herders make use of this opportunity to effect a genetic selection of their animals; so they place as sires only those males considered to be the strongest, largest, and displaying the preferred colour of the day, and which have abundant wool, this being of good quality.
>
> (ibid.: 222)

This statement implies that the selection is done only through the males, and that females with reproductive potential are not prevented from producing calves. Caro (1985: 251–5) reports two methods of llama and alpaca mating for Ulla Ulla in Bolivia. These take place between January and March, and the mating is known as *chaxruyapiña* ('to mix together'). The first method is very similar to that reported by Palacios Ríos, and it therefore involves choosing both male and female partners, but the account does not make it clear whether females with any qualities which may be regarded as undesirable are not mated. The second method is that of continuous association: in other words, selected male sires are separated from the male herd and are pastured and corralled with the females during the rutting period. This is said to be less productive, because after initial sexual activity the *jañachu* sires lose their sexual drive. In fact, Caro claims: 'The reproductive physiology and behavior of alpacas and llamas is quite complex and requires a great deal of human intervention. . . . The herders must guide and supervise the entire mating process' (ibid.: 251).

In Isluga, llamas and alpacas are left to mate without human supervision. The result is that, although herds are trained to act as discrete groups and should form endogamous breeding units, in practice the animals come into contact when grazing on the *bofedal*, and llamas and alpacas do on occasions mate with each other, hence the fairly frequent number of hybrids. When asked about mating matters, Enquelga herders refer to the marking ceremony, admitting that it is a matter of luck (*suerte*). Animal fertility is dependent on luck, and the multiplication of llamas, alpacas and sheep is ensured by the proper observation of the *wayñu*. However, one should never forget to recognize that the *uywiri* have the power to grant or withhold luck, hence *suerte* may be a limiting factor controlling equally human existence or the multiplication of one's animals.

In any case, camelid fertility is generally said to be low, and research on alpacas indicates a high incidence of embryonic mortality within the first thirty days of gestation (about 50 per cent) (Fernández Baca 1971: 19). Such problems are compounded by high rates of mortality in calves during their first three months, which in some years rises to 50 or 60 per cent (ibid.: 29). If half the impregnated females successfully give birth, and the mortality of the calves reaches another 50 per cent, the rate of herd increase is low indeed, given that only a little more than half the total numbers of females in a herd will be able to breed each year, since the remainder are too old, too young, barren, or gave birth too late during the season to mate the following year. Caro (1985: 255) points out that the growth rate for Ulla Ulla herds is higher for larger herds than for smaller ones, giving typical percentages of 13 per cent for a herd of 300 alpacas but only 5 per cent for a herd of 100 alpacas (these calculations assume that in each case twelve animals per year are slaughtered for consumption). This should be contrasted with an 18 per cent incremental increase reported in sheep flocks (Dahl and Hjort 1976: 231). In Enquelga, most herds number less than 100 head of llama, while alpaca herds are very small indeed. Since stud males are herded continuously with the females, and since males are sexually active all year round, calves are sometimes born outwith the main birth season, which coincides with the summer rains. These animals are especially vulnerable to cold night-time temperatures and shortages of pasture, which result in an insufficient quantity of milk in the mother. However, it should be pointed out that camelids have achieved a remarkable adaptation to the rigours of the Andean highlands, since the majority of calves are born during the morning. Research cited by Guerrero (1986: 29) in alpacas observed at a Peruvian research station indicates that most births occurred between 07.00 and 11.00 hours, with none before 05.00 or after 14.00 hours. This means that the coat of a newly born animal will dry long before nightfall, and therefore a camelid calf has a much greater advantage in this respect than a newly born lamb, which may be born at any time of the day or night.

Given the low incremental increase of camelid herds, it is in the herders' interest to promote the growth of their herds. However, any selection involving females would be to lessen the number of offspring produced. In Ingold's words, this would be 'contrary to the most basic premises of pastoral herd management' (Ingold 1980:1988 132), since to limit herd size in order to breed animals for quality would be to court disaster and increase the risks to which pastoralists are exposed.

Ingold develops his argument in reference to reindeer herds, but it is worth considering further. According to Ingold, the pastoralist seeks safety in numbers, and herders protect their herds by relaxing natural selective pressures that may operate on females and young, allowing animals to survive and reproduce when they might not have done so without protection. However, this is offset by artificial selection through the males, since the majority of reindeer bucks are removed by castration or slaughter. Thus any increase in variability which may occur in does and fawns would be counteracted by the greater stringency effected through the selection of the bucks (ibid.: 132). Ingold develops his argument to consider the evolution of domesticated species through the practice of artificial selection. 'Animals breed', he says, but humans can also 'breed animals' by establishing divergence (that is, different breeds) through artificial means in which a population of breeding stock is isolated from other animals of its species. Ingold argues that pastoralists are not necessarily ignorant of methods of artificial selection, but they cannot be used when pastoralists are locked in a subsistence economy based on their herding activities. They accumulate livestock in order to provide the household with security against environmental fluctuations. Therefore, Ingold suggests, 'pastoralists will only institute a policy of artificial selection designed to alter the hereditary constitution of their animals if the following two conditions are met: firstly the animal must cease to furnish all the essentials of their subsistence, and secondly, some alternative form of security to animal property must be available' (ibid.: 134–5).

He tests his thesis against archaeological evidence for ovine and caprine domestication during the seventh and sixth millennia BC when recognizably domesticated breeds of sheep and goats emerged against a background of land-intensive cereal agriculture and an expanding network of trading relations. He then considers evidence for the later domestication of cattle, the wild progenitor of which (the aurochs) was a woodland dweller, and he proposes that domestication would have been undertaken by already sedentary cultivators in the Middle East (ibid.: 136–41). Given the mixed subsistence economy, with agriculture, or agriculture and trading, as important activities, Ingold argues that those farmers were in a position to initiate policies of artificial selection, favouring innovative characteristics not generally shared by the wild form of the animal concerned.

Contemporary herders living at high altitudes in the Andes have severely restricted opportunities for diversifying their economic base. The evidence from Isluga supports Ingold's contention that herders will alter the hereditary constitution of their animal resource only if his two conditions are met. It is true that most families in Isluga also grow *quinua* and/or potatoes. People living in the villages of Manq"asaya are in a better position to do this, while the herders of those parts of Araxsaya where no cultivated crops may be grown only have access to fields they inherit at Islug marka. However, potatoes and *quinua* are susceptible to weather conditions, and such an unreliable source of production would not provide the necessary security for herders to constrain the reproductive potential of their animals. In fact, during periods of drought, camelids are much more reliable than agricultural crops, or even sheep, although in the short term, and under normal climatic conditions, sheep may be attractive, as their rates of flock growth are higher. It is therefore important for herders to attempt to maintain their camelid herd sizes at all costs, and any selection that does take place will only be done through the removal by castration or slaughter of some male animals of reproductive age.

As I have indicated above, Enquelga herders do not control the mating of their camelids, and beyond doing their utmost to provide access to good-quality pasture lands throughout the year (which, of course, is no simple task, and which would favour fibre production in all animals, whatever fibre quality the individual animal may have inherited), they do not actively try to promote through artificial selection the increase of individual animals with good-quality fibre.[1] A herder knows his or her animals as individuals, and each and every one of them is appreciated as such. The individual variation in fibre quality between different camelids means that some animals produce little, while others produce abundant but coarse fleece, and yet others produce abundant fine fleece. The herder values the animals for what they provide, and if a fleece is fine it will be converted into clothing, but if it is coarse it will be plaited into ropes and slings. This should not be taken to mean that Isluga herders are ignorant of the principles of animal breeding through artificial selection, or that a lack of interest in creating new distinct breeds implies a lack of intellectual sophistication on their part, which is implicit is Herre's statement (1963: 240) that 'selected breeds are a sign of a highly developed culture', and that 'selected breeds are "improved" towards higher performance when man's abilities increase'.

Unfortunately, this is a standpoint that seems to be shared by many outsiders visiting Isluga, including officials of the Instituto Nacional de Desarrollo Agro-Pecuario (INDAP) or intending purchasers of llamas. Many such visitors criticize the Isluga herds as being hopelessly 'inbred', listing as evidence traits such as blue eyes, short ears or deformed lips. One of these visitors even went on to explain, very seriously, that blue-eyed llamas cannot see in the sun and are more liable to fall in the water and drown. Isluga herders know that these are inherited traits, and they know that blue-eyed llamas are no less likely to fall and drown than their more numerous brown-eyed counterparts. In addition, llamas with short ears do not necessarily also inherit fleece of an inferior quality. It is true that traits such as a mouth deformity prevent the llamas and alpacas from eating properly, but such animals are slaughtered. In addressing the subject of improving body size and fleece yields in alpacas, Fernández Baca (1971: 36) points out that problems such as eye colour and the form of the ears complicate studies which seek to examine the effects of artificial selection, and that there is no evidence whatsoever that such characteristics have any influence on the productive capacity of the animals. The ulterior motives of outsiders who seek to 'improve' the camelid stocks of peasant producers in the Andes should be taken into consideration, since the outsider's perception of the standards to which llamas and alpacas should conform may be very different from the herders' own criteria.

If present-day herders in the Andes go no further in influencing the genetic make-up of their animals than to remove some of the males of reproductive age through castration or slaughter, it cannot be taken for granted that herders in the past would also have followed the same practices. Unlike the reindeer, the domesticated and wild forms of which display little morphological variation, there are clear differences in fleece structure and colouring between the domesticated and wild camelids. The fleeces of llamas and *wakaya* and *suri* alpacas may be said to lie at one end of a spectrum of continuous variation at the other end of which are to be found the fleeces of the wild camelids. However, my 'outsider' description of the characteristics of camelid fleece in Chapter 4 is influenced by Ryder's interpretation (Ryder 1969) of the evolution of sheep wool. Ryder's work shows that major changes have occurred in the structure of sheep wool over a period of several millennia. It is important to recognize that evolutionary schemes of this nature rest on two basic assumptions that have been pointed out by Payne (1968: 370). The first assumption is that it is reasonable to assume that the rate at which wild populations change under natural selection is slow or almost negligible during the past ten or twenty thousand years, in contrast to the great speed at which domestic populations change under human selection. Second, any interbreeding between wild and domestic stocks has not significantly altered the wild populations.[2]

Instead of arguing mainly from the present situation, using evidence from the past to confirm or amplify existing interpretations, Payne (1968: 371) believes that it is important 'to attempt to reconstruct past populations from internal evidence rather than to continue to identify single bones on the basis of modern data'. Payne is, of course, referring to faunal evidence consisting of animal bone remains, but it is also necessary to bear these points in mind when considering the archaeological fleece remains from the Tulan Quebrada. Fallet's conclusion that, on the basis of her examination of vertical sections of camelid hide, llama and alpaca are both the descendants of guanaco, and that the resemblance of alpaca fleece to that of vicuña is the result of convergent evolution, is an example of an evolutionary scheme derived from the present-day situation (Fallet 1961: 54–5). Her study is based on an examination of samples collected from a restricted number of live animals, and she did not examine archaeological samples of camelid fibre. Wheeler's interpretation of the faunal remains from the site of Telarmachay in the Puna de Junín, Peru, led her to propose an early emergence of domesticated alpacas in Level Lower V-1 from 6000 to 5500 BP (Wheeler 1984: 402). It is not known what the fleece of such animals would have looked like, but such information does not square easily with evolutionary schemes similar to the one proposed by Fallet, which is based on the assumption that alpacas were a later derivation from llamas, being intentionally bred by human beings as producers of fleece rather than of meat.

The Lupaqa herders of Chucuito

Much of the documentary evidence on camelid herding practices in the past derives from sources published on the densely populated western shore of Lake Titicaca. This area was that of the Lupaqa, an Aymara-speaking people who had been incorporated into the Inka Empire and at the centre of whose territory was the town of Chucuito.[3] In February 1567, the Spanish inspector Garci Diez de San Miguel was sent to Chucuito to investigate, among other things, the numbers of camelids owned by the Lupaqa people, the 'taxes' they had been required to contribute to the Inka Emperor and the taxes they were now paying to the Spanish crown. In addition, he was to enquire if any of the herds belonging to the Inka Emperor or to the state cults (for example, to the Sun) were still in existence. Chucuito was divided into the moieties of Alasaa and Maasaa (upper and lower, respectively, corresponding to the Araxsaya and Manq"asaya of Isluga). The cacique of Alasaa, Martín Cari, gave a very guarded reply when asked how many camelids the people possessed, saying that many Indians had cattle – he did not know how many – but they sold wool and made clothing from alpaca fleece (Diez de San Miguel 1964 [1567]: 18). However, when he was asked if there was a community chest, he replied in the negative:

> They do not collect things for the community because the indians pay the eighteen thousand pesos of tribute in Potosí when they go to work in the mines as previously stated and the thousand pieces of clothing they make and

> send to the royal officers of Potosí and the increase of the community livestock is used as already declared so that they do not have to put or remove [anything] from the community chest and when sometimes the day labour which the indians earn in Potosí is not enough to pay the eighteen thousand pesos, the shortfall is taken from the aforementioned community livestock. (ibid.: 24).

This statement suggests that it was important to maintain herd sizes, even though the herd referred to here was under communal ownership, belonging to Alasaa rather than to one family. One of the functions of this herd was to help feed the poor during religious festivals or at times of need (ibid.: 23). It was therefore a sort of 'bank' which obviated the need for storing money in the community chest, a concept which was echoed in Isluga when a herder declared with satisfaction during his family's *wayñu* ceremony that his llamas and alpacas were his 'bank'.

However, herd sizes were being steadily eroded in the years following the Spanish invasion. A previous cacique of Chucuito remembered that in Inka times, community herds were so numerous that there was a shortage of pasture (ibid.: 40), and a Spaniard in Chucuito declared that he had heard of a private individual who owned more than 50,000 head of cattle (ibid.: 50). No doubt, this was a highly exaggerated figure. The people themselves reported much more modest figures: 'generally all have native livestock [camelids] up to a hundred and more head and fifty and twenty and ten and three and two and of this order, and that some indians have no livestock, although very few' (ibid.: 97–8). Another census of herds was conducted a few years later, listing herds by household, lineage and moiety, with one household declaring 1,700 animals. This second inspection, conducted by Pedro Gutiérrez Flores, shows that people such as the priest Fray Agustín de Formicedo, against whom many complaints were registered, was actively promoting the importation of sheep from Spain. One way of doing this was to have the camelids of the 'community' herds sold off and replaced by sheep. These community herds originally belonged to the Inka Emperors and to the state religious cults, and they apparently had come under the control of different Aymara ethnic groups within a few years of the European invasion. Flores Ochoa argues that Europeans would have manipulated the replacement of camelids by sheep as a means of controlling the native populations of the *altiplano* (Flores Ochoa 1970: 67). In many cases, it seems that the Aymara people found it difficult to claim the community herds, and Martín Cari stated that there were no herds belonging to the Inka Emperor or to the Sun in Alasaa because the Spanish had taken everything many years previously (Gutiérrez Flores 1970 [1572]: 19, 44–5, 46). Similarly, Quechua speakers further north attempted to assert control over communally owned herds, which were designated as *sapsi*, a word that applied equally to animals and agricultural lands.[4]

It seems as though there was a need for such animals during colonial times. The documentation from Chucuito shows that there were some Aymara highland people without any animals, since herders were and always are subject to adverse weather conditions and diseases which may decimate their herds. Diez de San

Miguel (1964 [1567]: 99) was told that a thousand animals died in one year because of frost, and Garcilaso (1966 [1604]: 513–14) reports a disastrous epidemic of *karachi* (llama mange) between 1544 and 1545, saying that this highly contagious disease, which also affected wild camelids and foxes, was previously unknown.

As for more precise details of herding practices, we learn from Gutiérrez Flores (1970 [1572]: 43) that shearing took place, ideally, every three years. Most of the respondents, when questioned by Diez de San Miguel, replied that animals were shorn every three years (Diez de San Miguel 1964 [1567]: 152,154,157,161).[5] One respondent stated that it was better to shear every two years, since longer fleece is wasted due to the weathering effects of water and sun (ibid.: 164). However, these documents reveal little about herd management as practised by individual households of the seven towns in Lupaqa territory. Fortunately, further information can be gleaned from Bertonio's Aymara dictionary, which is based on Lupaqa Aymara and which was probably compiled not long after the crown inspections. It would seem that camelids were separated according to age and sex, the so-called *phatta* animals being separated into groups of older animals (*ancuta*) (Bertonio 1984 [1612], Bk II: 258). There is also an entry that refers to the marking of animals:

> To mark the sheep etc. with woollen yarn, or with some fringe. *Coyllutha, Puyllutha*, or to make some cut in the ear.
>
> (Bertonio 1984 [1612], Bk I: 429)

No indication is given as to the context of this marking, or whether it took place in a ceremony equivalent to the *wayñu* or *k'illpa* of present-day Aymara herders. The verb *puyllutha* cited by Bertonio derives from *p"illu*, a 'garland'. As a Jesuit priest, Bertonio often mentions whether the words he lists are related to 'pagan superstition', and in particular to rites of passage observed in connection with human beings, but in the case of animals he is silent on the matter.

Equally, there is no ceremony implicit in the word he lists for breeding animals:

> *Harccatha*; to put many stud animals in a flock of sheep
>
> (ibid., Bk II: 122)

This word is equivalent to *uywa jarq'aña* ('to pair off animals') in modern Aymara. However, there is more information supplied regarding the giving of animals to children, which coincided with a rite of passage, during the ceremony of the first hair cutting:

> *Jinchuma caura*. Sheep which they give to the children when they cut their hair at two, or three years of age. It seems to be a superstition.
>
> (ibid., Bk II: 134)

> Sheep which a father gives to his son or an uncle to his nephew. *Hinchuma*.
>
> (ibid. Bk I: 119)

Murra (1975a: 124–5) assumed that animals were also given to children at the time of marriage, although he recognized that other mechanisms involving the transmission of animal property from one person to another are not known.

Although the details are somewhat sketchy, it seems that herding in Lupaqa territory during the latter part of the sixteenth century was similar to recent pastoral practice. The main difference was the existence of the communally owned herds of camelids. Lupaqa herders were apparently anxious to maintain their herd sizes at all costs, in order to have live animals in reserve for times of hardship and climatic disaster, and to help cope with the new tax demands for which they were now liable. Data collected for the sixteenth-century censuses was arranged according to households, lineages and moieties, but Bertonio's dictionary shows that domesticated animals were given as personal property to individuals. This conforms to present-day Aymara practice in which animals are inherited by individuals and cared for by households. The Lupaqa herders used the successive generations of offspring of their animals as personal property, and if we knew more about the transmission of animals from one person to another we would undoubtedly understand networks of social relations among the herders themselves. There is no indication that herders purposefully sought to influence the genetic make-up of their herds, since Bertonio's translation of the Aymara *harccatha* implies that mating took place when the herders introduced stud animals into a herd of females: that is, in a manner similar to that practised by Chinchillapi and Ulla Ulla herders today. Thus, any selection would only take place through the males and would probably be cancelled out by a greater variety among young and female animals. It should also be pointed out that the household herds were left out on the same pasture grounds as those belonging to the caciques and the community, all 'mixed up' with the animals of the community authorities and lords (Diez de San Miguel 1964 [1567]: 78).

Camelids and the Inka Empire

Prior to the European invasion, the Lupaqa territories had come under Inka imperial rule (figure 6.1) and the Lupaqa had lost control of considerable numbers of camelids on a previous occasion. The Spanish administrator Hernando de Santillán wrote that after the Inkas conquered an area, a census was made of all the camelids and the Emperor distributed them among the Sun and certain state religious cults, to himself, and to the caciques and commoners, the number of animals given being allotted on a strictly hierarchical basis (Santillán 1968 [1563]: 108). Because the areas surrounding Lake Titicaca were the richest in camelids of all the territories conquered by the Inkas, they were an important source of wealth for the expanding empire, and the expropriation of their animals and the loss of autonomy suffered by the lakeside kingdoms were sources of tension in relations between these peoples and the new Cuzco-based administration. From an Inka point of view, the animals which belonged to the Emperor and those assigned to religious cults were called *qapaqllama*, which Murra translates as the 'herds of the mighty', but the herds belonging to the various ethnic groups were

Figure 6.1 Map of the Inka road network, showing the approximate extent of the Inka Empire (after Lumbreras 1974).

known as *waqchallama*, translated by Murra as 'the herds of the weak' (Murra 1975a: 135). According to Murra, the Inka élite manipulated this ideological stance to its own advantage at the expense of other ethnic peoples. Inka organization and management had some very important implications, at least for those herds that came under state control. As Inka influence grew, the Empire had access to increasingly varied resources, and the number of storehouses, which contained a variety of foodstuffs and other goods, increased. Llamas were necessary as beasts of burden to transport these goods, but they were also required in great numbers at times of war. Zárate, for example, reports that on one occasion, QuizQuiz, who was one of Ataw Wallpa's captains, had an army of more than 12,000 warriors, and that when he was suddenly forced to withdraw he left behind more than 15,000 llamas and 4,000 men and women (Zárate 1862 [1555]: 483).

In Inka times, before war was undertaken against another ethnic group, a camelid was sacrificed and its heart, lungs and entrails seized and examined for auguries and omens (Garcilaso 1966 [1604]: 360). Murra adds that black llamas and dogs were starved so that the enemy's heart would wither away as did the animals, and that hostile activities temporarily ceased to allow for the necessary offerings, which included the sacrifice of llamas, to honour and greet the new moon (Murra 1975a: 140).

According to Garcilaso, black camelids were preferred for Inka sacrifices. Black was regarded as being more divine than other colours, and Garcilaso explained this was because black animals were black all over, while white animals, in spite of having an all-white fleece, had a dark snout, which was held to be a defect (Garcilaso 1966 [1604]: 360). However, other accounts speak of a white male llama, elaborately dressed, which was accompanied by the royal insignia and which announced the coming of the Inka Emperor. Sarmiento described this *napa* llama as a white camelid wearing a red vest, gold earrings and a red shell necklace; in front of it a bearer carried a staff 'like a hafted feather cross' called *sauturpaucar* (Sarmiento 1942 [1572]: 40). Molina gave a similar but less full description of this llama, which he calls both the *Raymi napa* and *tupa guanaco*, in his description of the elaborate Qapaq Raymi festival that was celebrated in November (Molina 1947 [*ca* 1575]: 105).

There would seem to have been a symbolic colour coding in the use of certain camelids. Molina listed the animals sacrificed to the Sun and to other deities during the month of May:

> In this month they sacrificed to the Sun a great number of sheep [llamas] of all colours, some called *huacarpaña* which were white and woolly, and other sheep [llamas] called *huanacos*, and other white woolly alpacas, called *cuyllo*, and other alpacas called *paucarpaco*, which were woolly females of a bright reddish colour, and other alpacas called *oquipacos*, and other large sheep [llamas] called *Chumpi*, which were the colour of those that was almost tawny; and other sheep [llamas] called *llancallama*, which were black and woolly.
>
> (Molina 1947 [*ca* 1575]: 52–3)

The words used to describe these animals reveal an interest not only in colour, but also in other fleece qualities. Llamas known as *huarcarpaña* may well have been named after *warq'ar*, Quechua for 'white egret', combined with *arpay*, 'to sacrifice (to a divinity)' (González Holguín 1952 [1608]: 34; Dedenbach-Salazar Sáenz 1990: 193). This terminology seems to share an ideology also evident in Isluga, in the connection between camelids and birds associated with watery places. The word *cuyllo* may be the equivalent of the Quechua *qoyllur* ('shining'), another apparently favourable quality in the fleece of white and woolly alpacas. *Paucar*, as a description of a reddish brown, may well be the equivalent of the Isluga use of the word *chupika* for reddish-brown alpacas, since *chupika* is one of various Aymara terms for 'red'. *Llanco* seems to be the same as *llunka* in Isluga, for an all-black animal without a spot of another colour. The elaborate Quechua vocabulary of the Inka period for describing camelid fleece colours and qualities reveals that the real importance of llamas and alpacas in Inka times was to provide raw materials for weaving, a well-documented activity (Murra 1975b; Murra and Morris 1976).

In general, many of the chroniclers agree that brown (*pardo*) llamas were sacrificed to the deity Wiraqucha, white to the Sun, and piebald animals to the Thunder deity (Murra 1975a: 141; Rostworowski 1986: 34). Few accounts are as detailed as some of the descriptions given by Father Cobo, who reports the sacrifice of one hundred llamas during the month of Camay. These are described as having a light brown body colour, with white head and legs below the knees – 'if they were found', he added (Cobo 1956 [1653]: 212). Piebald animals were sacrificed in April. Cobo calls these camelids *murumuru* which is the Quechua equivalent of the Aymara *t'axllu*, while one hundred white and woolly llamas were sacrificed during the festival of Coya Raymi, the great festival sponsored by the Inka Empress (ibid.: 214, 217). Guaman Poma also lists certain occasions when llamas of different colours were sacrificed, but this author also hints at some regional variation, as he says that the Qulla people of Qullasuyu offered black camelids but the Puquina people of Qullasuyu white llamas (*'carneros blancos de cuyro'*) (Guaman Poma 1980 [1615]: 273).

The symbolic system(s) that seem to have lain behind the use of different camelid qualities in different contexts cannot be reconstructed completely. What is important here is the evidence for the exploitation of llamas and alpacas in a very rational way, often for religious purposes and to maintain the prestige of the Inka rulers, who claimed divine right for their legitimacy as Emperor and Empress. Large numbers of animals were regularly slaughtered and the meat shared out in what was designed to be a lavish display of 'generosity'. However, Garcilaso was anxious to point out that only male animals or barren females were sacrificed, never females with reproductive potential (Garcilaso 1966 [1604]: 360). This is supported by the oral Quechua traditions compiled in Huarochirí at the very end of the sixteenth century or the beginning of the seventeenth, which describe the sacrifice of a sterile female llama to a deity called Chuquihuampo (Taylor 1987: 159). Fray Martín de Morúa also observed that female camelids were not slaughtered because of their reproductive potential (Morúa 1946 [1590]: 170).

The Inkas were in a position to keep very detailed accounts of all the camelids under state ownership (for the religious cults) and also those owned by the various regional shrines throughout the Empire. Murra reports that the official in charge of the *quipu* (the knotted cords on which records were kept) in Qullasuyu told Polo de Ondegardo that he knew how many llamas had been raised in the area since it had formed part of the Inka Empire (Murra 1975a: 142). This being the case, the Inkas were well placed intentionally to breed animals of the colours they required,[6] to conduct other experiments and to keep records of the results. Certainly, the *suri* breed or variety of alpaca already existed in Inka-times, since there are Inka style silver models which represent a camelid with long fleece of slightly wavy locks reaching nearly to the animal's feet (Morris 1988: 252, fig. 8.11).

The faunal analysis undertaken by E.S. Wing (1988) of animal remains from Huánuco Pampa, an urban Inka period site located at nearly 4,000 m above sea level in the northern Central Andes, reveals an interesting situation. The animal remains consisted predominantly of camelid bones, identified by Wing as llamas and alpacas in approximately equal numbers. However, the remains included sixteen bones that were found to deviate significantly from the rest, being either more robust or more gracile than equivalent bones found in normal camelid populations. It is not known how many individual animals these bones represent, but the deviant bones are more commonly associated with certain walled compounds, and Wing suggests that these special animals may be the result of breeding experiments (Wing 1988: 170).

Therefore it can be seen that under the Inka Empire, the state was in a position to institute policies of purposeful artificial selection for qualities that were perceived as desirable in camelids, since camelids ceased to furnish all the essentials of their subsistence. Other forms of security to animal property were available, thus meeting Ingold's two essential conditions. This is not to deny that other peoples in other times and places might also have been in a position to attempt the breeding of camelids through artificial selection, but less coherent evidence is available. What cannot be denied is that although camelid herds were numerous and often large in those parts of the Empire suited to a pastoral economy, people maintained close control over herds belonging to the Inka rulers and to the state religious cults. Writing in 1563, Santillán (1968[1563]: 54) observed that the animals that previously belonged to the Inka Emperors were still known to the people: 'until today they are known to whom they belong by their signs'. It is not clear what he means by 'their signs', for the animals concerned were obviously the progeny of the animals that had belonged to the Emperors. Perhaps the people who managed to gain control over the animals not destroyed or expropriated by the Spanish used techniques analogous to those employed by Isluga herders discussed in Chapter 3, or those by the herders of Paratía (Flores Ochoa 1978) to enable them to recognize the animals and their offspring.

It is also apposite to remember that the Inka rulers claimed ownership over the wild guanaco and vicuña, which were designated *intipllama* ('the llamas of the Sun') (Guaman Poma 1980 [1615]: 288). As indicated in Chapter 2, the Inkas attempted to implement shearing strategies on vicuña and guanaco, which resembled those practised on domesticated animals. Additionally they culled animals that they considered to have undesirable qualities. It is not clear to what extent these strategies were actually practised, but there remains a distinct possibility that changes in the fleece structure in the wild animals might occur in wild populations over time. We should take heed of Payne's cautious remarks (1968: 370) referred to above, that the assumption that the rate by which changes in wild populations of animals occur under natural selection is slow is, in fact, only an assumption. Browman interprets Garcilaso's account of the capture of wild animals and the subsequent release of 'such males as were necessary as sires' (Garcilaso 1966 [1604]: 326) to mean that wild guanaco were infused into the potential domestic breeding pool in the form of newly domesticated cargo animals (Browman 1974: 194). However, Garcilaso himself does not state this as he refers to the release of a necessary number of sires (and he specifically mentions deer in his discussion of wild animals) to serve the wild populations of the species concerned. The conclusion that may be drawn is that the Inka rulers sought to apply similar management practices to wild and domesticated herds, but that the former were managed under conditions which in some senses resemble those of present-day ranching practices.

The exploitation of camelid fibre in pre-Inkaic times

Early iconographic representations of camelids exist that display an interest in the qualities of fibre, occurring in Wari, Tiwanaku and earlier contexts. The presence in iconographic imagery of fleece-bearing camelids provides a means for enquiring how people before the Inka period responded to fleece as a raw material. However, it is important to take into account that people did not always choose to express their responses to a particular material in visual form. In a fascinating article, George Miller and Richard Burger examined the subsistence system of the residential areas of Chavín de Huántar. One of their concerns was to address how ideological messages were encoded in Chavín imagery, which frequently depicts exotic animal species, such as caymans and felines. They examined how the exogenous iconographic themes were related to the social reality of the worshippers at the Chavín de Huántar temple (Miller and Burger 1995: 422). Their analysis of the faunal remains showed that hunted and herded camelids constituted an important component of the Chavín de Huántar subsistence economy, yet camelids do not feature in Chavín art. The people did not express their responses to materials such as fleece in the Chavín art that has survived.

There are Classic Tiwanaku ceramic effigy vessels with modelled camelid heads having a rope through the animal's ear. In addition, there are examples carved in stone, for example, on the back of the Bennett stela from Tiahuanaco itself (figure 6.2), and on a Tiwanaku-style stela from Huancane north of Lake Titicaca, and carved in wood on small snuff trays from San Pedro de Atacama cemeteries. The cemeteries of Quitor 5, Solcor 3, and Coyo Oriente have produced four such trays (figures 6.3 and 6.4; plates 6.1 and 6.2), which clearly show camelids with their characteristic dentition, long ears, a down-turning tail and feet with two digits. These depictions present elaborately 'dressed' animals decked with typical Tiwanaku symbols, and from which, in the case of one of the Solcor 3 trays, branching vegetal forms emerge. Torres points out that the motif born by this branching vegetation resembles an earlier motif, which has been tentatively identified as a seed head of cotton, and that this motif is probably also represented in a more stylized version on the snuff tray from Coyo Oriente, which has, in

Figure 6.2 Camelid carved on back of the Bennett stela from Tiahunaco (after Posnansky 1945).

Figure 6.3 Carved wooden trays from Quitor 5, Solcor 3 and Coyo Oriente cemeteries, San Pedro de Atacama. Upper row, from left to right, and lower left: tray No. 2235, Quitor 5; tray No. 1,874 from tomb No. 44, Solcor 3; tray No. 4,049–50, Coyo Oriente (after Torres 1987). Lower right: tray number 1,075, Solcor 3 (after Llagostera *et al.* 1988).

Figure 6.4 Map of the South-Central Andes.

addition, another plant motif which may represent maize (Torres 1987: 200). These two motifs also appear with the camelids depicted on the back of the Bennett stela. The Classic Tiwanaku phase (Tiwanaku 4) has been dated to AD 375–725 (Kolata 1983: 252), while the tombs which yielded the camelid design snuff trays at the site of Solcor 3 are assigned to phase B of the cemetery, which has evidence of Tiwanaku influence.[7] The tombs of Solcor 3 phase B overlap in time with Tiwanaku 4, the period that saw an expansion of influence during which Tiwanaku established what have been interpreted as administrative centres and economic colonies over the Bolivian highlands, eastwards into the jungle and westwards to the coast of southern Peru and northern Chile (ibid.: 252–3).

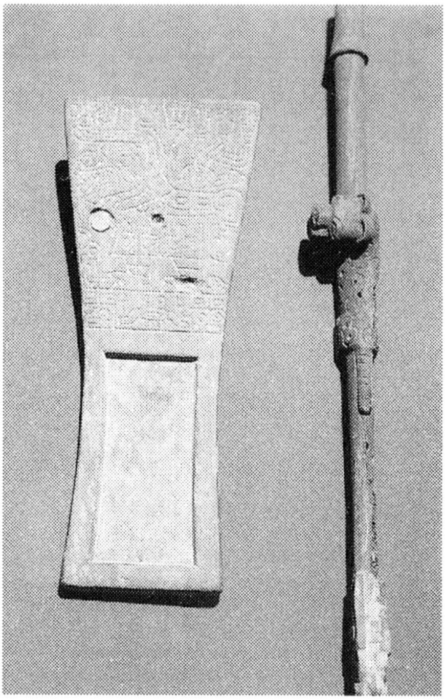

Plate 6.1 Wooden tray No. 1,874 and tube (note the head of a camelid carved at the bottom of the tube, seen from above) from Solcor 3 cemetery. By courtesy of the Museo Arqueológico R.P Gustavo Le Paige.

Plate 6.2 Wooden tray No. 1,075 from Solcor 3. By courtesy of the Museo Arqueológico R.P. Gustavo Le Paige.

The motifs associated with the camelids depicted on one of the Solcor 3 and the Coyo Oriente snuff trays show a fusion of the rigid, geometric Tiwanaku style with another more organic and descriptive style. Torres observed that the latter style is also displayed by the large Wari-style Robles Moqo ceramics from the site of Pacheco, near Nazca (Torres 1987: 201). The Pacheco vessels have been dated as belonging to the earliest Wari intrusion, that is, during Middle Horizon 1B (after AD 600) into the Nazca area (Menzel 1977: 53). Wari ceramics and textiles are often regarded as adopting and reinterpreting many of the symbols and stylistic conventions of Tiwanaku art, which were then taken further afield as the culture of Wari spread from the highlands in the Ayacucho area to coastal regions of Peru, although the precise nature of the relationship between Wari and Tiwanaku remains to be clarified (Menzel 1964; Kolata 1983: 253; Torres and Conklin 1995).

A fragmentary, but extraordinary Wari textile depicts a series of female camelids in the act of giving birth (Dedenbach-Salazar Sáenz 1990: illustration 7). This textile is highly unusual, but it shares a common theme that occurs in the Tiwanaku camelid-shaped incense burners and in the little camelid heads at the base of the elaborate tubes that accompany the snuff trays in the San Pedro de Atacama tombs: a rope pierces the ear of the camelid and passes round the neck. In the Wari textile, their backs are draped with stylized feathers, a feature that occurs in other Wari representations of camelids.[8]

Excavations at the site of Maymi in the lower Pisco valley at a distance of 13 km from the coast revealed a pit with more than fifty vessels in a mixture of forms and styles classified by the excavator, the late Martha Anders, as belonging to Middle Horizon epochs 1B and 2 (approximately AD 600–800). Among these vessels are forms and techniques of decoration not previously reported in Wari ceramics. These include vessel forms which are decorated in bas relief and with painting, or a mixture of bas relief, modelled appliqué and painting, with plants densely packed round a series of camelids and felines (Anders 1988: 5–6). The camelids have black fleece hanging down to the feet and the fleece is depicted in layers which resemble overlapping feathers (Anders personal communication December 1988). In Isluga, wild cats are considered to be the spiritual herders of domesticated camelids, and the Maymi pottery would seem to indicate that the association is a long-lived one, even though the social context for this association has changed dramatically over time. The Maymi camelids have red and white plied ropes hanging round their necks, from which a circular object is suspended, and these recall the undyed ropes which are used to hang the *wistalla* round the necks of Isluga camelids. Flores Ochoa (1968: 113) indicates that the Quechua herders of Paratía hang small bags of herbs around the necks of their animals. The ethnographic parallels support the idea that the Maymi ceramics depict concepts of abundance and fertility in the Wari religion. Previous interpretations have considered these concepts in reference to images of the human form,[9] and it is interesting to find camelid representations from the coast which appear to be alpacas, but the feather-like representation of the fleece staples evokes metaphors involving birds. A more typical Wari-style modelled vessel in the Museo Nacional de Arqueología, Antropología e Historia, Lima, takes the shape of a reclining camelid also displaying a similar feather-like treatment of fleece staples, which are drawn by means of incised lines (Lumbreras 1981: 138).

To summarize, during the seventh and eighth centuries AD, elaborate imagery was current in Tiwanaku and Wari art that can be interpreted as indicating the cultural importance of abundant fleece supplies. The imagery is given visual expression through representations that display an interest in fibre, whether of camelid fleece, cotton or the corn silk of a maize cob. This surely cannot be accidental in societies that were producing vast quantities of beautifully woven fabrics.

The theme can be traced even further back in time, for a mantle found in a tomb on the peninsular of Paracas in southern Peru is surrounded on all four sides with a frieze of ninety figures worked in a technique known as cross-knit looping, executed in camelid fibre yarns, with two of these figures representing camelids decked with vegetal forms and fruits (d'Harcourt 1974: 179; plates 93, No. 26, and 97, No. 50) (figure 6.5). Sawyer places this mantle early in the Nasca period (Sawyer 1962: 158), so it must have been made some time after AD 100. The Nasca culture flourished round the drainage of the Nazca valley in the south of Peru, and archaeologists using Rowe's chronological framework place it in the Early Intermediate period, but Menzel points out that the archaeological record indicates close contact between the south coast and the peoples of the Ayacucho area, which she says later developed into a 'special cultural relationship' between Nazca and the Wari capital (the type site of Huari itself), during the so-called Wari Empire in the succeeding Middle Horizon (Menzel 1977: 52–3).

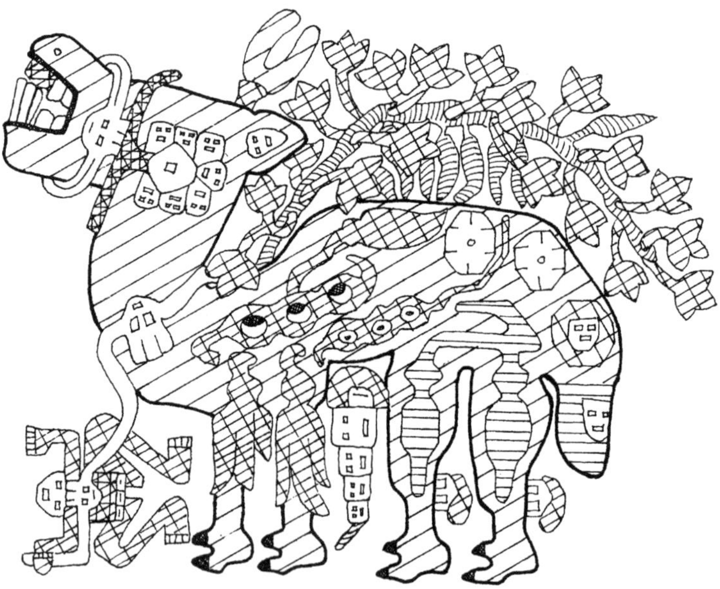

Figure 6.5 Camelid decked with plants and fruits from Paracas mantle (after Sawyer 1962, with corrections). The Brooklyn Museum, 38.121.

There were also close contacts between Nasca and the peoples of the highland areas south of Ayacucho, including Cuzco and the Titicaca basin, which of course was where the Tiwanaku sphere of influence was already expanding. The stylized representation of a camelid with abundant fleece and decorated with three geometric cruciform symbols in a tapestry from a site in the Azapa Valley near Arica in Chile[10] also belongs to this tradition (figure 6.6). According to Rivera, this textile dates from late in the Alto Ramírez series in the far north of Chile, the main period of which is said to have occurred between 500 BC and AD 300 (Rivera 1980: 90). However, it seems strange to trace back this particular type of imagery (camelids decked with vegetal motifs) to its earliest manifestations on the coast, rather than in the Andean highlands. The use of camelid fibre yarns can be traced back further on the Peruvian south coast (for example, Paracas), and Conklin and Moseley (1988: 155) point out that by the beginning of the Early Intermediate period, large quantities of camelid fibre yarn were being used in the numerous mortuary textiles in the Paracas cemeteries. Conklin (1985: 2) dates the first use of camelid fibre on the coast back to 900 BC (during the second half of the Early Horizon), but the situation is different if evidence from inland sites in the South-Central Andes is taken into consideration, as will be made clear in Chapter 8.

The presence of camelid fibre on the south coast of Peru is used to imply trading relations with the highlands from an early period. On the north coast of Peru, camelid fibre was not used in abundance until nearly a millennium later. Certain textiles from Paracas in the south of Peru, and from Alto Ramírez, on the coast of the far north of Chile, have been identified as belonging to the Pukara culture because of their similarity to the colours and designs used in Pukara ceramics (Conklin 1985). Novoa and Wheeler (1984: 123) mention that domestic camelids played a predominant role in the economy of the site of Pucará, and they suggest that perhaps specialized breeding for wool production occurred about 500 BC. Lumbreras (1974: 123) mentions that early Pukara ceramics are ornamented with compositions which include ducks and camelids, a significant combination in the light of the bird metaphors evoked by the Wari feather-like depiction of camelid fleece.

Of course, the weavings themselves constitute primary evidence for the fleece that was at the disposal of the weavers. The sophisticated textiles that occur frequently at Paracas and Nasca period sites on the coast demonstrate that dye technology was highly developed. The Paracas Cavernas tombs contained more cotton than camelid fibre in the textiles compared with the later Paracas Necropolis textiles, where the use of camelid fibre was common. O'Neale (1942) identified 190 different hues in the yarns used in this later period. Although uneven fading accounts for some of the tones, the number of different colours is impressive, and this also indicates that supplies of light-coloured camelid fleece suitable for dyeing were abundant. In Nasca period 4, the typical slit tapestries characteristic of this phase are executed in designs which are preferentially outlined with narrow black lines. In many instances, the black is missing in an otherwise well-conserved textile, as the black yarns have deteriorated over time,

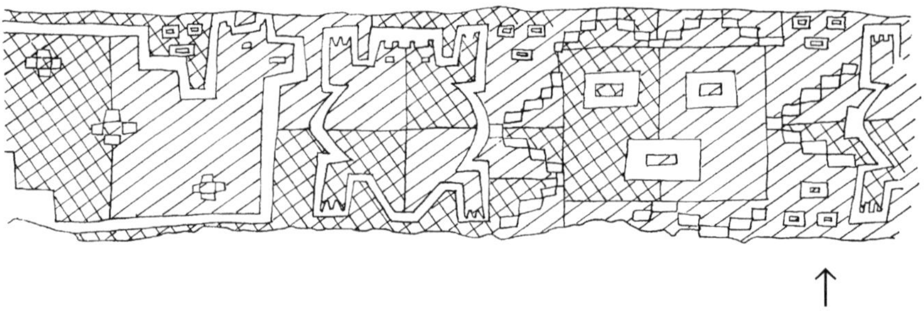

Figure 6.6 Alto Ramírez tapestry from the Azapa Valley, near Arica, Chile. The arrow shows the direction of the warp (after Rivera 1980).

presumably due to the corrosive effects of the mordant used to fix the black dye (plate 6.3). More research is needed on this topic, but there is a strong possibility that black fleece was not easily available during this period, and that the black camelids represented in the Maymi ceramics were a novelty on the coast in the subsequent Middle Horizon.

Black outlining is characteristic of both Nasca- and Wari-style painted ceramics. However, the outlining in Wari-style tapestry designs is frequently executed in white (plate 6.4), and it should be pointed out that in the earlier Pukara-style ceramics, designs were outlined by incisions which were later filled with white. Wari tapestries do contain areas of black (plate 6.5), but I am not aware if any dye analysis has been undertaken to check whether the black was dyed or the natural colour of fleece in these textiles. Some of the designs are outlined in black, like the ceramics. Both Wari and Tiwanaku tapestry tunics are noted for the exploitation of natural golden and tan brown colours, combined with the use of a few dyes to produce reds, blues and greens (Sawyer 1967; Stone 1986) (plates 6.6 and 6.7).[11] Thus there is ample evidence for camelid fleece occurring at the light brown end of the range of possible colours, with black less frequently represented.

It should also be noted that grey colours, in general, are very poorly represented in archaeological collections. I know of two four-pointed caps executed in a knotted technique, using undyed yarns, which are almost identical. One cap comes from the site of AZ 71 in the Azapa valley, near Arica in northern Chile, and the other is reported to have come from Chokari, between Potosí and Oruro in Bolivia (plate 6.8). The natural colours used are white, light brown, medium brown, grey and a very dark chocolate brown (practically black). These caps belong to a series associated with Tiwanaku contexts (Frame 1990: catalogue 7, plates 5 and 6). Another example of grey fleece occurs in a plaited rope (it is very similar in technique to those plaited today by the men of Isluga) which was used to apply a pad to the head of one of the skulls found in the cemetery of Quitor 6, San Pedro de Atacama, as part of an apparatus for artificially shaping the skull. The overall impression gained is that, as one goes back further in time, weavers had recourse to less variety in the natural colours available in camelid fleece.

Plate 6.3 Nasca-style tapestry band. The arrow indicates the direction of the warp. Museum of Mankind 1954 W Am: 5 552. By courtesy of the Trustees of the British Museum.

Plate 6.4 Fragment of a Wari tunic with white outlining. The arrow indicates the direction of the warp. Sawyer type IB (see Sawyer 1967). Museum of Mankind 1971 Am: 20. By courtesy of the Trustees of the British Museum.

Plate 6.5 Fragment of a Wari tunic which includes one of the side seams. The arrow indicates the direction of the warp. From the vicinity of Pampachiri, Department of Andahuaylas, Peru. Pampachiri school.

Plate 6.6 Fragment from a Wari textile found in the vicinity of Pampachiri, Department of Andahuaylas, Peru. The arrow indicates the direction of the warp. In the collection of Pampachiri school.

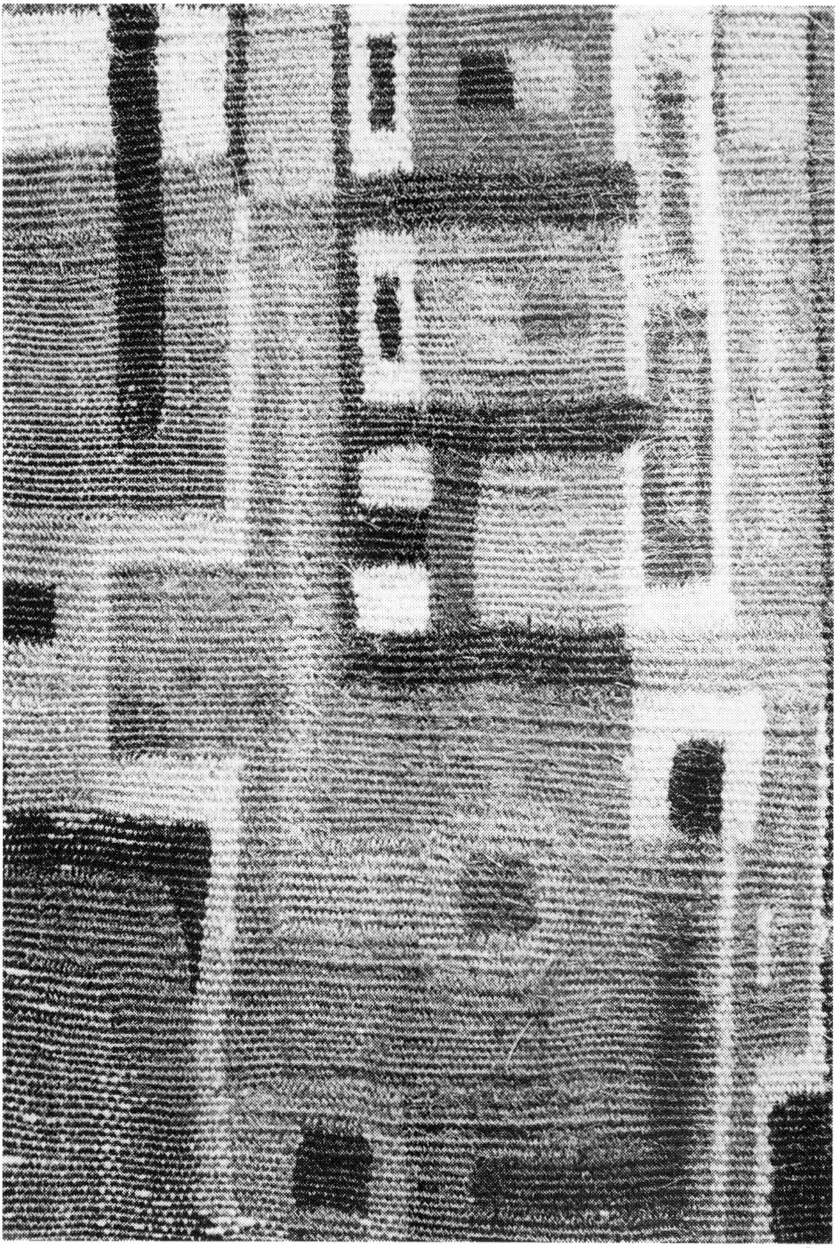

Plate 6.7 Detail of a tapestry fragment from a Wari-style tunic. Sawyer-type IIIB (see Sawyer 1967). The arrow indicates the direction of the warp. National Museum of Scotland 1954 1428. By courtesy of the National Museum of Scotland.

Plate 6.8 Tiwanaku-style four-pointed cap from Chokari, between Potosí and Oruro, Bolivia. Museo Antropológico de la Universidad de San Francisco Xavier de Chuquisaca, Sucre, Bolivia 3148 318–02–1674. By courtesy of the Museo Antropológico de la Universidad de San Francisco Xavier de Chuquisaca.

Most of the material culture referred to thus far in this search for an entry into the cultural responses to camelid fleeces of past societies, and the effect that differing responses would have on the manner of exploiting the fleece available, have been provided by societies that were growing in urban complexity: the Inka Empire, the expansive Wari and Tiwanaku cultures, and Pukara, with its centre at Pucará containing a stone walled court with semi-subterranean burial vaults built of carefully dressed stone blocks and slabs. The court is surrounded by constructions with doors facing the inner court (Bennett 1936). In the case of Wari expansion into the Jauja-Huancayo sector of the Mantaro drainage in the central Peruvian highlands during the sixth century AD, Browman states that the economic base of such societies was changed from pastoralism with secondary horticulture and hunting to agriculture with secondary herding, even though there were still large herds of camelids in the area at the time of the Spanish invasion (Browman 1974: 189). The above discussion on the use of different natural colours in Wari tapestries suggests that Wari people had an interest in increasing the range of these colours, particularly darker tones, and they might well have been in a position to implement selective breeding of camelids in order to produce different fleece colours. Mummified llamas and alpacas found at Moquegua, southern Peru, were identified by Wheeler *et al.* (1995) as having fine and coarse fleece types, indicating the existence of two breeds of llama and two breeds of alpaca during Wari times. Their research provides evidence for the different qualities of fleece available to Wari spinners.

The following chapter introduces the reader to the environment and archaeology of the Tulan Quebrada. Occupation of this region goes back to periods remote in time to the events of Wari and Tiwanaku. Moreover, this was an area that did not see the rise of urban centres and large-scale architecture. The archaeological setting constitutes the context within which the yarns and fleece remains will be discussed.

7 The emergence of herding societies in the Atacama

The study of the classification of fleece in Isluga as a technical activity demonstrates how herding technology and ideology influence the specific categories of yarn spun in the community and how natural fleece colours are deployed in weaving and plaiting. However, the spinning of yarn responds to a particular set of cultural circumstances involving human beings and herd animals. To examine further the historical particularities of a herding way of life and the implications for changes through time in fibre technology, I will review archaeological evidence from the past. The present chapter examines the emergence of herding societies in the Atacama, and it provides the context for the discussion in Chapter 8 of the yarns and fabrics they produced.

A series of *quebradas* – deep valleys with intermittent streams – intersect the arid rhyolitic plateau from the high-altitude terrain known as the *puna*. They descend to the Salar de Atacama, a salt basin with a few freshwater lakes seasonally occupied by flamingos. This is the environment that forms the setting for the archaeological sites that have produced the yarns and fabrics of camelid fibre analysed in the Appendix and Chapter 8. It is important to understand that this rugged and all too arid landscape that bears the imprint of human activity from the past is not an unchanging backdrop to those activities: in other words, they take place in a meaningful environment. However, detailed reconstructions of the Atacama environment cannot at present be securely correlated with changes in the archaeological record. The information currently available for reconstructing palaeoclimatic conditions in the region is somewhat contradictory, as will be made clear below.

The present chapter focuses on the Tulan and Purifica Quebradas[1] that have, for many millennia, formed part of a hunter-gatherer and herding way of life. Pastoral activities are a notable characteristic of these and other *quebradas* that drain into the Salar de Atacama. The people have added flocks of sheep and goats alongside their herds of llamas. Herders in the Atacama are even more constrained by temporal shortages of adequate pasture than are their counterparts in Isluga. This has given rise to a lifestyle with complex, nomadic, seasonal cycles of movement.

The streams in many of the *quebradas* are intermittent and they peter out before reaching the Salar. They have cut down into the sloping, rugged landscape from the high terrain, the *puna* proper, with its chain of extinct and active volcanoes and highland lakes that run along the length of the Chilean–Argentinian border. The *puna* is characterized by steppe-like vegetation dominated by the bunch grass *Paja brava*, while the land bordering the valleys tends to be sparsely covered by shrubs at about 3,000 m asl. However, the valley floors immediately adjacent to streams and watercourses do provide more abundant pasture of grasses and reeds. At a lower level, near the Salar, there are flat areas of vegetation known as *vegas*.[2]

The extreme dryness that characterizes the climate of this area is due to an almost total lack of precipitation from April to November. Summer rains and snow fall between December and March at altitudes above 3,000 m. Only occasionally does it rain over the Salar and town of San Pedro de Atacama. According to Popper (1977: 8), the aridity is due mainly to a stable subtropical high-pressure zone over the area. During the summer months, an equatorial front of low pressure moves in over the highlands, thus producing rain and snow sometimes called the Bolivian winter. Other factors involved in maintaining the aridity include the cooling effect of the Humboldt Current on the humid air masses of the Pacific Ocean. During the winter months, as temperatures drop, the air mass condenses and it forms thick fog on the coast, but any eastward movement of this moisture is obstructed by the coastal ranges of hills. Additionally, the chain of Andean mountains blocks any humid air masses from moving in from the Atlantic (ibid.: 9).

The low humidity level produces diurnal temperature changes, with a daily variation between high and low temperatures reaching up to 35°C on occasion, depending on altitude (ibid.: 9). Winter, the coldest period of the year, is between April and September, the cold being most intense from July to August. Snow may fall during this period at high altitudes.

Another feature of the area is the occurrence of strong winds from the south-west, especially in October and November, when they may reach more than 100 km per hour. Fine sediments whipped up from the Salar act as agents of erosion, and the winds also increase the high evaporation rates (ibid. 1977: 9).

The Tulan Quebrada is one of the most southerly of a number of valleys draining into the Salar. At its northern end, the Salar de Atacama is fed with water from the San Pedro and Vilama rivers. Further upstream, the Vilama river is joined by one of its tributaries, the Puritama, at Guatin, which is at 3,200 m asl, where irrigated fields are cultivated. The vegetational cover is more abundant here, and there is more pasture, which nowadays is exploited by donkeys, herds of llamas, and flocks of sheep and goats. The Puritama river is, in turn, joined by a tributary, the Purifica (or Puripica) river, which further upstream has cut down into a steep-sided gorge, and above which is located the site of PU 1. The landscape here is characterized by the frequent occurrence of very tall and imposing *cardon* cacti (*Trichocereus atacamensis*), which are exploited by the people of the area as a substitute for wood in roof construction and as planks for doors. This region is said to be more humid than the Tulan Quebrada. It certainly has the appearance of being moister, an impression also shared by T. Holden (personal communication) (plate 7.1). Despite the relative moistness of the Purifica Quebrada, few archaeological sites have been identified. In contrast, well over one hundred sites

Plate 7.1 The Purifica Quebrada, downstream from site PU 1; note the herd of llamas.

have been registered in the Tulan Quebrada. Since PU 1 is the site that evidently saw the domestication of camelids over 4,000 years ago, the question arises whether present-day climate is different from that which prevailed in the past. It is particularly important to explore this issue, given that both Browman (1974) and Hesse (1982b) have argued that increasing aridity encouraged hunters to start herding camelids as a source of 'tame meat'.

Climatic change in the Atacama

Unfortunately, there are no easy answers as to whether the climate has changed substantially over the past five millennia. Sources of information from which palaeoclimatic trends may be inferred are provided by pollen records (Heusser 1974; 1981; Markgraf 1989), sea-level fluctuations (Paskoff 1977), lake-level changes (Stine and Stine 1990) and recent glacial history (Clapperton and Sugden 1988; Rabassa and Clapperton 1990; Seltzer 1990). Much of this research focuses on areas further to the north or south of the San Pedro de Atacama area. In particular, many of the studies are of the southern Andes, at latitudes south of the subtropical anticyclonic region, where different conditions of atmospheric circulation prevail. As a consideration of palynological, climatic and glacial evidence demonstrates, it is not easy to combine this information to provide a coherent interpretation of palaeoclimatic conditions of the *quebradas* to the east of the Salar de Atacama.

The post-glacial desiccation described by Craig (1985: 27) as having taken place throughout most of the Late Pleistocene and the Historic period did not significantly affect the aridity of the coastal desert in the north of Chile. The lack of both human settlement and observed archaeological sites immediately to the west of the Salar de Atacama, at which point no river crosses the desert, supports such a view.

Glacial advances did, of course, occur in the Andes in the Late Pleistocene and during the Neoglacial interval of the Holocene (Clapperton 1990). The lack of glaciers in recent times in the San Pedro de Atacama area means that there were no large sources of water to contribute to the hydrology of the region, in contrast with the glaciated summits east of Santiago (Rabassa and Clapperton 1990: 155). Evidence for earlier Pleistocene glaciations on the summits of the high peaks of the Chilean Andes between 18° 10′ S and the geyser area of El Tatio (20° 20′ S), to the north of San Pedro de Atacama, has been listed by Paskoff (1977: 2–3). Paskoff also mentioned the alluvial fans and *quebradas* through which formerly active ephemeral streams passed in the coastal ranges of what is now the arid north of Chile. He attributed Pleistocene climatic changes to modifications of the atmospheric circulation. A displacement and a weakening of the South Pacific anticyclone would have resulted in more vigorous and far-reaching summer rains (the so-called Bolivian winter) (Paskoff 1977: 3–4).

Present-day factors important for producing glaciation in the Andes vary according to latitude. In the Cordillera Blanca of central Peru, snow turns to rain at approximately the same altitude all year round. The glaciers there are responsive to changes in temperature rather than in precipitation; thus snowlines occur between 4,800 m and 5,000 m, close to the annual 0° C isotherm (Seltzer 1990: 149). In contrast, further south in the Lake Titicaca area, the snowline responds to changes in precipitation rather than in temperature. Between latitudes 16° S and 21° S, the 0° C isotherm is said to drop from 4,800 m to 4,700 m asl. This change is accompanied by a decrease from 650 mm to 150 mm per year in precipitation; consequently, the snowline rises from 5,800 m to 6,300 m asl (ibid.: 149). Even further south, summits with altitudes greater than 6,000 m asl do not have glaciers, precisely because of the low precipitation totals, and the reduction in cloudiness (ibid.: 140). The Puripica and Tulan archaeological sites are located downslope from high mountain ranges that are subject to such arid regimes. True glaciers reappear much further south, from the San Juan-Mendoza area and southward, as the mean annual temperature decreases and precipitation increases (Rabassa and Clapperton 1990: 154).

Long ice cores taken from the summit of the Quelccaya ice cap in the Andes of southern Peru (13° 56′ S, 70° 50′ W) yielded a detailed 1,500-year record of precipitation history.[3] However, Quelccaya stands near the southernmost limits of the tropics, whereas the area under study here is much further south and is subject to subtropical conditions. In particular, there is a significant drop in precipitation, as witnessed by the higher snowline levels discussed above. Cores from the summit of Sajama (18° 06′ S, 68° 53′ W) at 6,542 m asl are of greater potential interest in the context of this book. A team of scientists headed by L.G.

Thompson reported elevated snowlines and a decreased net accumulation from the beginning of the Holocene until about 3,400 years ago. The team takes the presence of high concentrations of soluble species and dust in the cores to indicate that lake levels in the *altiplano* were low between 9000 and 3000 BP (Thompson, Davis, Mosley-Thompson, Sowers, Henderson, Zagorodonov, Lin, Mikhalenko, Campen, Bolzan, Cole-Dai and Francou 1998: 1863).

If the presently available data for reconstructing palaeoclimatic trends are ambiguous, then adequate short-term records collected over recent years are equally lacking. Popper (1977: 10) indicates that meteorological evidence that might support the proposition of a growing trend towards aridity in the region is at present inadequate, and that further research is necessary. Such studies are complicated by the diversion of water, especially from better-quality, less-saline sources near San Pedro de Atacama, in order to supply coastal cities such as Antofagasta. The desiccation of the Vega de Turi to the north in the upper Loa basin has provoked critical comment within Chile (Astorga 1986; Anon 1987) since subterranean and surface water is being extracted in large quantities to supply mines and urban centres in areas without locally available water sources.

Archaeologists' reconstructions of climatic conditions in the past are heterogeneous. Núñez (1983: 168) proposed a palaeoenvironmental scheme for the north of Chile, apparently based on the findings of research conducted in the semiarid area to the south of Copiapó. In it, he included a cooler and wetter period, with forest cover in interfluvial zones up to 2,000 m between 10,500 BP and 6500 BP. The available evidence in support of such a scheme is contradictory, since Llagostera (1979) used the presence of fish that inhabited warm waters in archaeological sites on the coast near Antofagasta (where these species are now locally extinct due to the coldness of the Humboldt Current) to indicate a thermal maximum between 8500 BP and 6500 BP. Aguerre *et al.* (1975) also argue for a climatic optimum in north-west Argentina between 8000 BP and 6000 BP.

The pollen record examined by Heusser (1974: 312) for the southern Chilean Lake District revealed a forest succession that included a warmer period, followed by a cooler and more humid one, between 6500 BP and 4500 BP. Heusser and Streeter's 1980 study of the fossil pollen in an 8.7-metre peat core, the chronology of which was controlled by twelve C-14 determinations, from Alerce, southern Chile, indicated a cold spell ending about 10,000 BP followed by a period of warmth between 9410 BP and 8600 BP. The Alerce sequence does not indicate a period of warm conditions before 6000 BP, but it does show a cooling trend, which set in with successive minima between 4950 BP and 3160 BP and between 3160 BP and 800 BP. This trend towards cooler and wetter conditions (the most pronounced peak in the precipitation levels is reported to have occurred between 4950 BP and 3160 BP) was interrupted twice, at about 3000 BP and 350 BP, by temperatures higher than at present in the Alerce region (Heusser and Streeter 1980: 1,346). The climatic conditions studied by Heusser and Streeter are located south of Latitude 30° S, and they are influenced by the anticyclone of the Pacific. However, detailed studies from the South-Central Andes are lacking, and a cooler and wetter trend might have been more widespread if one accepts Clapperton and

Sugden's evidence (1988: 186–8) for a Neoglacial advance from 5000 BP to 4000 BP, which indicates a cooler, moister period in the Andes. This interpretation is supported by studies of sediments in the lakes of the *altiplano* (Titicaca, Poopó and the salars of Uyuni and Coipasa), which indicate that all were larger and interconnected during the Late Pleistocene. Lake levels underwent a marked decrease during the Holocene. However, water levels fluctuated greatly between 5500–5000 BP and 4000 BP, and Lake Titicaca then gradually rose in two main steps, reaching its present-day level in *ca* 1600 BP (Wirrman and Mouguiart 1995: 353).

A great diversity in South American palaeoclimatic trends is reflected in the pollen record (Markgraf 1989). According to Markgraf, Compositae increased at the expense of Gramineae, but the latter were still dominant between 10,000 BP and 8000–7000 BP at high-altitude sites in Bolivia and north-west Argentina. She observes that this indicates increased temperatures and greater humidity. After 8000 or 7000 BP, herbaceous taxa were replaced by Compositae, *Ephedra* and Chenopodiaceae, indicating a greater *subpuna* (valley) component, or an up-slope shift of these taxa, and drier conditions until about 4000 BP. After this date, the high-altitude components increased again, and the presence of long-distance pollen suggests westerly or winter rains (Markgraf 1989: 14–15).

Clapperton (1990: 302) has proposed another scheme for main intervals of climatic change during the Holocene, deduced from glacier fluctuations and environmental indicators. The information relevant to the Andes is shown in table 7.1.

It is very difficult to tie these loose strands together to form a convincing palaeoclimatic reconstruction for the *quebradas* east of the Salar de Atacama. The cooler and wetter conditions that prevailed in the Andes during the period *ca* 4700 to 4200 BP in Clapperton's scheme (1983: 129) coincide with the period in which the inhabitants at Puripica adopted the herding of camelids. According to many authors, pastoralism is adopted as a subsistence economy in response to

Table 7.1 Main intervals of climatic change in the Andes.

Cold/wet interval *Glaciers advance* *(BP)*	*Warm/dry interval* *Glaciers retreat* *(BP)*	*Place*
	ca 9700 – 8600	Probably all parts of southern hemisphere
ca 8400–7500 (?)		Central and South Andes
ca 6300–6000		North Andes
ca 5200–5000		South Andes
ca 4700–3800		Central and South Andes
ca 2700–2000		All Andes
AD 1250–1930s		Most glaciers in the southern hemisphere have fluctuated during this interval

adversely changing climatic conditions. Increasing aridity is often seen as reducing the availability of pasture, a trend that is accompanied by reduced populations of wild animals. Pastoralism, according to Browman (1974) and Hesse (1982b), enables people to maintain supplies of meat in the face of such adverse conditions. However, cooler and wetter conditions have the result of depressing the snowline on the summits to the east of the Salar de Atacama, and also of masking lower-altitude pasture grounds with seasonal snowfall. However, the increased moisture would have increased the amount of pasture near the Salar, producing favourable conditions for more abundant wild camelid populations. It is probable that if there was a moister period at this time, then the change was only a relative one. Human coprolites and flotation samples from sites in the Tulan Quebrada dating from or after the fifth millennium BP analysed by T. Holden (1990: 280) contain plant remains that suggest that the environment two or three millennia ago was not significantly different from that of the present.

In recent times, the pasture grounds in the Tulan Quebrada have been heavily grazed. Where overgrazing occurs, there are changes in the vegetation, since *Brea* (*Tessaria absinthioides*), a resinous plant inedible to sheep, is replacing more palatable forage species. This leads Holden to comment that the Vega de Tilopozo, which is more isolated and further to the west, is probably more characteristic of less heavily grazed pasture ground in a similar environment, since the vegetation cover is more varied. Sheep and goats crop grasses and plants much more closely to the ground than camelids do. The overgrazing that has occurred since the introduction of sheep and goats into the area, following the Spanish invasion of the sixteenth century, means that present-day pasture availability is probably not comparable with the situation in pre-Hispanic times, even though the area may not have undergone dramatic changes over the past two millennia.[4] In high-altitude pasture grounds, Popper noted signs of an ecological imbalance with an increase of herbivore rodents that feed on the roots of grasses. This increase may be due to the local extinction of a feline predator, or to the fact that human beings are no longer hunting the rodents as a source of food (Popper 1977: 16). In sum, it is not safe to assume that present-day herding cycles necessarily resemble those operative in the past. However, before proceeding to describe the archaeological patterning in this area, I will briefly present two examples of herding regimes in the San Pedro de Atacama area.

Present-day herding strategies

Earlier ethnographic accounts of herding in the highlands of the Atacama include Boman (1908) and, in the communities bordering the Salar de Atacama itself, that of Bowman (1924). Here, I will briefly give an indication of the extent of land used in two present-day cases, one in the Purifica Quebrada and the other in the Tulan Quebrada (figure 7.1).

Guatin, at 25 km to the north-east of San Pedro de Atacama, is situated at the confluence of the Puritama and Vilama rivers, as mentioned above. In the 1980s it was the home of an old man, Juan Vilca, who employed an assistant to help him

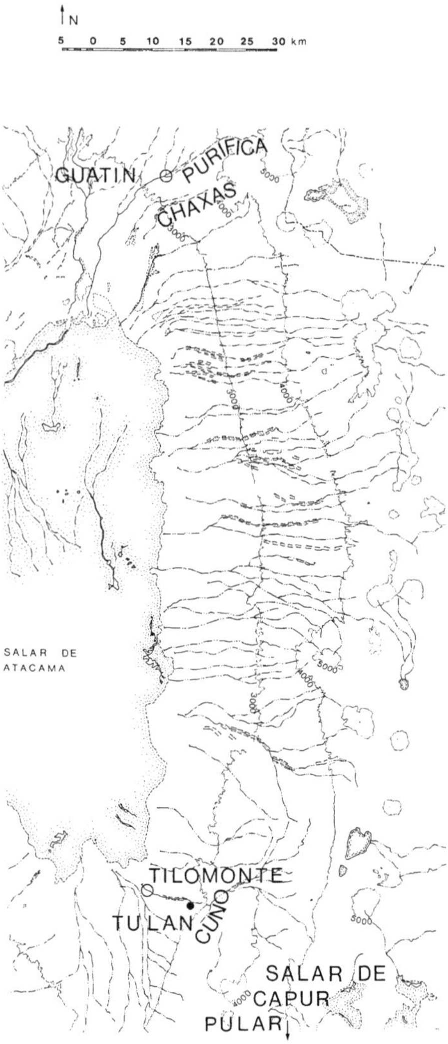

Figure 7.1 Map showing the Purifica and Tulan Quebradas.

herd his animals and whose way of life was studied by Serracino and Stehberg (1975). They reported that Vilca's most pressing task was that of finding adequate pasture for his animals (ibid.: 88). Every morning, the sheep and goats were taken from their corral and later brought back at sunset. Vilca owned at least three residences.[5] Depending on the quantity of the pasture available at his main residence at Guatin, on climatic factors, whether or not he had a paid assistant, and on the state of his health, Vilca took his flocks to one of his other *estancias* and remained there for weeks or several months.

Juan Vilca inherited land for herding his animals from his family, and Serracino and Stehberg (ibid.: 89) defined the area as stretching from the river Puritama along the length of the Purifica (on both sides) and as far south as the Chaxas river. This area included two *estancias*: Fundiciones, some 7 km upstream the Purifica river from Guatin, and Iglesias, some 15 km to the east of Guatin. He tended to use the higher-altitude grazing grounds in dry periods, since there is sufficient pasture at Guatin in the rainy season from January to March. While sheep and goats were always watched over by a herder during the daytime, the herd of llamas belonging to Vilca was left on its own, often at levels of about 4,000 m (ibid.: 90). However, these llamas often followed the Purifica river downstream to lower altitudes.

Currently, the Tulan Quebrada is used by more families than is the Purifica Quebrada. Tilomonte, situated some 16 km south of the village of San Roque de Peine, is but one of the areas belonging to the inhabitants of Peine. In Peine, a narrow river makes possible the cultivation of maize, wheat and alfalfa in about three hectares of land, and it also supports an abundance of chañar and algarrobo trees (Mostny *et al.* 1954: 11). The residents have access to various lands for cultivation or for herding. According to Mostny, the most important *vegas* used by the people are those of Púlar, said to be one-and-a-half days' walk southwards towards the Argentinian border (ibid.: 11). She mentions that they also have access to the *vegas* of Tambillo, which are, in addition, exploited by the residents of Socaire and Toconao. Most of these pasture lands are communally owned, but Mostny says that some people regard grazing lands as their own private property by virtue of prolonged use by one family. The situation is therefore somewhat similar to that in Isluga, since houses and cultivated land constitute private and heritable property (ibid.: 11 and 79). It is not clear from this account how regularly people from Peine herd animals in Tambillo, which is over 60 km distant, or if access to these *vegas* is due to a spouse retaining rights to pasture animals there. Mostny states that fiscal territory belonging to the Comuna of Peine consists of the oases of Peine and Tilomonte and the Vega de Tilopozo; in fact, all the vegas and water holes are said to be fiscal property (ibid.: 11 and 160).

During my fieldwork in the Tulan Quebrada in October 1987, I observed two different herds of llamas, which were at that time unsupervised by herders. Early in the morning, the llamas descended to the banks of the Tulan river, and late in the afternoon they ascended to the dry land above the canyon, each herd setting off in different directions. One of the families using this valley was a married couple whose main residence was stated by Núñez (1988: 453) to be Tilomonte, although they never remained in one place for long. The pasture grounds used by the Chaile family range from Tilomonte via Tulan as far south as Púlar, a distance of about 50 km. In the late 1980s they did not use the pastures at Tilocalar or Tilopozo. According to Núñez, they owned twelve llamas of their own, but also in the same herd were llamas belonging to other members of the family resident in Peine and Socaire. This herd was left to fend for itself while the Chaile couple herded the sheep and goats. During the winter months from May to August, they remained below altitudes of 3,000 m, moving to higher grounds in September or October.

They stayed at high-altitude places of residence in November and December, returning to lower altitudes some time between January and April, depending on the weather conditions (ibid.: 452–3). When moving from one pasture ground to another, the herders followed the Tulan river and they spent the night in rock shelters surrounded by dry-stone walling that divides the shelter for the human beings from the adjacent corrals for the animals (ibid.: 455). Having followed the river upstream, the herders then moved southwards towards the high-altitude *vegas* of Púlar or Capur, following one of the valleys such as Cuno. Another reason given by Núñez (ibid.: 453) to encourage both human beings and animals to migrate to higher zones is the high incidence of troublesome midges (*jerjele*), small mosquitos found in the lower and warmer oases.

In sum, the herding cycles observed along the Purifica Quebrada and along the Tulan Quebrada are similar, in that flocks of sheep and goats receive more constant supervision than do herds of llamas, while all animals move from one level to another according to season and the availability of pasture. The main differences observed are due to the availability of pasture, since the apparently more humid environment along the Purifica Quebrada supports fewer herd animals. Thus the territory occupied by Juan Vilca is less extensive than that utilized by the Chaile couple in their nomadic cycle of movement from Tilomonte to Púlar.

On the use of ethnographic analogies

Present-day llama herding practices in the Atacama have been used as analogies for the herding of camelids in the past. Núñez (1988: 456) has claimed that the low density of camelid herding, with little human intervention, as practised today by the Chaile family, may be used as a model to account for the archaeological record of site TU 54, located near the spring of Tulan.[6] Similarly, Rabey (1989: 269) maintained that highland Argentinian llama herding practices, unmixed as they are with alpacas, are of great theoretical interest because they provide a heuristic model for understanding the domestication process, and the relationship between human beings and animals in that process. The neglect typified by Rabey of the high-altitude herders of north-west Argentina towards their llamas may be more apparent than real, and the apparent lack of supervision of domesticated camelids in the present requires some comment.

In Isluga, herders train their animals to act as a unit and they frequently accompany their animals to encourage them to follow habitually a sequence of movements within a relatively circumscribed area. Stated in other words, animal and human interests coincide, and the human owners encourage their animals to move from higher-ground, dry sleeping places to wet *bofedales* during the daytime until sunset. This is a cycle of events the animals follow anyway, but herders take them to places where they can be easily found, and where the human owners have rights to allow their animals to graze. In the valleys surrounding the Salar de Atacama, human control over llamas is exerted in a more subtle manner. Not only are water resources more limited, but the llamas must be kept apart from the cultivated lands at Guatin and at Peine and, to a lesser extent, at Tilomonte. This means that their owners have a good idea where their animals will be, even though they only supervise their herds at intervals of one to two weeks, since llamas descend to water on a daily basis. Herders ensure that sufficient adult animals remain in the herd to guide the younger animals and to maintain good herd discipline. However, the less vigilant control apparently exerted by Atacama herders can be made possible only in the absence both of predators and also of large numbers of wild camelids. Nowadays, pumas are very rare indeed, and are all but absent in the hills. Foxes may prey on young llamas, and herders often construct traps of stone to deal with this danger. In addition, female and newly born llamas are separated from the males during the first part of the year and are kept under closer supervision. Thus losses due to predation are minimized. Another important factor, the scarcity of large herds of wild camelids in the area, means that llamas have little chance of going feral and joining their wild counterparts. Indeed, wild camelids are more likely to join the company of herds of llamas. On 6 October 1987, I observed a vicuña in one of the herds of llamas grazing on the bank of the Tulan river (plate 7.2). The vicuña was nervous as it was aware of human beings on the other side of the *quebrada*, but it did not run off and it remained with the herd as the animals made their way downstream. This vicuña was possibly a solitary male displaced from a family group.

The type of pastoral control over llama herds described above is also practised in the Vega de Turi, further north in the upper Loa basin, and it has been reported

Plate 7.2 Llamas accompanied by a solitary vicuña, immediately downstream from the spring at Tulan. The photograph was taken from near site TU 57.

in north-west Argentina (Merlino and Rabey 1978; 1985; Rabey 1989). This characteristic style of camelid herding in the highland pasture areas of the Atacama in territory on each side of the present-day international frontier between Chile and Argentina is the result of many factors. Following the Spanish invasion, human settlement patterns changed as the dispersed population was settled into more centralized villages by the Spanish, a process particularly occurring with the reforms of Viceroy Toledo in Peru after 1572. In addition, sheep and goats were introduced into the Andes. Another introduction was that of alfalfa, a source of fodder which, unlike native Andean plants, may be stored. In the intervening centuries, the exploitation of various mines in the area attracted migrant (especially male) labour, taking more men than women away from communities. This meant that women became largely responsible for maintaining the homestead, while male members of the family tended to spend longer periods working in mining centres such as Chuquicamata (Valdés *et al.* 1983: 24, 48–9).[7] Hence, few families possess enough members to supervise closely sheep, goats and llamas. Since llamas can be trained and are predictable in their movements, more constant supervision is given to the sheep and goats. Nevertheless, Merlino and Rabey (1978: 52–8, photo 3) reported practices comparable to those in Isluga, including the irrigation of pasture lands by canals and the celebration of the Pachamama in a particular form of cultural adaptation in an annual ritual herding cycle to which they ascribed some antiquity.

Rabey (1989) concluded that the herding system he described for highland north-west Argentina does not constitute a pastoral system. However, he did not study the present-day situation to observe whether a structural transformation has taken place that would denote a shift from one dominant mode of resource exploitation to another, and it is not clear how his material may be used to help understand the domestication process, as he claimed.

A study of the material record from late archaic period sites in the region indicates that hunter-gatherer peoples added the herding of camelids to their economy. The first site in the study area at which this transition has been detected is PU 1. However, as pastoralists, one of their most important concerns was surely to keep their own animals alive for as long as possible. Hence the bulk of their diet probably derived from their hunting and gathering activities. The herders of PU 1 and of later sites in the Tulan Quebrada must have kept their herds of domesticated camelids in a context where herds of wild camelids were commonplace. Presumably, the herds with human owners temporarily displaced the wild herds in the narrow Tulan canyon when site TU 54 was occupied. Hesse (1982b: 12) claims that pastoralism established in the Atacama an adaptation which ensured its success by destroying the potential for other systems. Presumably he is referring to the hunting of wild camelids.

However, Chapter 8 will present evidence for the continuing exploitation of raw materials provided by wild animals. It should also be remembered that the eighteenth-century evidence for vicuña hunting cited in Chapter 2 suggests that herds of vicuña and guanaco were not scarce. Thus the pastoralist societies using the Purifica and Tulan rivers and the surrounding pastures, should be seen against a background which was populated by far greater numbers of guanaco and vicuña herds than at present. Because present-day conditions are markedly changed, it is my opinion that ethnographic analogies based on the present-day exploitation of pasture in the Atacama should be used with great caution.

Archaeological sites in the Tulan and Purifica Quebradas

More than 108 sites have been registered in the Tulan Quebrada, and only those relevant to the concerns of this study are considered here. Some of the sites have an occupation that continues into the present and it should be emphasized that the Tulan Quebrada is not an isolated unit of space. This drainage system is part of an extensive network of intermittent river valleys to the east of the Salar de Atacama, and reference is made to other sites where necessary. In particular, the site of Puripica 1 (PU 1), where analysis suggests that the domestication of camelids took place, is discussed in greater detail.

An attempt is made also to place these sites in a chronological context. Archaeology in the north of Chile has been plagued by the shortage of reliable dates, and chronological frameworks proposed to account for the patterning observed in the archaeological record often rest on shaky foundations. The site that has produced the most numerous items analysed in the Appendix, the site of Tulan 54 (TU 54), is securely dated by a series of six C-14 dates. However, problems remain in establishing temporal correlations with other sites. The discussion here attempts to place this site in a chronological context by considering changes in the patterning observed in certain cultural practices. Changes in house construction through time are contrasted with other activities, which are seen to demonstrate more continuity over time and which give the sites of this area its particular character. The use of red pigment, the making of beads and the use of a tool type (microperforator) are considered here in more detail. These all constitute activities that left a durable imprint at various points of the landscape. The important activity of making yarn is discussed in Chapter 8. Finally, the evidence described here for the archaeological sites is considered in the context of published frameworks for chronology and the evolution of resource exploitation.

To date, the site with the greatest antiquity of all the valleys draining into the Salar de Atacama is the rock shelter of San Lorenzo (figure 7.2). However, evidence for the early occupation of the area is slight. After about 5000 BP, the evidence becomes more abundant in both rock shelters and open-air sites. All the sites that have yielded the animal fibre remains and the yarns and fabrics studied in this book have been located in the Tulan Quebrada (figs 7.3–7.5).[8]

Tulan 67

TU 67 is a small rock shelter in the district of Tilomonte. Inside the shelter, the rock wall bears linear designs painted in red, including a human figure, apparently wearing some form of a pubic covering, outlines of camelids and geometric designs. The rock shelter is situated underneath an enormous overhanging rock, and

Figure 7.2 Map showing the location of Archaic period sites in the South-Central Andes discussed in the text.

modern herders have constructed dry-stone walling under and around this rock (figure 7.6). Beyond the shelter of the overhang, and on the surface of the ground, is an arc of small upright stones. A large scatter of surface material, mostly lithic, covers the ground on both sides of the arc. Previous excavations in the rock shelter of a trench 3 m long and 1 m wide have not been published.[9] Between 5 and 7 September 1987, a grid square measuring 1 m by 1.5 m was marked out at the western end of the trench and excavated in seven stratigraphic layers.

All layers in this grid square (number 4) contained animal bones, lithic material, feathers and vegetal material; bird beaks were present in most. Small sticks, painted red or pink, were present in layers III, IV, V and VII, as well as two pieces of shaped wood, also painted red, in layer IV. Animal bones, painted red, were encountered in layer VII. Camelid dung was found in all layers, apart from IV, where a small

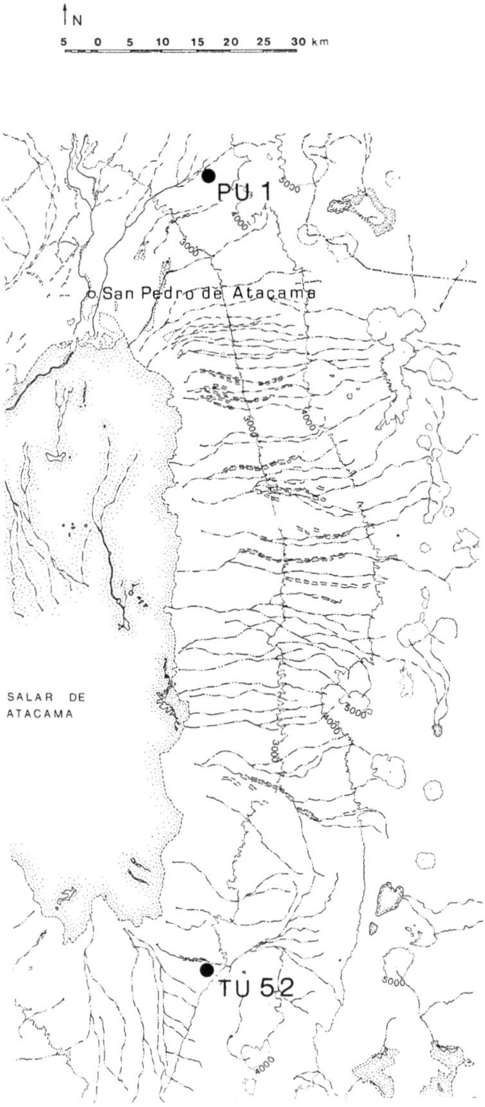

Figure 7.3 Map showing the location of sites PU 1 and TU 52.

amount of rodent droppings were observed. Sherds of undiagnostic pottery were found in layers I and III. A snail shell and two pieces of shell appeared in the layer III deposit, as well as a shell bead similar to those from TU 52 and PU 1; another bead of this type was found in layer IV. Oval shell beads and a broken piece of worked and pierced mussel shell were located in layer V.

The stone tools present on the site include typical Tulan-type blades and scrapers. Layer IV contained the scalloped base of a Miscanti-type point with

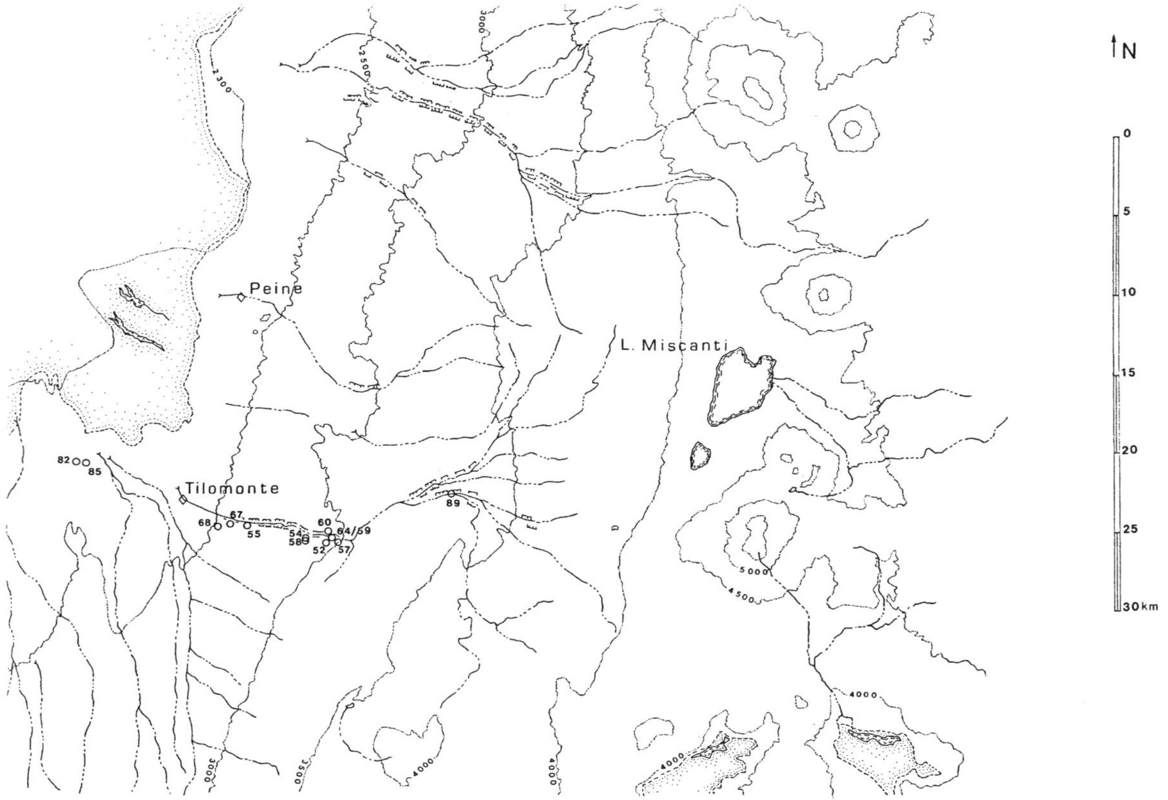

Figure 7.4 Map showing the location of sites in the Tulan Quebrada discussed in the text.

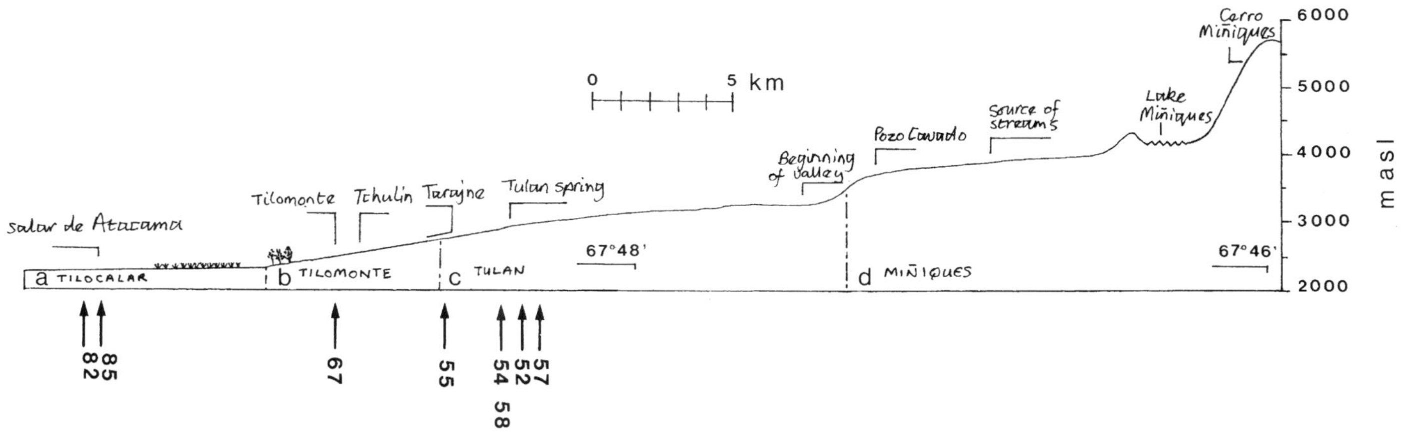

Figure 7.5 Schematic cross-section of the Tulan Quebrada (adapted from Núñez and Santoro 1988).

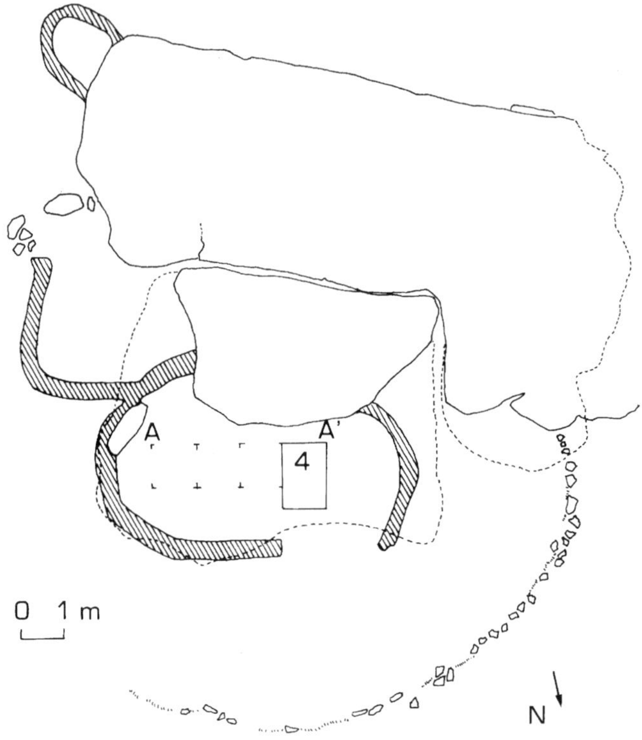

Figure 7.6 Plan of site TU 67.

denticulate edges, and a type of perforator with a wide body, which is said to be characteristic of the Tambillo industry. Thick leaf-shaped points of chert or tufa and triangular points of volcanic glass were found in layers V to VII. Layer V also contained a chert microperforator with a twisted outline. Núñez (1988: 207) considered layer V to be preceramic and of the Tambillo pattern as it contained a heart-shaped point with a slightly concave base of grey volcanic glass that characterizes the pattern. However, blades and flakes of the later Tulan industry as described for sites TU 52 and 51 below were also present in the lowest layer. A piece of tabular stone in layer III was used as a palette to grind red pigment (plate 7.3), and a small piece of red mineral was found in layer IV.

Yarns and animal fibres were found in layers I to V. Layer VII contained a piece of textile and also a piece of basketry. The textile, at least, would seem to have dropped down from an upper layer, since the C-14 date on a charcoal sample from the context in which it was found produced a result of 8190 ± 120 BP (Beta 25535). In northern Chile, this is far too early for a loom product that was woven perhaps with the aid of heddles. The two AMS dates obtained from samples of camelid fibre demonstrate that disturbance did occur. They are inverted: layer V yielded a date of 5320 ± 90 (OxA 1842)and layer VII, 4870 ± 65 BP (OxA

Plate 7.3 TU 67, grid square 4, layer III. A selection of lithic material, including a broken tabular stone palette with traces of red pigment. By courtesy of the Museo R.P. Gustavo Le Paige S.J.

1843). In the upper layers the deposits are also disturbed. Holden (personal communication) reports the presence of alfalfa seed in layer II and a large beetle was found still alive in the layer IV material. The passage of such an insect through the stratigraphy would allow smaller items to drop to lower layers. It should also be remembered that rodent droppings were found in layer IV, and this is also an indication of disruptive animal activity in the deposits. Thus no layer may be assumed to be intact.

The yarns and the textile fragment are analysed in the Appendix. Although the deposits cover a long timespan, the archaeological material is relatively slight, but analysis of the yarns reveals some differences when compared with those excavated from open-air sites such as TU 54. This theme is discussed further in the following chapter.

Tulan 52

This is an open-air site with circular structures located near Tulan itself, which is at Latitude 23° 48′ S and Longitude 68° 1′ W, on high ground immediately south of the Quebrada, at an altitude of about 2,925 m. An excavation of 28 sq. m at this site revealed four habitation structures. The structures consisted of a hollowed-out ground area around which were placed upright stone slabs. Deposits up to 1.35 m in depth containing archaeological material which covered these structures are described by Núñez (1981: 143), with the majority of the lithic and bone remains occurring in the intermediate stratigraphic zone. Charcoal from a hearth in the lowest stratigraphic zone yielded a date of 4270 ± 80 BP (N 2488), while a second sample from the uppermost zone yielded a date of 4340 ± 95 BP (N 2487). Within the three stratigraphic zones are various microstrata, and some lenses of aeolic sediments, which suggest seasonality in the occupation of the site.

The lithic assemblage at TU 52 includes lanceolate points, triangular points and stemmed points, with the first of these three categories being represented in far greater numbers than the other two. The most abundant type of knife present is an unmodified blade, although another type, that of a bifacial flake with pressure flaking, is also common. There are scrapers in various forms; awls or perforators and microperforators; hammer stones; grinding stones; and miscellaneous items, including striated pieces of pumice, beads of copper ore, stone and shell (ibid.: 145–7). It has been suggested that the predominance of lanceolate points and cutting implements indicates hunting and butchering activities (ibid.: 147). The lithic assemblage of TU 52 resembles that of the nearby lithic workshop of TU 51, for which a C-14 date of 4990 ± 100 BP (N 2486) was obtained. Both these assemblages rely heavily on the exploitation of a sedimentary rock[10] which is locally available, and in many cases this was taken into the site in the form of blades, which were worked further on site (ibid.: 144). Quartz, calcedony, chert and volcanic glass were also used to a lesser extent, and presumably these materials were found in higher-altitude locations.

The most abundant category of bone fragments from TU 52 represents the remains of camelids, although Hesse (1982a: 203) also identified rodent bones, of which 75 per cent were chinchilla (table 7.2). Bird bones are much less frequent and include flamingo remains, which to Hesse (ibid.) suggests a seasonal occupation of the site during summer when these birds migrate to the highland lakes.

In October 1987, a column 25 sq. cm was excavated at the eastern end of the previous excavation. Sterile deposits were encountered at a maximum depth of

Table 7.2 Distribution of bone fragments in samples from TA 1, TU 52 and PU 1.

	TA 1	*TU 52*	*PU 1*
Camelids	1,047	12,096	3,426
Rodents	1,085	2,067	825
Birds	76	101	239
Percentages of rodent species			
Ctenomys	98	20	6
Lagidium	2	5	37
Chinchilla	0	75	50
Other	0.2	0	7

Source: After Hesse 1982a

81 cm, above which lay a hearth marked by small fire-blackened stones and containing pieces of charcoal. At TU 52, hearths are located between, not within, the circular structures (Núñez 1981: 144). The upper layers of this excavation were found to contain small pieces of camelid fibre, larger quantities of rodent fur and yarns spun from camelid fibre. Beads of shell also were encountered.

Puripica 1

This site is located on high ground which dominates the Purifica or Puripica river at an altitude of about 3,500 m asl, some 5 km upstream from Guatin. Locally, the site is known as Los Morteros because of the large numbers of broken and perforated mortars with a conical hollow which are scattered over the site, which is estimated by Núñez (1988: 415) as having a visible occupation of 400 sq. m. In 1956, Father Le Paige dug into the site and reported the existence of hearth and animal bones (Le Paige MS 217). More recent, systematic excavations covering 27 sq. m with deposits 70 to 80 cm deep were reported by Núñez (1981). Circular house structures similar to those at TU 52 were revealed. Four stratigraphic layers were described by Núñez (ibid.: 150), of which the lowest, III and IV, consist of similar deposits arbitrarily divided at a depth of 55 cm. Two charcoal samples from the bottom layer, but from different parts of the excavation, produced C-14 dates of 4050 ± 95 BP (N 2360) and 4815 ± 70 BP (SI 3113).

The stone tool assemblage of PU 1 has been characterized as being based on the use of flakes rather than blades, and it includes blanks and finished tools. Basalt, from locally available sources, was frequently used, while chert, quartz and volcanic glasses were used to a lesser extent. Projectile points, most of which are lanceolate in shape, occur in lower frequencies relative to cutting tools than is the case with the TU 52 lithic assemblage (figure 7.7). However, microlithic perforators are slightly more abundant than scrapers at PU 1. Núñez (ibid.: 152) considers that both the lithic assemblages (TU 52 and PU 1) would have been suitable for skinning animals and working hide, but that the higher incidence of knives over projectile points at Puripica might indicate that hides from tamed animals were available. Broken grinding stones (*morteros*) and rubbing stones (*manos*) occur in the stratigraphy. Plate 7.4 shows a broken *mano*, used for grinding red pigment, from layer III of grid square AA7.

Faunal analysis of the bone remains from PU 1 shows a predominance of camelid bones, while the presence of rodents is also attested: chinchilla represent 50 per cent and viscacha 37 per cent of the rodent bones (Hesse 1982a: 203). Hesse (ibid.) reported a greater variety of birds at PU 1. Examination of the camelid bone remains from TU 52, TA 1 (Tambillo) and PU 1 enabled Hesse (ibid.: 203–6) to demonstrate a bimodal distribution in the measurements on foot bones, taking it to indicate the presence of a larger form of camelid, which he equated with guanaco, and a smaller form, which he equated with vicuña. He then proceeded to examine the harvest profiles for the two different groups by attempting to establish the rates at which animals of different ages died (ibid.: 206–9).[11] Hesse established these profiles (figure 7.8) by calculating the proportions of fused

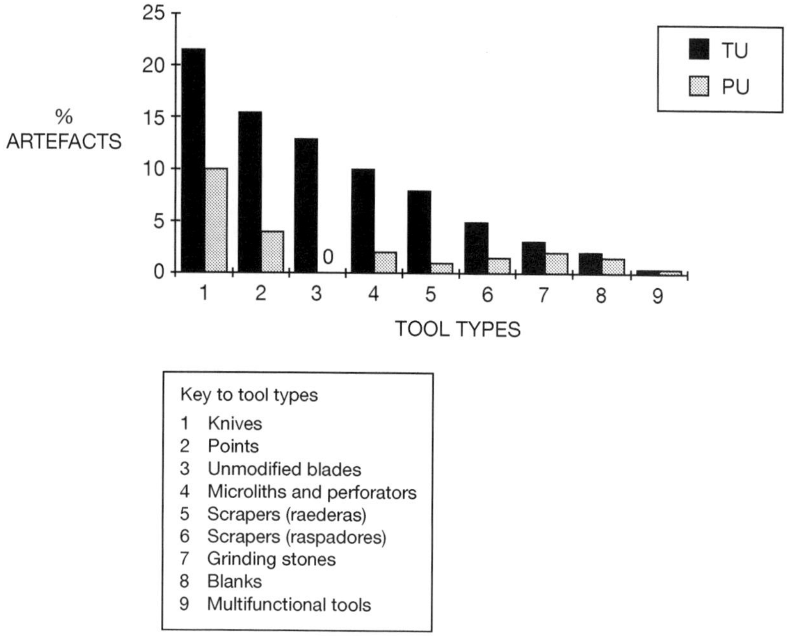

Figure 7.7 Bar diagram showing frequencies of projectile points and cutting tools at TU 52 and PU 1. TU 52: 2,706 classified artefacts (78.6%); PU 1: 734 classified artefacts (21.4 %).

epiphyses of the different bone types according to eight sequential age stages proposed by Elizabeth Wing (1972: 330, table 2). The results showed clear differences between the three sites. About 75 per cent of the camelids were found to be skeletally mature at death in the TA 1 sample. At PU 1, the majority of the animals died young, and more than 50 per cent were dead before the end of age stage one. The TU 52 sample showed a more even distribution: about 20 per cent of the camelids died before the end of age stage one, 50 per cent between age stages one and eight, and 30 per cent at age stages greater than eight (Hesse 1982a: 206). Separate harvest profiles for each of the two forms (large and small) revealed anomalies, but the TA 1 sample was too small to be analysed by the methods employed by Hesse. In the case of the small camelids, the percentage which survived long enough to have fused epiphyses in their feet is 63 per cent for TU 52 and 71 per cent for PU 1 but, in the case of the large camelids, 59 per cent of the camelids were mature at TU 52 but only 42 per cent survived to a mature age at PU 1. Another anomaly observed by Hesse (ibid.: 206, 209) is the greater proportion of toe bones from the small camelids at PU 1, which he interpreted as a bias caused by the practice of two different carcass processing patterns. The entire skeleton of the large camelids is represented, but the small form was evidently butchered away from the site where the long bones remained, with the toe bones, presumably with meat and hide, having been carried to the site of PU 1. Hesse

Plate 7.4 PU 1, grid square AA7, layer III. Part of an upper grinding stone (*mano*) with traces of red pigment. By courtesy of the Museo R.P. Gustavo Le Paige S.J.

concludes that the harvest profile for PU 1 probably indicates that the large camelids were being domesticated, since the 50 to 60 per cent loss of the large camelids before they reached the end of age stage one may have been caused by death due to a disease such as enterotoxaemia, which may occur if animals are kept in restricted and crowded conditions (Hesse 1982a: 210; 1982b: 11; Novoa and Wheeler 1984: 124). Hesse pinpoints this change as occurring between layers III and II.

The proposition that domestication occurred at PU 1 is supported by the fact that projectile points are less abundant at this site and by the presence of blocks incised with depictions of camelids. Six such incised boulders are known from the site. Some of them were located during excavation, including a small block (figure 7.9) that Núñez says was associated with the sample, which yielded the C-14 date of 4815 ± 70 BP (SI 3113). Another boulder formed part of an upper section of walling in grid square 7, while some of the upright slabs in the lowest course of the walling in the same house display long striations of an unknown function. Among the incised blocks found on the surface of the site is one depicting two camelids, one above the other (figure 7.10) and yet another displays two camelids, one in front of the other, above four long, wide incised lines (figure 7.11). All the

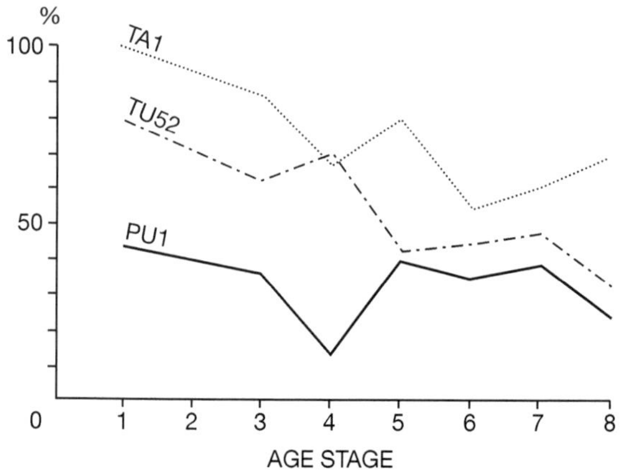

Figure 7.8 Harvest profiles for sites TA 1, TU 52 and PU 1 (after Hesse 1982a). The curves represent the survivorship of camelids relative to a series of age stages proposed by Wing (1972).

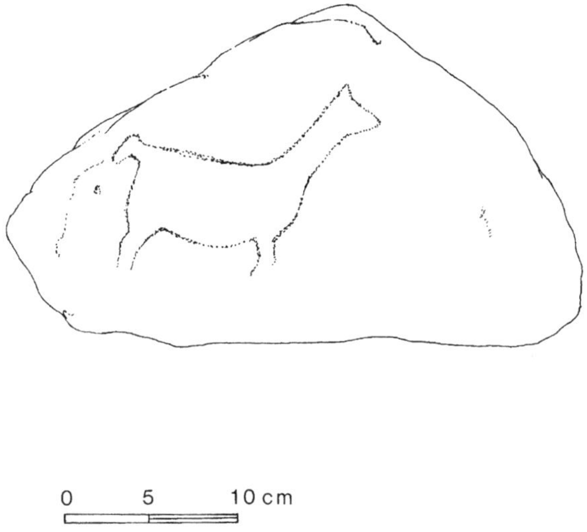

Figure 7.9 Block incised with drawing of a camelid from PU 1, associated with a date of 4815 ± 70 BP.

camelid depictions mentioned thus far face towards the viewer's right. Stylistically, they resemble the camelids incised on a rock face at Kalina on the upper Loa. The panel is numbered 101-UR-I and it is assigned to the Kalina phase by Berenguer *et al.* (1985: 91–2) (figure 7.12).

At both Kalina and Puripica, the camelids are drawn with large bodies and schematic heads, and the feet are not drawn in detail, since the lower limbs are indicated by two lines which do not meet. Berenguer *et al.* argue that the dating

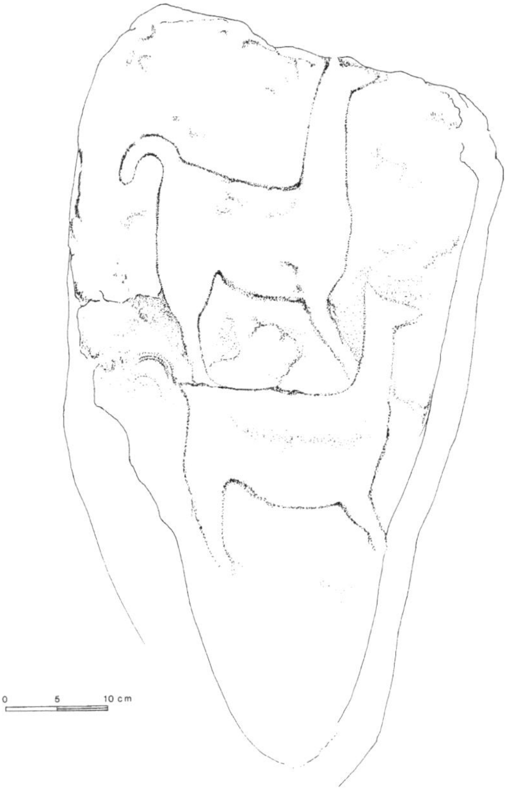

Figure 7.10 Block incised with drawing of two camelids from PU 1.

of the Kalina panel is supported by the stylistic similarity to the Puripica art and by the proximity of the former to a circular structure (site SBa 101) at which perforated and broken mortars with a conical hollow are present. A test excavation at this site revealed a stone tool assemblage similar to that of the Chiu Chiu Complex, while a charcoal sample from layer 2 produced a C-14 date of 3950 ± 50 BP (Beta 6844) (Berenguer *et al.* 1985: 91). Like the Puripica camelids, the majority of the animals face towards the viewer's right, which at Kalina means upstream. However, my own work at PU 1 in 1987 produced another small block when grid square BB6 was excavated, the base lying at a depth of 35 cm and the upper face at 20 cm below the surface (figure 7.13). The camelid depicted in this case faces the viewer's left. This stone, which was partially overlain by a broken piece of a mortar with a conical hollow, formed part of a layer of stones placed over one side of the house beneath, without respect for the underlying architecture (plate 7.5). By the time this layer of stones was laid down, the deposits had virtually covered the upright slabs of the circular structure. The excavation of 8 sq. m in October 1987, further to those reported by Núñez (1981), followed this stone alignment, which incorporated some very large upstanding blocks. The alignment

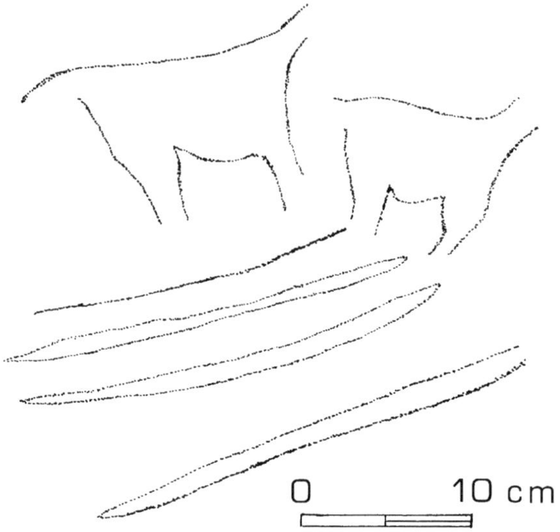

Figure 7.11 Block incised with drawing of two camelids and wide striations from PU 1.

Figure 7.12 Rock face incised with drawings of camelids, panel 101-UR-I, at Kalina (after Berenguer *et al.* 1985).

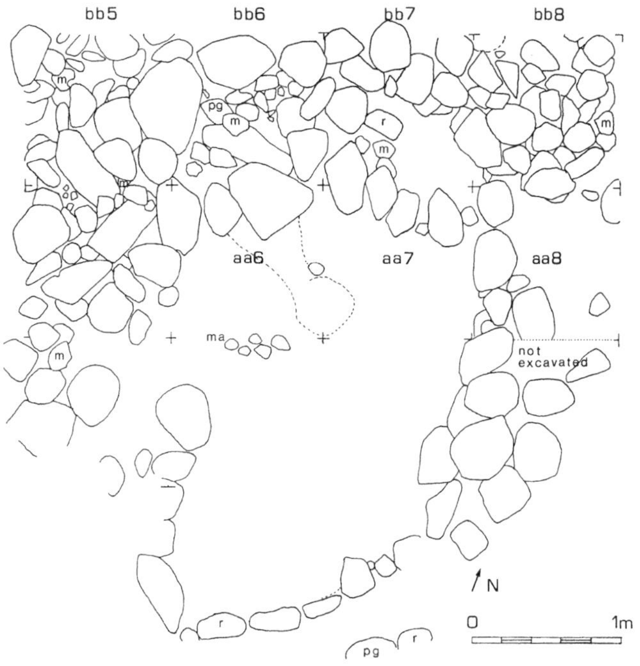

Figure 7.13 Plan of 1987 excavation at PU 1.

lay in the uppermost layer (I). At the southernmost extreme, the stone alignment was found to be cut by the pit excavated by Le Paige in 1956. Loose charcoal collected from the middle zone of layer II in grid square AA7 produced a C-14 date of 4290 ± 60 BP (Beta 32390), so the covering of the adjacent grid squares with the stones must have been done some time after this date.

Most of the animal bones and lithic material came from inside the circular structure. Since the deposits are moist and full of modern roots, the variety of organic remains found at some of the Tulan sites is not present here. Much of the bone material is in a very fragmentary state, but a complete camelid rib was found in layer II in grid square AA6. Most of the lithic material and animal bones came from the upper part of layer III and the lower middle part of layer II. The very bottom of layer II produced a scarce amount of material, and this represents a slight hiatus in the occupation of the site. It should be remembered that Hesse noted changes in the harvest profile for the large form of camelid as taking place between layers III and II.

It is a matter for regret that such an important site for understanding the domestication of camelids in the Atacama should not yield direct evidence for camelid fibre. More spectacular is the detection of a puma (*Felis concolor*) bone in the sample examined by Hesse (1984: 57), a species he did not identify in the other samples he examined. The presence of this bone is significant and suggests that the human group was finding it necessary to protect their animals from a feared predator.

Plate 7.5 PU 1, grid square BB6. Layer of stones, which includes a stone incised with a camelid overlain by a broken grinding stone.

As an archaeological site, PU 1 is relatively isolated, unlike the clusters of closely spaced Tulan sites. However, Le Paige (1965: 9–11) reported the existence of two extensive and dense surface scatters of lithic material near the lava flow of Las Fundiciones in the vicinity.

Tulan 85

On the edge of the *vega* of Tilocalar at the foot of a steep spur of land at the edge of one of the southern tips of the Salar de Atacama is the site of TU 85, at an altitude of just over 2,300 m asl. The site is a broad, seemingly low mound, which is partially overlapped by extensive modern corrals, but excavation revealed archaeological deposits up to 2 m in depth. Much of this material consisted of compact layers of ash, at the margins of which and between which organic remains were preserved. Some 11 sq. m have been excavated, including grid square 11, which was excavated in December 1989. In grid square 11, yarns and animal fibres were found in layers IV to VIII and in layers XIII to XV (out of sixteen, a total depth of about 1.75 m). Layer VIII also contained a fragmentary textile with pure black warp and weft, which is analysed in the Appendix. Unfortunately, provision was not made to do this fieldwork for the museum in San Pedro de Atacama until the last week of my visit. Consequently, there was no time to examine the yarns and animal fibres excavated from grid square 11, and the yarns are not included in the Appendix. The lithic material from this grid square includes stemmed points with denticulate edges and various microperforators. Small shell beads similar to those of TU 52 and PU 1 were also found, as well as larger, more sequin-like beads of shell.

According to Núñez (1988: 272), the imbricated pottery sherds, microperforators and stemmed points of volcanic glass with denticulate edges from this site are similar to those found in TU 54. Two C-14 dates are available: 3140 ± 70 BP (Beta 25508) and 2660 ± 80 BP (Beta 32388), and this evidence suggests that TU 85 and TU 54 were at some stage contemporary with each other, although the former must have spanned a longer time scale than the latter. However, the archaeological material seems to be more sparsely represented than at TU 54, and much of the surviving animal bone is in a charred and very fragmentary condition. Grid square 2 contained some yarns and animal fibres in layers II, III and VI.

The excavation of grid square 7, towards the eastern end of the main trench, revealed some human burials at a depth of approximately 1 m from the surface. Body III, that of a child, still had the remains of a turban in place, although conditions of preservation did not permit the survival of other items of clothing. The significance of this find is discussed in further detail in the following chapter.

The disposition of large boulders in the surface of the mound, and also at the bottom of the excavated trench, suggests that some architectural features are present.

Tulan 54

TU 54 is the site that has produced the most organic remains of all the sites discussed in this chapter. The foundations of a loosely clustered group of oval structures are surrounded by dense deposits of midden at the site, which is about 500 m downstream from the shepherd's house of Tulan. TU 54 is on the southern side of the Quebrada, directly overlooking the river in the canyon below, and it is not far from sites TU 52 and 51 (plate 7.6). It is located approximately at 2,900 m asl. The results of previous excavations in the 1970s have not been published, and most of the yarns and fabrics from this site listed in the Appendix come from those excavations. Núñez mentioned in an interim report (1976b: 148) that TU 54 is a campsite with a mixture of preceramic and ceramic components. However, excavations have proved that ceramics are present throughout the TU 54 midden, and the site is now securely dated by a series of radiocarbon determinations, which demonstrate that the site was occupied about three thousand years ago.

Grid square 1 is an isolated excavation of 1 sq. m, to a depth of some 65 cm in 5-cm spits. The excavation was abandoned before reaching sterile deposits. Grid squares 2 to 6 form a trench in the midden immediately to the south-east of a structure with an oval ground plan. This structure measures 2.62 m by 2.10 m, which means that it is too small to be regarded as a house. Perhaps temporary shelters were erected on the site before it went out of use. Further excavations have revealed irregularly shaped houses incorporating stones incised with drawings of llama heads, and a boulder with a camelid (Núñez 1994: 91).

Plate 7.6 Site TU 54, with part of one of the oval structures in the foreground. Note the herd of llamas in the Tulan Quebrada.

Grid square 2 was excavated in spits of 5 cm, but the excavation of the remaining squares followed stratigraphic layers. In October 1987, grid square B3 was excavated on the north side of grid square 3. The lower part of layer 11 contained a small piece of sheet metal, perhaps copper, weighing less than 0.05 grams. An AMS date of 2840 ± 60 (OxA 1839) BP is available for this layer, while the upper part of layer 11 was dated at 2940 ± 60 BP (OxA 1838). Layer 17 was dated at 3080 ± 70 (BP (OxA 1840), and layer 27, at 3000 ± 65 BP (OxA 1841). These AMS dates on samples of yarn spun from camelid fibre (all of which are from grid square B3) are supported by two C-14 dates, one of 2900 ± 70 BP (Beta 18197) for grid square 2 on a combined sample of charcoal from a depth of 85–90 cms and of animal dung from 75–80 cms, and another date of 3030 ± 70 BP (Beta 25506) on a sample of charcoal from the bottom of grid square 3.

In general, the midden deposits are built up from many layers of vegetal debris and sediments, including a thin lens of small pieces of red ochre (layer 12). The mounds of vegetal matter were evidently set on fire frequently, as layers of ash and charred material occur throughout the stratigraphy. The upper layers near the present-day land surface contain much ash, and this seems to have prevented the percolation of rainwater to lower layers, thus allowing the unusual conditions for the preservation of organic remains at this site. A rapid build-up of midden deposits would also have helped protect these organic remains, and the C-14 dates argue for a fairly short-term occupation of the site.

Among the most characteristic items from the excavation are stemmed points of volcanic glass with denticulate edges, tools made from *toba Tulan*, micro-perforators or awls made from chert, and a profusion of shell and stone beads. Most of these beads resemble the plain type also found at PU 1, TU 52 and TU 85, but also present are sequin-like beads and two shell pendants with ornithomorphic ornament (figure 7.14; plate 7.7). Pottery also occurs sparsely throughout the stratigraphy from the earliest layers. It appears in the form of sherds, some of which were repaired in antiquity. Ceramic types include coarse, brown burnished ware and black sherds with corrugated and incised decoration. Surface finds include mortars with a wide, shallow hollow (unlike the conical hollow of mortars at the sites of TA 1, TU 52 and PU 1).

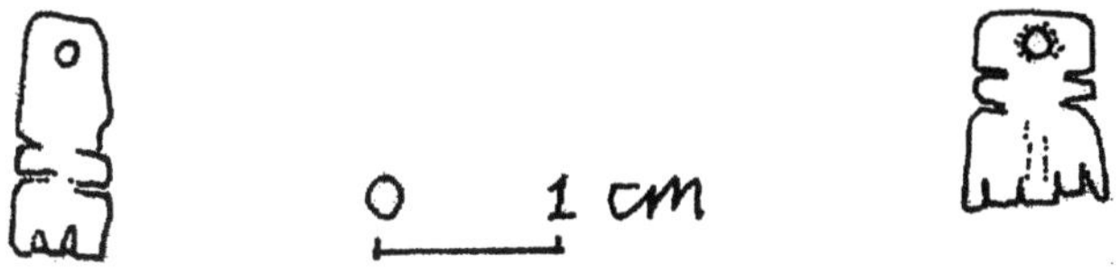

Figure 7.14 Shell beads with ornithomorphic ornament, TU 54. The left-hand one is from grid square 6, layer a, and the one on the right is from grid square 3, layer k.

Hesse examined a sample of 577 animal bones from TU 54, but he does not say from which grid squares this sample was collected. However, the distribution of the identifications is shown in table 7.3. *Metriopelia* is a ground dove. The rodents were identified as *Lagidium* or *Chinchilla* (that is, viscacha or chinchilla, a total of sixteen specimens), *Ctenomys* (cholulo, one specimen), *Phylottis* (leaf-eared mouse, two specimens) and one unidentified fragment of a rodent bone. Detailed analyses of the camelid faunal material are not reported. However, it should be mentioned that the inventory of finds includes camelid bones painted red (plate 7.8).

Various items from the TU 54 material indicate an interest in the colour red. Small pieces of red ochre are abundant, and they include the previously mentioned lens, which consists of a thin, even scattering in grid square 3 (layer 12). Tabular stone palettes and stone tools with traces of red also appear. Twigs painted red have been noted and, finally, grid square B3 yielded an unidentified object which consisted of feathers (some of which were dyed red) bound to a piece of cane (figure 7.15).

T. Holden examined human coprolites from throughout the site. Two types of material dominated the contents: *Opuntia* cactus seed, known locally as *cume*, and the fibre of *Schoenoplectus* rushes, known locally as *junquillo*. Meat, bone and hair were also present, but in very much reduced quantities. Holden concludes that most of the foodstuffs were collected and processed close to the site itself, an observation that is also valid for the flotation samples from site TU 85 and from the gut content of human burials from site TU 58 which he also analysed (Holden 1990: 296).

Plate 7.7 Shell and stone beads from TU 54, grid square B3. The broken bead, second from the right in the lower row, is the flaked-off section of an unfinished stone bead. By courtesy of the Museo R.P. Gustavo Le Paige S.J.

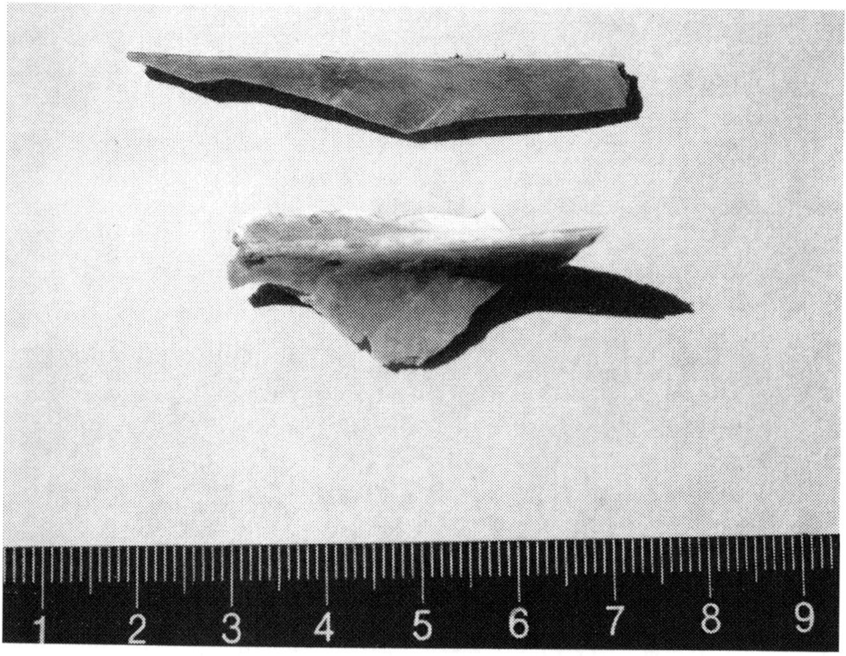

Plate 7.8 TU 54, grid square 3, layer g. Camelid bones painted red. By courtesy of the Museo R.P. Gustavo Le Paige S.J.

Table 7.3 Animal bones identified at TU 54.

Taxon.	*No.*
Mammals	
Camelid	141
Rodent	20
Bird	
(*Metriopelia sp.*)	1
Bone scrap	415

Source: Hesse 1984

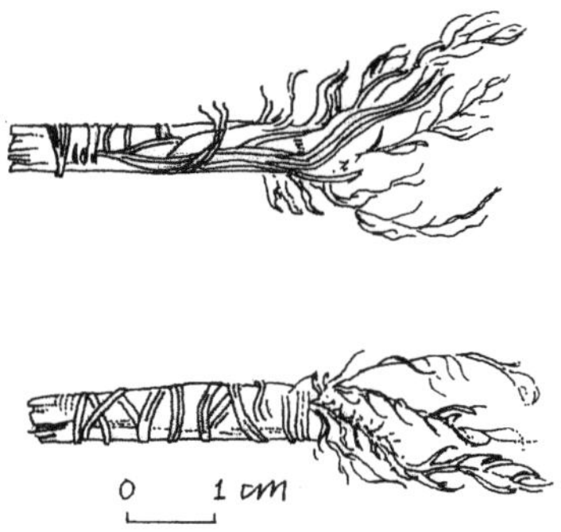

Figure 7.15 Unidentified object consisting of feathers, dyed red, attached to a piece of cane, from TU 54, grid square B3, layer 22.

The yarns and fabrics from grid squares 1 to 6 of TU 54 are listed in detail in the Appendix. Rodent skins and pieces of camelid hide were also found. This material is discussed in more detail in the following chapter.

Tulan 55

A small cave on the southern bank of the Tulan river at about 2,800 m asl (Latitude 23° 48′ S, Longitude 68° 2′ W), close to the floor of the valley, was originally prospected and surveyed by Father Le Paige. The walls of the cave and a large horizontal block, which fell from the roof at some stage, are covered with incised petroglyphs, mostly of camelids. Le Paige located his excavation near the fallen block of rock, and he defined three major strata, the first to a depth of 15 cm from the surface, the second between 15 cm and 63 cm, and the third from 63 cm to a maximum depth of 118 cm. A charcoal sample from the bottom layer is said to have produced a date of 3710 years BP (Madrid), but no further details regarding this C-14 determination are available (Núñez 1976a: 108). The sample was associated with vegetal material, feathers, camelid bones and black pottery, and the result obtained has been used to date the first use of ceramics in the area.

However, supporting data has not been forthcoming from other sites to confirm such an association. Rodent droppings appear in all occupation layers of the site, and they may well have been disturbed by rodent activity. It is evident that this cave has been used over a long period of time. Núñez conducted further excavations, in a trench measuring 4 m by 1 m following the north–south axis of the cave. Nine layers were defined, the fifth being a sterile deposit. A sample of charcoal and vegetal material produced a C-14 date of 2700 ± 100 BP (Beta 16981) for layer IX, which is more coherent in the context of the initial use of pottery in the area. A pit in grid square 2 included a fragment of pink silk material and a cactus spine needle, while just above this was a tiny piece of paper with old Spanish writing. Thus the occupation covers a long timespan, and the archaeological content of the layers is slight.

V. Popper studied the botanical remains from the deposits excavated by Núñez. She identified grasses, sedges, maize, chañar and algarrobo, with greatest concentrations in the upper layers and in the large pit. In particular, the maize remains come from the upper parts of the stratigraphy, but concentrations of sedges were found in the lowermost layer with cultural material, and algarrobo samples were found throughout the layers (Popper 1977: 33–6). Popper also identified a fragment of a gourd shell (*Lagenaria siceraria*) in layer IX, which she considers to be a domesticate (ibid.: 47). She also identified *Opuntia* cactus seeds in coprolites in layer IX and concentrations of seeds in the grid square 2 pit and in layer II. The botanical remains indicate that plants were taken to the cave from different microenvironments, since chañar and algarrobo grow in the low oases near the Salar de Atacama, but the deposits also contained *ephedra* branch fragments, a shrub known locally as *pingo pingo*, which grows at high altitude (up to 4,500 m).

In 1986, I examined some of the material from the TU 55 excavations, and the yarns are listed in the Appendix. However, not all of the material was available for study, thus the inventory is incomplete. Popper confirmed the presence of a large number of yarns in the excavated material, and that a random testing found that none were of cotton (Popper 1977: 46). She mentioned that eight yarns were made from plant fibres, one of these being tied round a piece of wood, and that a small fragment of netting was present in layer IX of grid square 3 (ibid.: 46).

Tulan 58

This site is a cemetery situated on the southern bank of a shallow, dry *quebrada* immediately to the south of TU 54. Núñez (1988: 169) reported the existence of two layers in the excavated structures: the upper contains tombs, which have been exposed to the elements, and the lower, intact tombs underneath piles of selected slabs. Tombs one and two are apart from the main concentration, they are eroded and/or disturbed. Tomb three was also disturbed, but it contained 446 shell beads (figure 7.16; plate 7.9), some stone tools and pottery fragments. Tomb four was underneath tomb three and Núñez described its contents as having the remains of a possible adult along with what is described as twisted vegetal fibre, fine yarns, a fragment of basketry, a thick piece of wood (said to be the remains of a marking post), a bifacial stemmed point with denticulate edges of volcanic glass, and two microperforators of quartz. In addition, a 'woollen' weaving of sophisticated technique is said to have come from the tomb (Núñez 1988: 170). However, these yarns, basketry and fabric have not been made available for study, and the only item included in the Appendix is an object of unknown use (perhaps the handle of a basket?), which had deteriorated and fallen into several pieces. Tomb five contained a neonatal child and it is said to be associated with tomb four. The child had a stone-carved quadruped (a feline?) placed inside the mouth, and green stone beads in various shapes and sizes at the level of the chest (figure 7.16). An adjacent tomb, also in the lower layer, is said to have contained the remains of an adult. Tomb seven contained the remains of an adult with what is described as a head ornament underneath the skull, with beads and sticks stuck together. The inventory of tomb seven also included a grinding stone (*mano*) and ochre next to the chest, and a scraping tool. Another skeleton was said to be beneath. Tomb eight contained human remains, but no inventory items are listed (ibid.: 171). On the surface of the site, near the tombs, are three mortars with a shallow, wide hollow.

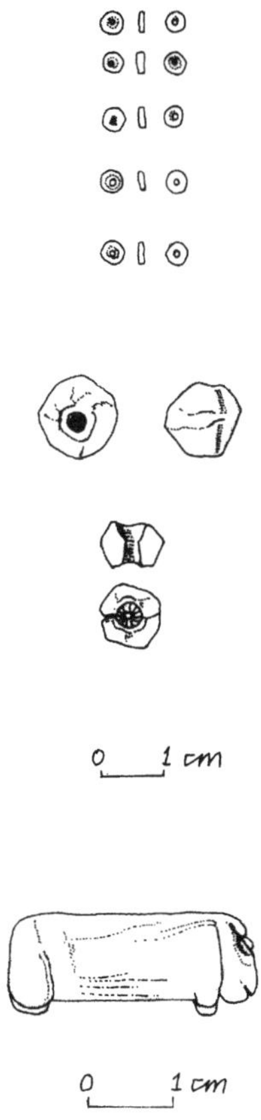

Figure 7.16 TU 58: small shell disc beads from tomb 3; large green stone beads and carved stone quadruped from tomb 4.

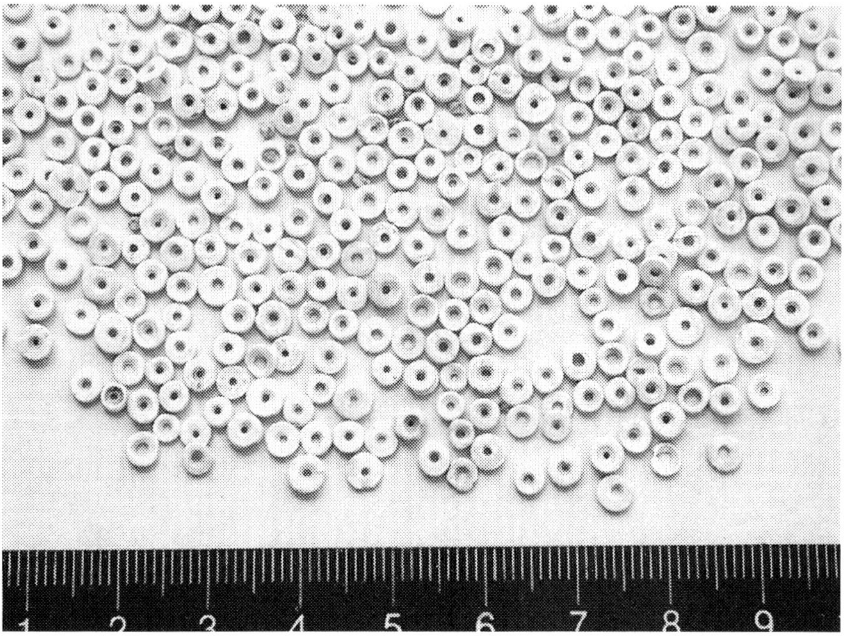

Plate 7.9 TU 58, tomb 3. Beads and carved stone quadruped (see also figure 7.16). By courtesy of the Museo R.P. Gustavo Le Paige S.J.

To date, pottery has not been reported in intact tombs, but the presence of shell beads, microperforators and other stone tools has been used to suggest that sites 58 and 54 are contemporary (ibid.: 170). A C-14 date on a sample of dung is available for tomb six, in the lower layer: 2240 ± 50 BP (Beta 32387) (Núñez personal communication). This date suggests that this tomb at TU 58 was in fact more recent than TU 54, although at the moment there is no way of knowing which tombs were contemporary, or for how long a period the cemetery was in use. The stone and shell beads at TU 58 are distinctive, since the large green stone beads have not yet been found in other Tulan sites. Both these beads and many of the thin, flat, disc-like shell beads have an hour glass perforation (figure 7.16; plate 7.9), thus the manufacture of the beads at TU 58 is different from that at TU 54, where the disc-like shell beads have a cylindrical perforation, as indeed do those of TU 52 and PU 1.

Tulan 82

This is the most recent open-air site to be considered in this section that has yielded yarn, fabric and animal fibre remains. Le Paige (MS) first reported its existence. It is situated on high ground on a spur of land jutting out to the west of site TU 85, and it overlooks the pasture grounds of the Vega de Tilocalar. TU 82 consists of closely spaced, large and small stone-built, circular structures. The walls are incurving, and the smaller structures are roofed with a false vault. These buildings are aligned along a north–south axis, in accordance with the layout of the land. To the south of the site is a large arc of standing and fallen stones of unknown function. Excavation of a 1 sq. m grid square in one of the circular structures revealed four stratigraphic layers, to a maximum depth of 60 cm (Núñez 1988: 259). Núñez mentions that various types of pottery were found, including classic fine polished ware of the San Pedro de Atacama culture, in the stratigraphy and also on the surface of the site. In addition, a dense scatter of copper mineral was found on the surface inside one of the circular structures at the north-eastern side of the site. A charcoal sample yielded a C-14 date of 1610 ± 60 BP (Beta 25529).

Other sites further upstream at Tulan follow a similar architectural pattern, in that they have densely clustered buildings with a circular groundplan (TU 57) and, in the case of TU 59, incurving walls, where the preservation of the architecture allows such an observation to be made.

Tulan 64

This site is a shallow rock shelter situated at the back of a rectangular corral. At the foot of the rock face are large blocks of fallen rock, the various faces of which have been incised with drawings. The rock art was evidently done at different times and periods, for it includes presumably pre-Hispanic panels of camelids (figure 7.17) and also a post-Hispanic human figure on a quadruped with reins, as well as incised crosses.

Excavation of 1 sq. m took place on 7 and 8 October 1987, in an area free of fallen rock. The grid square was excavated in 5-cm spits. Ceramic sherds were found in nearly all layers, including fine polished wares and coarser burnished pottery, but llama dung was concentrated in upper layers. A barbed and tanged arrowhead came from layer 10–15 cm, another from 15–20 cm and three from 25–30 cm (plate 7.10). All are broken. Also in the deposits was a scant amount of lithic waste. As the excavation progressed, a standing stone (the very top of

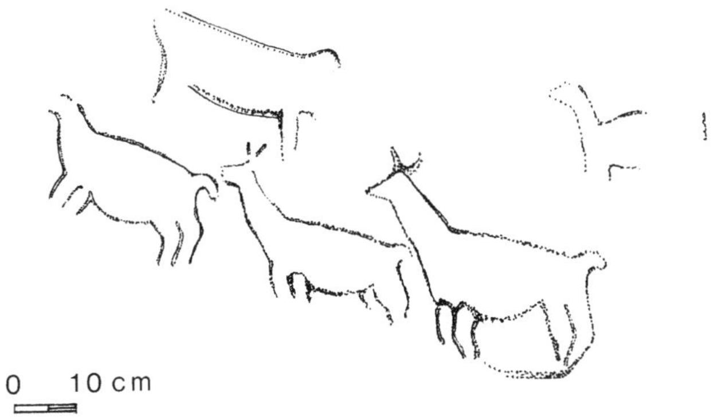

Figure 7.17 Panel of rock art, TU 64.

Plate 7.10 TU 64. Barbed and tanged arrowheads. By courtesy of the Museo R.P. Gustavo Le Paige S.J.

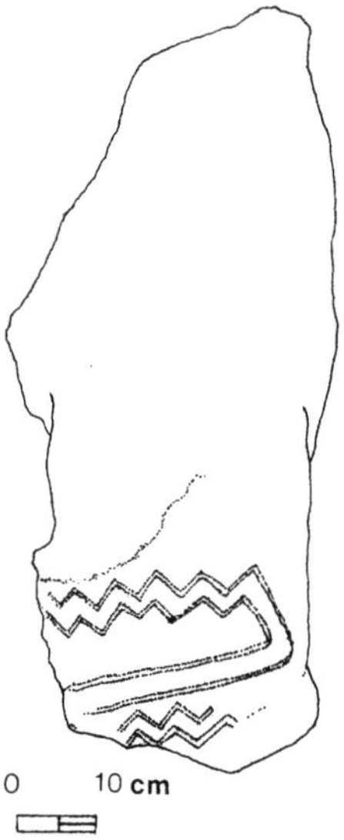

Figure 7.18 Re-used standing stone with geometric design, TU 64.

which emerged above the surface of the ground before excavation began) was uncovered. This stone was inserted into a socket dug into sterile soil and it was held in place by a small mound of stones round the eastern side of the socket. Sterile soil was encountered at 65 cm. The standing stone had evidently been re-used. When it was removed, it was found to bear an incised geometric design at the end that had been placed downward into the socket (figure 7.18); this means that the design was no longer visible when the stone was inserted into the socket. A C-14 date on a sample of charcoal produced a result of 1540 ± 80 BP (Beta 25530) (Núñez 1988: 193).

Tulan 60

The footpath leading from the shepherd's house at Tulan drops into the valley near site TU 64, leading to the spring of rather saline water which gives rise to the perennial river which flows as far as Tilomonte. It ascends steeply on the northern side of the Quebrada, passing close to a great panel of incised drawings on a vertical slab of cliff face overlooking the spring. This panel is known as site TU 60, and it includes drawings of life-sized camelids, which face a more or less vertical fissure in the rock face (figure 7.19). Immediately to the left of this crack, a life-sized puma confronts the camelids and faces the viewer's right. It surmounts another camelid, which also faces right. These main figures are flanked by other camelids, most of which face inward to the vertical axis formed by the fissure in the rock. The panel is extensive and many animals have been superimposed on top of others, but not always respecting the original orientation. Camelids are depicted with llama-like tails, upright ears and bifurcated toes. One of these confronts the puma, and it is represented as having a bulging belly, as if pregnant. Beneath the main panel is a series of small animals, including camelids, puma and goose-like birds. The conjunction of camelids, puma and aquatic birds in a panel of rock art close to a spring cannot be accidental in the light of the bird associations discussed in the previous chapters, and of the repeated combination through time of pumas and camelids in both artistic iconography and ritual practice. The superimposition of camelids on top of camelids suggests that the constant re-drawing of the images was important at this significant spot in the landscape. However, the panel cannot be reliably dated at present.

Berenguer and Martínez (1986) have discussed a comparable sequence of rock art located at Taira in the middle Loa and the theme of camelids and springs in the light of a Quechua myth collected at the end of the sixteenth or early seventeenth century (one of the Huarochirí narratives). The dating of the Taira panels is equally problematic.[12] A large, solitary camelid depicted as though drinking water (figure 7.20) appears on a large expanse of rock above the river through the Quebrada de Jerez, which flows past Toconao (south of San Pedro de Atacama) and which drains into the Salar. This camelid is similar in style to the Tulan and Taira ones. Because it is sited immediately below a late pre-Inkaic fortress, it may well date from late pre-Hispanic times.

Archaeological sites and modern corrals jostle together and overlie each other in the Tulan Quebrada, a situation in contrast with the apparently more favoured conditions at Puripica, with its more abundant pasture and fewer sites. The Tulan sites examined here are concentrated between Tilocalar and Tulan. Other sites are known at higher altitudes, for example, Pozo Cavado (TU 89), at an altitude of about 3,500 m asl, a rock shelter site, and at Estancia La Cueva (TU 88), a campsite, both of which are situated in *quebradas* tributary to the Tarajne/Tulan drainage system. Niemeyer and Schiappacasse (1968; 1976) studied sites bordering the lakes of Miscanti and Miñiques at even higher altitudes. Humid conditions have not favoured the preservation of organic remains from such sites, and inevitably the present study is based on evidence from the lower to middle sections of the Tarajne/Tulan drainage.

Figure 7.19 Panel of rock art, TU 60.

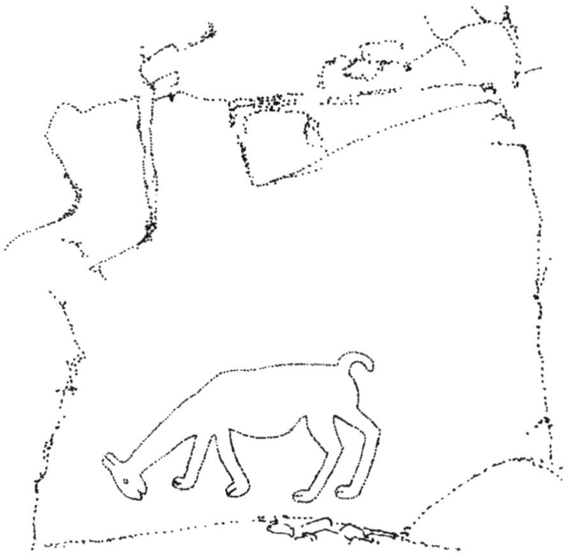

Figure 7.20 Panel of rock art in the Quebrada de Jerez, Toconao, San Pedro de Atacama, with incised life-sized camelid.

Time and chronology in the Tulan Quebrada

Although some of the sites mentioned above have been used over long periods of time, in particular the rock shelter of TU 67 and the cave of TU 55, the individual occupations by human groups have left fairly scant artefactual remains. This situation can be contrasted with open-air sites such as TU 54, a site occupied for a short timespan but where an abundance of objects were deposited.

During the tenth and ninth millennia BP, evidence is slight for human activity in the San Pedro de Atacama area, in places such as Tambillo (TA 1) and the rock shelters of TU 67 and TU 68. The AMS dates for TU 67 present tenuous evidence for continued human activity in the Tulan Quebrada. More substantial evidence for human occupation is available in the area after about five thousand years ago; it is attested at sites such as TU 51 and TU 52, and further to the north, at Puripica. Site SBa 101 at Kalina and sites of the Complejo Chiu Chiu also correspond to this period. During this time, human activity left more endurable evidence in the form of circular buildings with substantial foundations of upright vertical slabs set round a hollowed-out floor. The next period for which there is substantial evidence comes from sites with pottery, about three thousand years ago (sites TU 85, TU 54 and TU 55).

Following another hiatus, there is a series of sites at Tulan, with circular houses located on the south bank of the Quebrada above the spring. The high-altitude rock shelter, TU 89, at Pozo Cavado, and the lower-level agglutinated village of TU 82 at Tilocalar may also belong to this period, during the first millennium AD.

Fine polished black pottery and other ceramic types characteristic of what is known as the San Pedro de Atacama culture were used at least at some of these sites.[13]

A fragment of polished black pottery has been reported from the site of Tulor 1, an extensive agglutinated village in one of the oases of San Pedro de Atacama, as well as abundant sherds of a grey burnished ware in the second (last) occupation layer of the site (Llagostera *et al.* 1984: 137). The houses have a circular ground plan and incurving walls, built of mud brick. A series of post holes indicate that the roof was supported by posts (ibid.: 147). The excavators considered the architecture of Tulor to derive from a foreign pattern, possibly from north-west Argentina, where similar sites with agglutinated circular houses surrounded by an enclosing wall are found, even though the Argentinian sites are less extensive than Tulor. The incised pottery and other ceramic wares found in the lower layer at Tulor are also said to correspond to Argentinian forms, but in the second layer the variety in pottery is replaced by two basic types (ibid.: 140–2).

The circular houses at Tulan have incurving walls (sites TU 82, TU 57 and TU 59). However, the methods of construction were different: dry-stone slabs at TU 82, walling with outer and inner faces and rubble infill at TU 57, and mud and stone at TU 59. Thus patterns of construction display some variety.

The patchiness of the evidence for pre-Hispanic occupation in the Tulan Quebrada is highlighted by the very few sites that can be reliably placed in later pre-Hispanic times. Evidently, the cave at TU 55 was still being used, and Le Paige (MS Tilomonte Sur) reported a badly sacked cemetery at Tilomonte, numbered TU 77 in Núñez (1988).

Despite the changes in house building techniques over time, other material practices demonstrate some continuity through time. The use of red pigment is an example of such continuity of practice, although it is not always evident to what surfaces the pigment was intended to be applied. A grinding stone with traces of red on the grinding surface was found in grid square F 5/6 at TA 1, near a hearth, and also a piece of red mineral. The rear wall of the rock shelter TU 67 was painted with red. Elsewhere, Núñez (1988: 293) reports the existence of stylized camelids painted in red on a small block next to site TU 89 at Pozo Cavado. As mentioned above, the deposits at TU 67 contained bones and sticks painted red, a palette used for grinding red in layer III, and a piece of red mineral in layer IV. Red mineral and camelid bones painted red were deposited three thousand years ago at TU 54. An early interest in the colour red is significant as it is related to the development of red dye technology. Some of the yarns listed in the Appendix were dyed red, and they are discussed further in the following chapter.

Bead making was an activity which was developed over time but which shows continuity in the persistence of flat, small, disc-like beads of stone, shell or bone that occur on many sites, including TU 67, TU 52, PU 1, TU 85, TU 54, TU 58 and TU 57. A black stone bead from TU 54 (plate 7.7, see p. 186) obviously broke during manufacture, and the striations caused by the perforating and the working of the bead were not sanded smooth. This evidence shows that beads were made on site, and the sheer numbers in which they sometimes occur indicate that it was an important activity. The frequent presence of shell demonstrates that contact in some form or another was made with the coast, from where mussels and other shells were taken. It would seem that supplies of shell were more frequently obtained in later periods, from the time of TU 54 onwards. Certainly, the evidence from TU 58 indicates that an abundant supply of what was evidently an important but exotic resource was exploited by the peoples of the valleys draining into the Salar de Atacama. The spinning of fine yarns on which to thread the beads is an activity that was also undertaken at these sites.

Microperforators or awls made from some form of chert are another example of an artefact found in different Tulan sites over long periods of time. Examples of this tool type occur at earlier sites such as TU 52 and PU 1, they are still found at TU 54, and they occur again at TU 57. Elsewhere in northern Chile they are associated with sites of the Complejo Chiu Chiu (2700–1675 BC), and they are long-lived in the sequence proposed by Druss (1978: 42–68), surviving into the beginning of the subsequent Vega Alta period sites in the middle Loa region. Núñez (1976b: 148) mentioned a mixture of preceramic and early agropastoral components at TU 54. Subsequent excavation did not support his suggestion that preceramic lithics were present among the surface material of the site, and perhaps the abundant presence of microperforators helped form the initial impression of such a mixture of tools from different periods. Following the excavations reported by Núñez (1981), he argued for a certain degree of continuity in the stone tool tradition of the Tulan Quebrada from that of site TU 52 to the lithic assemblage of TU 54, which includes microperforators of chert, stemmed projectile points and Tulan-type blade tools (Núñez 1981: 149).

This situation is in contrast to that reported by Benavente from her work at the site of Chiu Chiu 200, a site with habitation structures assigned to the Vega Alta phase. Benavente argued for the superposition of a new tradition in the middle Loa in which the microliths were replaced by larger, parrot-beak piercing tools made from andesite, and stemmed points by projectile points with a concave base. In addition, she relates the appearance of corrugated pottery at Chiu Chiu 200 to similar pottery from San Francisco, north-west Argentina and from Bolivia. A series of thermoluminescence dates on pottery from Chiu Chiu 200 demonstrates, with a high degree of probability, that this site was more or less contemporary with TU 54.[14] While broken sherds of coarse, brown burnished ware and black sherds with corrugated and incised decoration are found at both sites, Benavente's analysis shows that modelled and painted ceramics are represented at Chiu Chiu 200 (Benavente 1982: 88). Fabrics and yarns are also present, and they will be compared with the TU 54 material in the following chapter.

Seasonal occupation and seasonal movement in the Atacama

The presence of lenses of windblown aeolic sediments within the upper and lower stratigraphic zones at TU 52 suggests that the site was periodically abandoned and reoccupied. According to the model presented by Núñez (1981: 153), site TU 52 was used by hunters and gatherers at a time when radical changes were taking place in the subsistence economy at PU 1. He allied the domestication of camelids with a growing trend towards sedentism. Núñez thought that 'the development of early preceramic herding would have stimulated a sedentary lifestyle with greater sociocultural, productive and demographic complexity of a proto-Formative nature' (ibid.: 158).

Núñez gave his 1981 article the subtitle of 'towards sedentarization', but he did not define what he means by the term. Implicit in his argument is the idea that 'food production' is combined with a more sedentary lifestyle, while hunting and gathering is dependent on nomadic movement. However, if the form of 'food production' under consideration here is to be understood in the specific sense that the people of Puripica began to herd camelids and that they came to regard certain living animals as the private property of individual people, it should be stressed that this particular form of establishing property rights is different to that of claiming ownership to plots of land for the cultivation of crops. In the case of pastoralists, social relationships of ownership are not concerned with immobile property such as land, but with movable property in the form of camelid herds. The foregoing chapters have demonstrated that pastoralists in the Andes are constantly preoccupied with the need to find adequate pasture for their animals, and that they take their animals to different pasture grounds on a seasonal basis. At site PU 1 itself there are hints of periodic absence of occupation, for between layers III and II the deposition of lithic tools and animal bones was markedly reduced, while Núñez (ibid.: 150) mentions the presence of aeolic sediments in layer II. The seasons at which PU 1 was occupied still remain to be clarified. It is likely that the occupants did, in fact, follow nomadic cycles of movement to gain access to fresh pasturage and to hunting grounds.

In the chronology offered by Núñez, sites TU 52 and PU 1 are late Archaic period occupations, but sites TU 54, TU 85 and TU 82 belong to the Formative period. Núñez explained that the concept of Archaic is used to designate transitional or 'experimental' societies leading to different forms of sedentary life, and that it is to be regarded as an evolutionary stage without chronological implications (Núñez 1983: 172). In his 1981 article, he recognized that, with the gradual changes caused by the beginnings of the cultivation of plants and the herding of animals, not all human groups dedicated themselves to sedentism: hunters and gatherers retained their mobile lifestyle, as did some herders (Núñez 1989: 81). However, when considering the site of Tulor 1, Núñez interpreted the evidence from this extensive agglutinated village as representing the achievement of fully fledged sedentism in an area suited to the undertaking of irrigation for agricultural purposes, and the fulfilment of a process that purportedly enabled sedentary life to provide more food and support a larger population (Núñez 1989: 97). An examination of the stratigraphy of house 2 at Tulor 1 shows a thick accumulation of sand between the two occupation layers (Llagostera *et al.* 1984: 146). This suggests that the house was abandoned for an as yet undefined period of time, and that seasonal or periodic occupation occurred, even though at first sight this site may seem characteristic of the permanent settlement that Núñez regards as the hallmark of the Formative period.

Sedentism and nomadism should perhaps be regarded as the opposite extremes of a continuum between which human activity may shift. Human groups using Formative period sites in the Tulan Quebrada probably resided in a series of settlements. Their nomadic movements did not necessarily involve going back and forth only in the Tulan/Tarajne drainage system. The fact that the archaeological investigations reported here have concentrated in one valley should not be used to imply that neighbouring valleys were not also used in nomadic cycles of movement.

On the definition of subsistence economies

To date, little or no evidence has been detected for the practice of agriculture at the sites described above in the Tulan Quebrada. The presence of some cultigens has been noted. Popper (1977: 47) identified remains of a gourd shell (Lagenaria siceraria) in layer IX at TU 55. Holden identified some examples of a pale seeded *Chenopodium* sp. in four of the TU 54 coprolites and, in flotation samples from the same site, a single seed that compares favourably with *Capsicum* pepper (Holden 1990: 301). However, Holden points out that the pepper seed is very small for a cultivated variety and that the *Chenopodium* seeds occur in such small amounts that they cannot have been important in the diet of the people using site TU 54. In present times, the cultivation of crops takes place at Guatin and, to a limited extent, at Tilomonte, thus the majority of the Tulan sites and PU 1 are not immediately close to agricultural land. It seems that subsistence activities in these two areas have been based on hunting and gathering, which was later, over several millennia, combined with herding.

The faunal analysis of the PU 1 material suggests that the domestication of camelids took place between layers III and II of the site. This interpretation has been taken to permit the proposition that the camelid bones at later sites such as TU 54, TU 82 and TU 57 represent domesticated llamas, and that sites belonging to chronologically later stages are pastoral or part of an agropastoral economy. However, it must be recognized that hunting and gathering remained important activities for which there is archaeological evidence at the sites listed above. Holden's analysis of the coprolites indicates that the gathering of locally available plant resources provided the bulk of the food consumed, and Hesse's work demonstrates the importance of the hunting of rodents, a theme further explored by Olmos (1980) at sites TU 52 and PU 1. Snares, which probably served to trap birds, occur in the TU 54 midden deposits, and these are described in the following chapter. In addition, site TU 64, with its enigmatic standing stone, seems to be related to hunting activities because of the presence of barbed and tanged arrowheads. Therefore, hunting and gathering activities continued to be practised into the Formative period and beyond; residual hunting for rodents continues to this day.

Ingold (1984: 5) points out that hunter-gatherer societies are conventionally understood to consist of people who do nothing else in the way of obtaining food than gather and hunt, and that as soon as they begin to herd or to farm in addition to their hunting and gathering activities, they are classed as pastoralists or cultivators. He objects to the classification of hunters, trappers and fishers of northern Eurasia as food-producing pastoralists on the basis that they kept small herds of domestic reindeer for their transport needs. Ingold argues that we should conceptualize social structure as being a composite entity, and that different practical activities such as hunting and herding correspond to the various means by which society exploits natural resources. He recommends that when people keep their own herds in the context of a hunting economy, the important question is 'to what extent do relations established in respect of the control and use of animal property constitute the social matrix within which the activities of hunting are carried on?' (ibid.: 5). If the herding of animals owned by individuals is seen to constitute the dominant matrix, then the society in question may be characterized as pastoral, but Ingold warns that this structural dominance may have nothing to do with the proportions of time and energy devoted to the activities involved in such tasks as hunting and herding. Pastoralists do not necessarily spend most of their time herding, and their herds may not provide the bulk of consumable produce. The balance of the activities undertaken by human groups may change during the lifetime of an individual person or, in the longer term, over several generations, setting up fluctuations caused by a society's response to changing conditions with which it may be confronted. According to Ingold, such changes do not amount to structural transformation unless a shift of dominance takes place from one mode of resource exploitation to another.

At present, it is not possible to specify when the dominant structural matrix of a hunting and gathering society shifted to that of a pastoral society in the Tulan Quebrada, although the implication from PU 1 is that between layers III and II such a shift began to occur. However, at the moment, there is no means of quantifying to what extent the herding of camelids was practised, and until supporting evidence from other sites is available, the evidence from PU 1 remains isolated. In the Chiu Chiu area, Pollard and Drew suggest (1975: 297, 303) that domesticated camelids were not introduced until *ca* 500–200 BC, during the phase Pollard calls Vega Alta II. However, Benavente disagrees with Pollard's interpretation (Pollard 1971) that the Vega Alta sites of the middle Loa represent variation through time (Pollard's Vega Alta phases I and II), and she proposes that domesticated camelids were used as beasts of burden at site Chiu Chiu 200 about 800 BC. In her view, this site would have served as a central village and the other Vega Alta sites represent functional variation, being used for different purposes during the same period (Benavente 1981; 1982). Detailed faunal analyses of bone remains from the Complejo Chiu Chiu sites and the subsequent Vega Alta sites would probably help clarify the situation.

The next chapter examines the different categories of animal fibres, yarns and fabrics that have been excavated from some of the sites considered above. Thus Chapter 8 provides the necessary evidence for the concluding chapter, which will examine the question of what difference owning one's animals should have on the exploitation of raw materials, particularly as regards the production of yarns and fabrics.

8 The yarns and fabrics of Tulan societies

The archaeological setting discussed in the previous chapter constituted the context in which Tulan people transformed camelid fibre into yarn and fabric. In the present chapter, I consider this aspect of the material culture made and used by societies in the Tulan Quebrada.

Excavation at sites TU 52, TU 67, TU 85, TU 54, TU 55 and TU 82 produced animal hair remains, pieces of hide and yarns. Some of these sites contained fragments of fabrics, and a few of them, plus another site – TU 58 – also contained fragments of basketry. This range of organic material offers an insight into the choices and preferences people made in selecting fibre, whether animal or human hair, or plant fibre.

It is frequently said that Andean fabric-making techniques are based on the initial exploitation of vegetal fibres. Junius Bird stated that ancient Peruvian fibre technology was based on cotton rather than wool or any other fibre (Bird 1954: 13; Bennett and Bird 1949: 258). Franquemont (1986: 82) repeated this view, claiming that until the middle of the first millennium BC, cloth in the Andes was made 'almost exclusively' of cotton. I regard such opinions to be derived from a more northerly coastal perspective, due to the favourable preservation of organic remains in the dry desert conditions of the coast in the Central Andes. In the South-Central Andes, camelid fibre was predominant in the yarns dating from the eleventh to the tenth millennium BP at the site of Inca Cueva cave 4 in north-west Argentina (Aschero 1984: 68).

One of the most important characteristics reported here for the yarn and fabric assemblages[1] from the Tulan Quebrada is the reliance on camelid fibres. However, few detailed, published reports exist of the yarn and fabric assemblages from Andean sites within a comparable chronological framework to the Tulan sites discussed here. To date, the study of the yarns from Guitarrero Cave in the north of Peru is one of the few published accounts that considers yarns (as a class of artefact) as an assemblage (Adovasio and Lynch 1973; Adovasio and Maslowski 1980). However, Guitarrero Cave is in an area where spinners selected vegetal fibres as their main raw material, and the published reports are couched in terms that do not make it easy to set the yarns of those assemblages alongside those listed in the Appendix.

For the purposes of this book, the animal hair remains are divided into two broad categories: unmodified remains of hair and skins with the fleece still adhering, showing no obvious signs of having been prepared, and modified remains, a category that includes yarns, fabrics and textiles. This second category also contains less securely identified items which may be fragments of yarns or rovings, or small remnants of yarn of which the spin has slackened.

The site that yielded the greatest profusion of such remains in both categories was TU 54. It produced the bulk of the material described in the inventory in the Appendix, and it inevitably provides the focal point of this chapter. The sheer quantity of the remains demonstrates the importance of camelids as a source of fleece three thousand years ago in the Tulan Quebrada.

An analysis of the yarns, fabrics and basketry indicates that these sub-divisions are not always discrete and mutually exclusive. In the Appendix I have included all elongate structures, and not just spun threads, in the category 'yarns'. Therefore, examples of three-strand plaiting (Nos 652, 653 and 654) are included with the yarns. In Irene Emery's classification of the primary structures of fabrics, three-strand plaiting is a simple version of a fabric structure called oblique interlacing. However, I decided to group these items with the yarns, since the interlacing was evidently intended to convert the fibres into a broad and long strand. Emery (1966: 10) defined 'yarn' as the general term for assembling fibre or filaments in a continuous strand, which may be used in fabric construction. Her book is enormously useful for identifying fabric structures in museum artefacts. However, since her book consists of a classification of fabrics, there is a tendency in it to assume that spinners produce yarn in order to weave, loop or interlace it so that it can be converted into a fabric. My work indicates that not all of the yarns produced on Late Archaic and Early Formative sites of the Atacama were intended to be employed in the making of fabrics. The large number of beads made from shell and other materials at sites in the Atacama and the 'turban' of yarns wrapped round the child's skull at TU 85 indicate that yarn was often intended to remain as such. It was not subsequently used in fabric-making activities.

Irene Emery (1966) distinguished between textiles – a two-element structure in which the weft interlaces with the warp – and other fabric-making techniques, such as felt and looping. Apart from the anomalous fragment of textile encountered in the lowermost layer of the rockshelter TU 67, most of the early fabrics from the Tulan sites were made by variants of looping techniques. Emery made a further distinction between linking, looping and interloooping. Maureen MacKenzie has examined the process of making looped fabric structures in more detail in her 1991 study of string bags made by the Telefol people in Papua New Guinea. She observes that the basic difference between looping (including all the string bag making techniques in Papua New Guinea) and interlooping (knitting and crochet) is that in looping techniques, the worker uses the end of the yarn. To create a new loop, she pulls the working end of the thread through the previously created loop. In interlooping, a continuous thread is used; a loop is pulled through and held in place by the loops above and below it. MacKenzie comments: 'thus knitting and crochet use a theoretically endless thread which does not pass through the body of the fabric; while looping, by its nature, must make use of a limited length which cannot be longer than is reeled easily across the palm' (1991: 210). She reports that 5 m is the ideal length of thread for looping purposes, and that in Papua New Guinea looping is interconnected with the spinning of workable lengths of yarn on the spinner's thigh, the method used in that part of the world. Unfortunately, it is not possible to reconstruct the optimum working lengths of yarn favoured by the Tulan spinners of the Archaic period, nor can we be certain of the methods of spinning employed. However, MacKenzie's insights into the interrelated processes of producing yarn and looping are of great interest, because spinning and looping are characteristic of the sites under discussion here.

The camelid fleeces used to spin the majority of the yarns listed in the Appendix are of a light brown or mid-brown colour. White appears fairly frequently, while dark brown, black and grey are very rare indeed in the Tulan sample. With archaeological yarns and fabrics, defining colour adequately is problematic. Rebecca Stone (1986: 138) regards colour as being 'capricious' on both a perceptual and physical level because of its 'inherently changeable nature', which makes it difficult to record and evaluate. She recommends the *Munsell Book of Colour* as an aid to recording colour in a standardized manner. However, many of the hues with which she was dealing were dyed. In the Tulan sample, only two dyed colours are represented – red and blue – and the latter is present only at site TU 82. Variation occurring at the light brown end of the natural brown spectrum is extremely subtle in the yarns and fleeces examined for this work. In addition, natural daylight was the only type of lighting available in the museum at San Pedro de Atacama when the material was analysed, and it varied greatly in quality throughout the course of a working day. I therefore decided not to use a standardized colour chart, and I assigned the observed colours to conventional categories such as 'white', 'very light brown', 'light brown' and 'mid brown'.

A further problem with archaeological samples is that animal fibres often undergo colour change following centuries of deposition in soil. Human hair, for example, has a tendency to turn a reddish colour in burial sites in the South-Central Andes, a tendency attributed by Virginia Ines Fortich Baca (1980: 130) to the presence of nitrates in the deposits. The camelid fleeces and yarns observed in the Tulan sample sometimes tend towards a yellow hue. Not only does the white often become yellowish, but also there are pronounced beige and yellowish brown colours present among the archaeological items that are not comparable with samples collected from live camelids.

Analysis of the Tulan yarns and fabrics

Tulan 67

The fibre material excavated from grid square 4 included forty-nine undyed yarns of camelid fibre, four camelid yarns dyed red, one roving of camelid fibre, one piece of fur dyed red, one fragment of textile, one piece of basketry and loose pieces of animal hair. Layers I to V produced the yarns, while the textile and basketry came from layer VII. These remains were possibly deposited over a long period of time but, as discussed in Chapter 7, the stratigraphy is disturbed and it is not possible to distinguish between the yarns from the preceramic and ceramic occupations. The fifty-five yarns listed in the inventory include only one of vegetal fibre (layer III), and the remainder (fifty-three yarns and one roving) are of camelid fibre. Given the emphasis on the use of red at this site (see pp. 168–72), it is surprising that only four of the camelid yarns should show signs of having been dyed red. This is, in fact, a similar proportion to the fifty-seven dyed red yarns out of the total sample of 702 yarns from grid squares 1 to 6 at TU 54. However, the proportion of yarns spun in the opposite direction (that is, S spun) is greater at TU 67, since thirty-three out of the total of fifty-five yarns have an initial S spin, but at TU 54, only forty-seven out of 702 yarns have an initial S spin. In addition, the TU 67 yarns tend to be somewhat finer in diameter than those from grid squares 1 to 6 of TU 54. Approximately half of the TU 67 yarns (twenty-seven out of fifty-five) have a diameter of 1 mm or less, but only 100 out of a total of 702 yarns from TU 54 have a similar fine diameter. It is possible that the yarn-making activities carried out at the rock shelter may have served a more specialized purpose than at the habitation sites. The yarns are mostly two-ply or single, unplied yarns, and only one has a Z4S construction.

Among the painted designs on the rear wall of the rock shelter is a stylized human figure, which appears to wear a fringed pubic covering (plate 8.1). While it is not possible to correlate the painting with any of the archaeological layers, it is interesting to compare TU 67 with the coastal cave site of La Capilla, immediately south of Arica, where twenty-three fringed pubic coverings were found, folded and tied, in the archaeological deposits (Muñoz and Chacama 1982: plate 8).[2] There are two panels of painted geometric, zoomorphic and solar designs on the walls of the cave, and the excavators of La Capilla considered that, in the absence of hearths, the site served as a cult centre where rites of passage were enacted, during which youths probably put on pubic coverings (ibid.: 42–5). Although the La Capilla garments were made of vegetal fibre, it is more likely that camelid fibre would have been used for such garments in the Tulan Quebrada, if indeed they were worn as suggested by the TU 67 rock painting. Three human figures wearing fringed pubic coverings are painted on rock near the confluence of the Salado and Caspana rivers (Le Paige 1965: plate 17b), and camelid fibre was used to make fringed pubic coverings excavated from a cemetery at site Tarapacá 40.[3] Additionally, Santoro (1981: 41) reported the presence of 'hanks of yarn for pubic coverings' in the Azapa phase section of the cemetery at site AZ 71 near Arica.

The fragment of textile found in layer VII of TU 67 appears to be anomalous. In the far north of Chile, true weaving makes its first appearance in narrow woven bands, such as those wrapped round the heads of two children in the funerary site Tiliviche 2 (tombs 18 and 19). According to the description in Standen and Núñez (1984: 142), the width of the band in tomb 18 consists of four warp threads and that in tomb 19 consists of seven warp threads; in both cases warp and weft

Plate 8.1 Stylized human figure wearing what appears to be a fringed pubic covering, painted on the southern wall of rock shelter TU 67.

are of camelid fibre. The excavators report a date of 3780 ± 100 BP (N 3772) for sector A of the cemetery, but tombs 18 and 19 are in sector B, which they suspect may be some two or three hundred years younger than sector A (ibid.: 147).[4] These narrow bands are unlike the TU 67 fragment, which appears to have come from a larger piece of cloth.

Tulan 52

If the dates reported for the lower stratigraphic zone (4270 ± 80 BP [N 2488]) and the uppermost zone (4340 ± 95 BP [N 2487]) of the main excavation may be used to argue for a fairly short-term occupation of the site towards the end of the fifth millennium BP, then the camelid fibre used in the nine yarns excavated from a column at the eastern end of the main excavation would have come from the hides of hunted animals. The small staples of fleece in layers VI, IX and Xa are of light natural brown colours, which have turned somewhat yellowish in the deposits. Seven of the yarns were evenly spun and plied from mid-brown fleece (also yellowish in colour). Two, unplied fragments in layer VII were dyed red; perhaps these represent one yarn broken into two pieces. This colour is now an even, light pink, and very light-coloured fleece was evidently selected for the yarn. The yarn sample from this site is too small to be significant.

However, Zlatar (1983: 22) reported 'woollen' (presumably camelid fibre) yarns and fabrics from the Late Archaic coastal site of Caleta Huelén 42. In north-west Argentina, the cave site of Inca Cueva 7, with one C-14 measurement of 4080 ± 80 BP (T 1773), contained yarns and fabrics, about 90 per cent of which were of vegetal fibre (sisal: *Agave sisalana*), the remainder being made of camelid fibre, human hair and sinews (Aguerre *et al.* 1973: 205; 1975: 213). Similarly, excavations at the earlier Middle Archaic coastal site of Caleta Abtao (Abtao 1) revealed ten 'woollen' yarns and a fragment of netting made from camelid fibre among more numerous yarns and fabrics made from vegetal fibre (Boisset *et al.* 1969: 101). Indeed, references to the existence of camelid yarns or fabrics are fairly common at coastal sites, where the conditions for the preservation of organic remains are better than in the Tulan Quebrada; such sites include Camarones 14 (Niemeyer and Schiappacasse 1977), Conanoxa W(a) (Niemeyer and Schiappacasse 1963), Tiliviche 2 (Standen and Núñez 1984) and Quiani 7 (Dauelsberg 1974). At these sites, vegetal fibre is reported as being more common than camelid, but camelid fibre is said to be slightly more predominant than vegetal in sites of the Los Morrillos Culture in the Cuyo region of Argentina, with dates of between 5900 and 2500 BC (Micheli 1979: 1–3).

Tulan 85

Yarns and animal fibres from TU 85 are few in number and small in size. The camelid fibre yarns and pieces of camelid fleece tend to be of light yellowish-brown colours. However, two other important finds are known from the site. The first is a textile from grid square 11, with both warp and weft made from black camelid fibre (fabric inventory No. 9 in the Appendix). Second, a skeleton of a child was found in a group of human burials in grid square 7. The remains of a 'turban' were still in place round the head of the child. It consisted of a textile (fabric inventory No. 8) which covered the hair and which was, in turn, wrapped by a considerable quantity of fine, evenly spun and plied camelid fibre yarns (plate 8.2). At least two yarns were used; one was undyed and the other was dyed red (yarn inventory No. 752). 'Turbans' are better documented for the Arica area on the coast of the far north of Chile, where cemeteries of the Late Archaic and Formative periods have been found to contain corpses with the head wrapped in ever more elaborate 'turbans' (Muñoz 1982: 135–6; Ulloa 1985: 20; Agüero Piwonka 1993: 74). At TU 85 the use of a textile between the hair and the yarn 'turban' is unusual, although Carolina Agüero Piwonka reported a group of seven 'turbans' from the site of PLM 7, Arica, which were wrapped round a textile placed over the head (type No. 13 in Agüero Piwonka 1995: 105, plate 2). 'Turbans' with red or black yarns wrapped round the head have been reported from cemeteries near the site

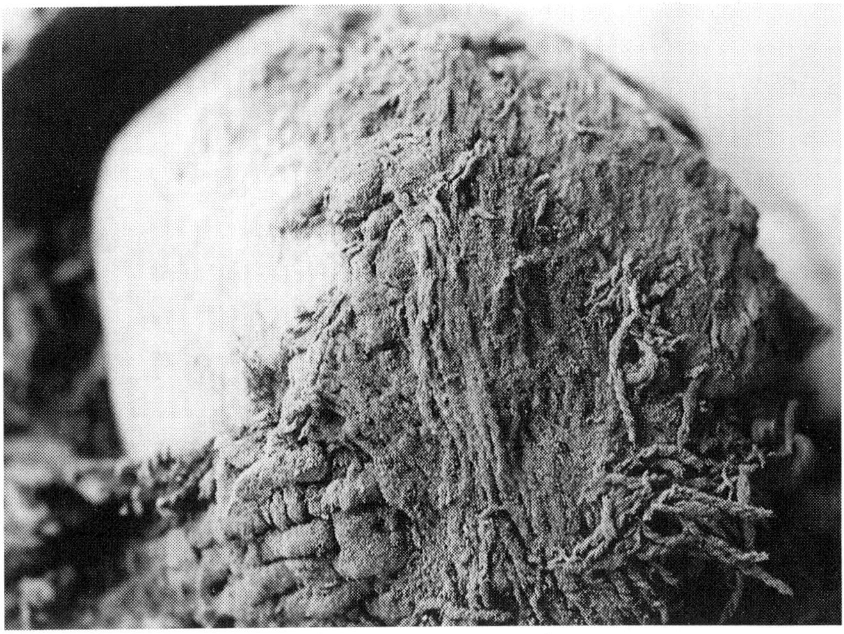

Plate 8.2 Remains of textile (fabric No. 8) and yarn turban (yarn No. 752) on the skull of a child at site TU 85.

of Guatacondo on the River Loa (Meighan 1980: 123). Evidence was detected at some of these burials for the removal of the head from the body: in some cases the head was replaced and in other instances it was missing. Meighan associated such practices with a 'skull cult' (ibid.: 121).

Tulan 54

The yarn assemblage

As previously noted, TU 54 has produced a profusion of organic remains. The large amount of animal fibre and yarn remains, and some fabrics, indicate the excellent, but unusual conditions that permitted the preservation of material not normally found at an open-air site at the same altitude. In this section, I examine the structure of the yarns, and the raw materials used in their construction. Then I discuss how different fleece colours were used, and the evidence for dyes, and, finally, the possible functions of the yarns.

The yarn assemblage from TU 54 is particularly impressive, including balls of yarn and a vast number of fragmentary pieces. Grid squares 1 to 6 contained 666 items listed as yarns (including rovings) listed in the yarn inventory. Some of these items are units of two or more yarns knotted together. There are thirty-six such units and, since yarns with a different construction and/or fibre content were often tied together to form these units, there are, in total, 702 yarns from grid squares

Plate 8.3 Small hank of yarn from layer 27 of grid square B3, TU 54. The yarn is Z2S, tight ply, 1 mm diam. This yarn was dated and produced a C-14 determination of 3000 ± 65 BP (OxA-1841).

1 to 6. The spun yarns range in diameter from less than 1 mm to a maximum of 7 mm, but some of the rovings are wider, and the plaited lengths of vegetal fibre reach up to 1.6 cm in width. Between one and ten strands are combined to make the yarns. Thus there is a great deal more variety displayed by the TU 54 yarn assemblage than that of TU 67. Although 100 out of the overall total of 702 yarns have a diameter of 1 mm or less, the average yarn diameter of the yarns from grid squares 1 to 6 is 2 mm (plate 8.3). A wide variety of yarn structures are also present, and there are some wider yarns constructed from as many as ten fine strands. Table 8.1 lists the different structures employed in the TU 54 yarns, and the frequencies with which they occur.

In addition to the more regular range of combinations of plying and re-plying, there are four yarns that cannot be represented by the normal notation employed in this book, since they have an asymmetrical construction. In other words, they consist of a combination of re-plied yarns that do not contain a balanced number of constituent single strands. The details are given in table 8.2 on page 210.

While the first three of these anomalous yarns are fairly short (between 65 mm and 110 mm in length), the last one listed is 330 mm long. Yarn No. 388 follows a more normal pattern of plying and re-plying, and it may simply be the result of re-plying two yarns which happen to be composed of a different number of strands, in which case the asymmetry may be fortuitous. However, the other three are distinctly unusual in their construction. The inclusion of a dyed red, plied yarn in No. 185, and the combination of a cotton yarn with two camelid yarns, one of which was dyed red, in yarn No. 186 makes one suspect that such yarns were, in fact, special.

Other yarns may be picked out for their special effects. Yarn No. 184 is a 50-mm length of plaited cord, but it is not possible to discern the construction of this piece without undoing it, as it has an applied pile of vegetal fibres that obscures the structure. Both plaited cord and pile are dyed red, and the effect is of a 'furred' cord. This 'furred' cord does not in any way resemble those analysed by d'Harcourt

Table 8.1 Frequencies of the different constructions employed in TU 54 yarns (grid squares 1–6).

Structure of yarn	*Number of yarns*
Z	127
S	11
I	13
?	5
Z2S	378
S2Z	28
I2Z	4
Z2I	5
Z2Z	3
S2S	2
Z3S	33
S3Z	5
Z4S	35
S4Z	1
Z5S	2
Z6S	2
Z7S	2
Z8S	3
Z10S	1
Z2S2Z	25
S2Z2Z	3
Z2S3Z	2
S2Z3S	1
Z2S5Z	1
Z2S2Z2S	1
Z2S2Z2Z	1
Three-strand plaiting	3
Atypical (asymmetrical)	5
Total	702

Key: Z = Z spun; S = S spun; I = no appreciable spin detected; ? = fragmentary condition, spin direction unclear.

Table 8.2 TU 54 yarns with asymmetric construction.

Inv. No.	*Construction*	*No. of strands*	*Final ply angle*	*Fibre*
185	Z S Z Z S Z Z Z	6	S	Camelid
186	Z S) cotton S Z Z S) dyed red Z Z	5	S	Camelid and cotton
388	Z Z S Z Z Z Z S Z Z	6	Z	Camelid
549	Z Z Z Z S Z S Z Z Z S Z Z	8	S	Camelid

(1974: 110–12; plates 61A, 74C, 76B–D), made by means of knots into which cut camelid fibre pile was introduced, or by wrapping a short camelid fibre fringe round a central cord. A yarn notable for the perfection of its spinning is No. 255: two light brown strands and one white strand are plied together (Z3S), but one of the light brown strands is more tightly spun and is finer in diameter than the other, and the former always overlies the latter. Because of the different tension in the spinning of the two light brown strands, the light is reflected in a different manner and a subtle colour difference is produced. Yet another unusual yarn is No. 512, which has a commonplace Z2S construction, but it is distinguished by the fact that both strands alternate from fine to coarse in regular succession. Since the change in diameter occurs simultaneously in both the strands over 235 mm out of a total yarn length of 421 mm, the special effect might have been intentional and not merely accidental. Finally, it is difficult to explain the presence of a very fine Z2S yarn, dyed red, about 115 mm long, which was incorporated in one of the strands of yarn, No. 508 (plate 8.4), which would otherwise be an unremarkable Z2S yarn of mid-brown camelid fibre.

The four anomalous yarns with their asymmetric structure and five 'special effect' yarns exploit camelid fibre and demonstrate the skilful use of this raw material on the part of the spinners. It is not known why they were spun in a manner that departs so radically from the norm for TU 54, but they demonstrate

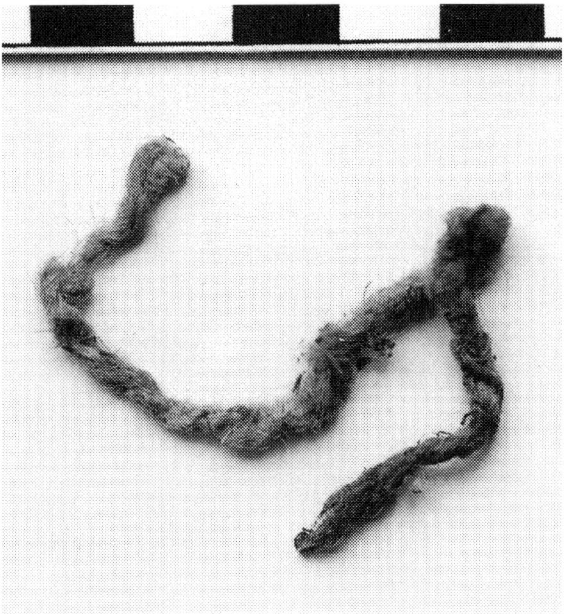

Plate 8.4 Yarn No. 508 from grid square 3, layer n, TU 54. The yarn is Z2S, medium ply, 3 mm diameter. One of the two strands incorporates a very fine yarn, 11.5 cm long, which has traces of red dye.

that the spinners were fully in control of their medium. As will be made clear below, yarns that are incorporated into cloth are not remarkable for any special effect or deviance from 'normal' spinning standards.

Camelid fibre was the preferred raw material for spinning yarn at TU 54. Of the total count of 702 individual yarns, there are five yarns of bast fibre, ten yarns of hard leaf vegetal fibre, six yarns of human hair, and two of cotton. In other words, only twenty-three yarns do not contain camelid fibre at all, and even the two cotton yarns were tied together with a slip knot (yarn inventory No. 555), so they might formerly have been part of the same yarn. A further twelve yarns combine camelid fibre and bast fibre, one strange yarn (No. 188) combines camelid fibre and unshredded hard leaf (a camelid fibre yarn was joined on to a forked twig), and one yarn (the previously discussed No. 186) combines a cotton two-ply yarn with three strands of camelid fibre. Yarns made from non-camelid fibres and yarns that combine different fibres are listed in table 8.3. Given the large size of the yarn sample, the reliance on the use of camelid fibre at TU 54 is truly remarkable. To date, no other site of the same period has been published with details of similar material. However, this should not be taken to suggest that the people who used site TU 54 were in any way unusual in their preference for spinning camelid fibre while their coastal contemporaries preferred to exploit bast and cotton. In all probability, the picture has been skewed by the preferential survival of organic

Table 8.3 Non-camelid fibre in the yarn assemblage of TU 54, grid squares 1–6 inclusive: composition of the fibre content of the yarns spun from non-camelid fibre, and non-camelid fibres combined with camelid fibres.

Grid square	*Layer*	*Item No.*	*Part*	*fibre*
3	O	555	a	Cotton
3	O	555	b	Cotton
2	P	186		Camelid/cotton
2	P	187		Bast
3	G	235		Bast
3	G	242		Bast
3	K	320		Bast
4	L	705		Bast
2	O	162		Camelid/bast
2	P	184		Camelid/bast
2	R	210		Camelid/bast
3	J	264		Camelid/bast
3	L	393		Camelid/bast
3	L	434		Camelid/bast
3	L	435		Camelid/bast
3	N	506		Camelid/bast
3	P	573		Camelid/bast
4	J	669		Camelid/bast
5	A	728		Camelid/bast
5	A	729		Camelid/bast
3	G	238		Hard leaf vegetal
3	G	241		Hard leaf vegetal
3	L	410		Hard leaf vegetal
3	Q	641		Hard leaf vegetal
3	Q	642		Hard leaf vegetal
3	R	652		Hard leaf vegetal
3	R	653		Hard leaf vegetal
3	R	654		Hard leaf vegetal
4	L	703		Hard leaf vegetal
4	L	704		Hard leaf vegetal
2	P	188		Camelid/hard leaf
1	B	73		Human hair
1	C	76		Human hair
1	C	77		Human hair
2	J	119		Human hair
3	O	544		Human hair
3	R	651		Human hair

remains along the coast, where vegetal fibres were more frequently used, and by the lack of surviving archaeological materials from highland sites, where it was readily available.

Most fleece colours in the TU 54 camelid fibre yarns are at the very light to mid-brown end of the natural brown spectrum. White camelid fleece occurs as one or more strand in forty-four of the 679 camelid fibre yarns. There are thirty-two yarns that are completely white (fourteen unplied yarns, seventeen two-ply yarns and one Z4S yarn). It seems that the TU 54 spinners had some white fleece at their disposal. In contrast, only one out fifty-four camelid fibre yarns at TU 67 is white.

At the other end of the natural brown spectrum, there is an isolated occurrence of a dark brown yarn in grid square 3 (yarn inventory No. 254). This yarn is in a fragile condition, but the colour of the fibre is uniform: the yarn has a S2Z construction, and an overhand knot was tied at each end (to prevent the ply from unravelling?). The only other camelid yarn from grid squares 1 to 6 that contains some fairly dark camelid fibres is yarn No. 675, but the fleece used to spin this yarn was predominantly mid-brown, and the slightly darker than normal fibres may well have been the guard fibres growing on the withers or along the top of the animal's back. Thus the one dark brown camelid yarn is highly unusual in the TU 54 assemblage. However, a tiny piece of black camelid fleece turned up in the deposits from the bottom of grid square B3. Since light to mid-brown natural colours are overwhelmingly predominant in the camelid fibre yarns at this site, it is hard to explain the presence of dark brown and such a small quantity of pure black, unless colour variation was already underway in the coats of domesticated stock. Whether the camelids owned by the people of TU 54 grew these unusual colours or whether the fleeces were brought in from somewhere else is another matter.

Grid squares 1 to 6 yielded fifty-seven yarns with traces of having been dyed red. In some cases, the dye has all but disappeared, probably due to bacterial action, and the colour only remains between plied strands or caught in a knot. Only a few of the yarns retain any degree of saturation in the red hue and these are short in length. Perhaps faded yarns were thrown away and found their way into the midden deposits, but their owners kept yarns that retained their strong colour. Since the colour has faded in most of the dyed yarns, it is possible to observe the natural colour of the fleece. Yarns of very light brown, light brown, yellow brown, and mid-brown were selected for dyeing. In no observed instance was white chosen, and this suggests that white fleece was not abundantly available and that it was retained as such. Where the dye remains sufficiently strong in colour, it may be noted that the yarn was dyed after the plying or re-plying stage of construction. In addition, some single strands show that they were formerly plied. It is not known whether fleece was also dyed before being spun and plied.

In this chapter, the term 'red' is used as a colour category encompassing a maroon-like red at one extreme to pale orange reds and pinks at the other. Two TU 54 samples were submitted for dye analysis to Dr Jan Wouters of the Koninklijk Instituut voor het Kunstpatrimonium, Brussels. Samples from yarn Nos 228 and 559 were analysed for dyes present using HPLC (high performance liquid chromatography), a technique said to enable both separation and clear identification of all the dye components present (Wouters and Verhecken 1989a; 1989b). The results are as follows:

Yarn 562: 2% pseudopurpurin
98% purpurin
Vegetal dye source is a *Relbunium* species.

Yarn 231: Only traces of purpurin were found, since most dye had disappeared from the fibre.
Vegetal dye source is a *Relbunium* species.
(Wouters personal communication of 15 March 1990)

It was not possible to distinguish the exact *Relbunium* species involved and, to date, such species have not yet been observed in the Tulan Quebrada (Holden personal communication). Twenty-five species of *Relbunium* occur widely from Mexico to the south of Chile, and the roots of some of them were used for obtaining red dye (Fester and Lexow 1973; Donkin 1977).[5]

Further systematic analyses of both the dyes used in archaeological yarns and of the sources of botanical and other natural dye substances are required. However, it seems that red dye was widely used from a fairly early period. A red yarn (cotton) sample that Grieder had analysed from material excavated at La Galgada, northern Peru, proved to have *Relbunium* as the dye source. It was combined with alum as a mordant, and he argued that this dye technology was already established when the earliest La Galgada textiles were made (Grieder *et al.* 1988: 180–1). The other dyed colour reported for the La Galgada preceramic textiles is black, but a more varied colour repertory comes in the ceramic phases (bright yellow, dark brown, and rare cases of blue and pale purple).[6]

At the very end of the Late Archaic in the far north of Chile, a simple, looped fabric (single element) from Camarones 15 is reported as having stripes of yellow, dark red, dark brown and reddish yellow (Rivera *et al.* 1974: 100). However it is not clear in the report whether the looping is of camelid or vegetal fibre, and samples were not submitted for dye analysis. As noted above, red camelid fibre yarns are present at TU 52, another Late Archaic period site. Thus it would seem that red dye technology was already well established by the time site TU 54 was occupied. To date, I have not recognized other dyed colours in the material from the site.

Twenty-four out of the fifty-seven red-dyed yarns in the yarn assemblage from grid squares 1 to 6 have a diameter of 1 mm or less, and sixteen out of the total fifty-seven have an initial S spin. In other words, camelid fibre yarns that were dyed are also more likely to be finer, or to be spun and plied in the opposite direction to the normal manner for TU 54. Given the presence of red in the atypical yarns discussed above, the implication is that red yarns were, in some sense, 'special'.

A 'typical' yarn from the TU 54 yarn assemblage is of light or mid-brown camelid fibre, with a Z2S construction, a medium-ply angle, and a diameter of 2 mm. However, as table 8.1 on p. 209 demonstrates, yarn structures do display some variety. The combinations of plying and re-plying impart strength and a greater robustness to the final yarn, making it more cord-like in nature, but there are virtually no yarns that might be classified as 'rope'. This study of the yarn assemblage from TU 54 gives no indication that llamas were being used as beasts of burden, since heavy-duty ropes would be necessary to hold items (such as firewood) in place on the backs of pack animals.

The constituent yarns in the fabrics from TU 54 were made almost exclusively of 'typical' yarns. Since a good number of yarns deviate from the 'normal' Z2S pattern, it would seem that they were not intended for making into cloth. It is not clear for what purpose many of the yarns listed in the inventory were intended. Some of them seem not to have reached the final stage of construction. Item No. 496, for example, is a ball wound from two different single Z-spun strands, perhaps in preparation for plying.

In contrast, other yarns were evidently intended to secure a small number of items in a bundle. They take the form of a tied loop, often secured with a granny or reef knot. The frequencies of the various knots detected in the material from TU 54, grid squares 1 to 6, are listed in table 8.4. Thirty-six items in the inventory are of two or more yarns tied together (plate 8.5), although many more yarns had the ends tied together, or one or more simple or overhand knots along the length. Some of these knots were presumably made to unite the ends of a broken yarn: for example, yarn No. 675. Yarn No. 115 is re-plied, with two overhand knots at one end, and the knots may have served to prevent the yarn from unravelling (as suggested above in the case of the dark brown yarn) (figure 8.1). Niemeyer and Schiappacasse (1963: plate XVI, fig. 5) offer a diagram of the heading of a simple looped bag, in which the working thread was looped round a camelid cord with a Z2S6Z construction, from the preceramic site of Conanoxa W(a). It is likely that yarn No. 115 served a similar function.

A recognizable item in the TU 54 assemblage is made from rigid, hard-leaf material, consisting of two unspun strands twisted together in the Z direction, in the form of a loop secured with a slip knot (Nos 241 and 703 in the yarn inventory; No. 704 probably belongs to the same category, but the piece is incomplete). These are evidently small snares. Probably they were used for trapping birds, since a rodent would have easily nibbled such a device, enabling it to escape (figure 8.2). The presence of snares indicates that the trapping of wild animals would have complemented the pastoral economy of the herders, perhaps enabling them to avoid slaughtering their own llamas.

Given the abundant presence of beads at site TU 54, the finer yarns may have been intended for threading beads. Of the beads examined, most are pierced with a hole, which is about 1 mm in diameter, and the finer yarns would have been suitable for such a purpose. To date, no lengths of sinew have been observed in the TU 54 midden deposits, thus camelid fibre, finely spun and plied, and perhaps dyed red, was probably used. Five extremely small pieces of pilled camelid fibre, some 1 mm to 1.5 mm in diameter, were found in grid square B3. Four of these are a light brown, one was dyed red. Pilling occurs when fleece has undergone rubbing or friction, and in this case it might have been caused by threading beads on yarn. The presence of these items supports the suggestion that fine yarns of camelid fibre were used for stringing beads together.

Table 8.4 Frequencies of knots in the yarn assemblage of TU 54, grid squares 1–6 inclusive.

*Type of knot**	*No. of knots*	*Total no. of yarns*	*No. of camelid fibre yarns*	*No. of non-cotton vegetal fibre yarns*	*No. of cotton yarns*	*No. of human hair yarns*
Reef	10	9	9	–	–	–
Granny	27	22	22	–	–	–
Overhand	46	37	32	4	–	1
Slip	7	7	3	2	1	1
Figure-of-eight	1	1	1	–	–	–
Larkshead	2	2	2**	–	–	–
Half-hitch	13	8	7	1	–	–
Unidentified	36	32	31	–	–	1
Totals	142	118***	107	7	1	3

Key

* The observed 'set' of the knot is recorded here. Emery (1966: 34) warns that although the interlacings cannot be changed if the knot is loosened, the original set of the knot may be lost. The interlacings may be the same in two different knots, but the conformation can differ.

** One of these yarns (item No. 393) combined one strand camelid and one strand bast fibre.

*** This total includes ten yarns (item Nos 146, 241, 390, 478, 556, 591, 599, 643, 649, 684) which have been counted twice because they incorporate two types of knot.

Plate 8.5 Unit of three yarns tied by means of three granny knots into a circle. TU 54, grid square 3, layer q. Yarn inventory No. 644.

The fabrics

Fragments of fabrics from TU 54 are few in number and they were located in grid squares 2, 3 and B3. They were found in a limited area within the stratigraphy at a depth of about 70 cm from the surface. The largest piece comes from layer 17 of grid square B3 and it has an associated AMS date of 3080 ± 70 BP (OxA 1840) (plate 8.6).

Four out of five fabrics listed in the fabric inventory (fabric Nos 2, 4, 5 and 6) were formed by simple looping on a foundation element crossed right-over-left with interlocked stitches (figure 8.3). In all four cases, the looping is compact. Fabric No. 2 (figure 8.4) has a single (unplied) foundation element, of the same natural yellowish-brown colour as the two-ply looping element. This fabric has a

Figure 8.1 Yarn No. 115, grid square 2, 40–45 cm, layer i, TU 54, tied with two overhand knots at one end. Z2S5Z; five yarns (tight ply, each with a diameter of 1 mm) have been re-plied Z. Mid-brown, with traces of red dye.

Figure 8.2 Items identified as parts of snares: yarn No. 241, grid square 3, layer g, and yarn Nos 703 and 704, grid square 4, layer l, TU 54. Made of Z-plied strips of vegetal material, loose ply angle. Items 241 and 703 are tied with a slip knot.

series of two-ply yarns doubled over the lags between the looped stitches of the bottom row. Three of these additional yarns are worn and fragmentary, but the remaining three are longer and they tend to twist back on themselves in the Z direction. The purpose of this fringe is unknown, but it may be compared to the edging of a fabric excavated from the Argentinian site of Inca Cueva cave 7 (figure 8.5).[7]

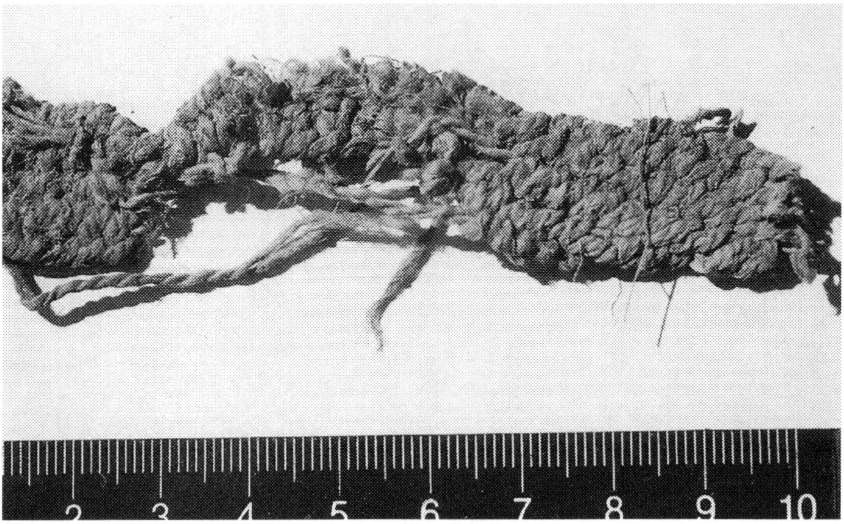

Plate 8.6 Detail of looped fabric No. 6, grid square B3, layer 17, TU 54. Simple looping on a foundation element crossed right-over-left (interlocked stitches). Compact fabric.

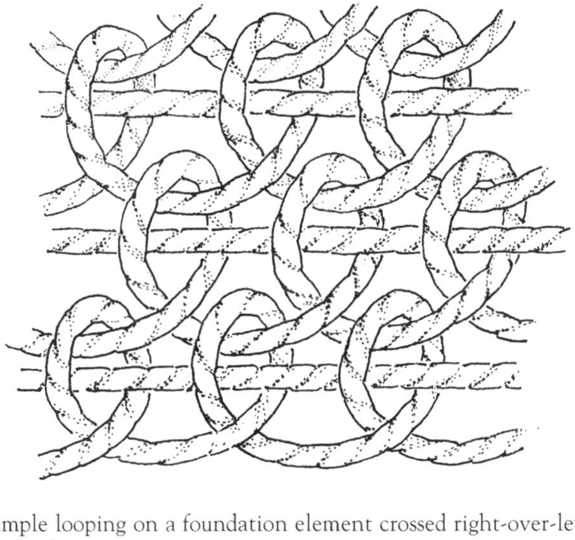

Figure 8.3 Simple looping on a foundation element crossed right-over-left (interlocked stitches).

Fabrics Nos 2, 5 and 6 are of natural brown colours, but fabric No. 4 incorporates a dyed Z2S yarn, the colour of which is now an orange-pink. The fragment is small in size, but the design seems to have been of alternating light brown and dyed horizontal bands (figure 8.6), effected by changing the active looping element, while the foundation yarn is undyed light brown throughout the piece. Caps bearing characteristic stepped designs using various dyed colours executed in

Figure 8.4 Fabric No. 2, grid square 2, from a depth of 65–70 cm (layer l), TU 54. Simple looping on a foundation element, crossed right-over-left (interlocked stitches). Compact fabric.

Figure 8.5 Looped fabric from Inca Cueva cave 7, Argentina (after Aguerre *et al.* 1973).

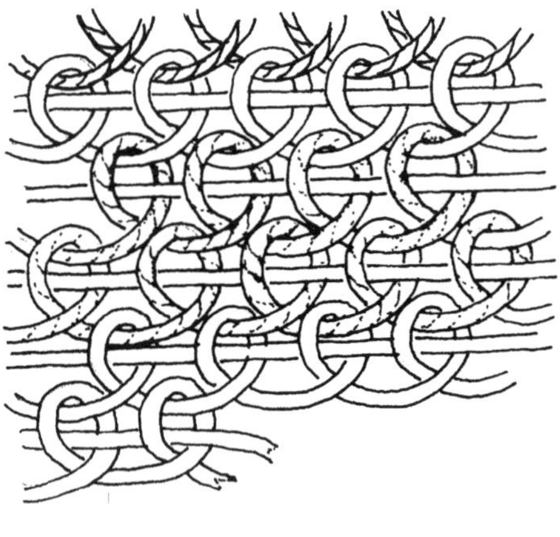

Figure 8.6 Fabric No. 4, grid square 3, layer p, TU 54. Simple looping on a foundation element crossed right-over-left (interlocked stitches). Compact fabric. The foundation and looping yarns are of a light brown colour, except for two rows and the uppermost fragmentary row of looping, which were dyed. This colour is now an orange-pink.

single-element simple looping are known from the later Alto Ramírez sites near Arica, but there is a precedent which predates both TU 54 and Alto Ramírez, since a fragment of simple looping (single element) made from camelid fibre yarns is known from grave number 14 at Tiliviche 2: the design is of stepped horizontals (Standen and Núñez 1984: plate 5f). A compact piece of simple looping with stripes of different colours is also represented by a fragment from Camarones 15 (Rivera *et al.* 1974: 100). These examples from other sites do not have the foundation element that is present in the TU 54 fragment with dyed striping.

The remaining fabric from TU 54 is No. 3 (figure 8.7). Unlike the other four fragments, the structure of this piece is single-element simple looping crossed right-over-left. The initial row of looping is of a light beige natural-coloured yarn, which is supported on an inactive yarn. The passive yarn is two-ply and of a mid-brown colour, like the three subsequent rows of looping. It should be noted that the second row of looping does not form interlocked stitches round the initial inactive supporting element: the active yarn of the second row makes loops only round the lags between each stitch (not stitch and foundation) of the previous row.

Looped fabric structures are ideally suited for making bags and pouches, as the fabric may be slightly elastic, with a certain amount of stretch. If the piece is worked spirally from the base, there is no need for side seams. These qualities also make such fabric structures suitable for headgear. Both bags and caps were identified at Inca Cueva cave 7, the latter being more rounded than the former (Aguerre *et al.* 1973: figures 5, 6 and 20). At site AZ 70 in the Azapa valley, a human head was found wrapped in a cloth of simple looping crossed right-over-left (Mujica 1985: plate 6.13), as well as two caps with the characteristic stepped design (ibid.: plate 6.2). These items came from the Alto Ramírez section of the site.

Looping may also be worked in the form of a flat piece of cloth. Although most of the TU 54 pieces are small, the stitches lie in horizontal lines and do not have the appearance of having been worked in a spiral. The width of fabric No. 6 suggests that this fragment was part of a larger, flat fabric (see plate 8.6). It may have been part of a garment such as a mantle or tunic. Compactly worked, simple looped tunics of camelid fibre with so-called guilloche designs resembling stylized plied yarns, interlocked snakes heads and other stylized motifs are known from Ocucaje at the end of the Early Horizon in Peru.[8]

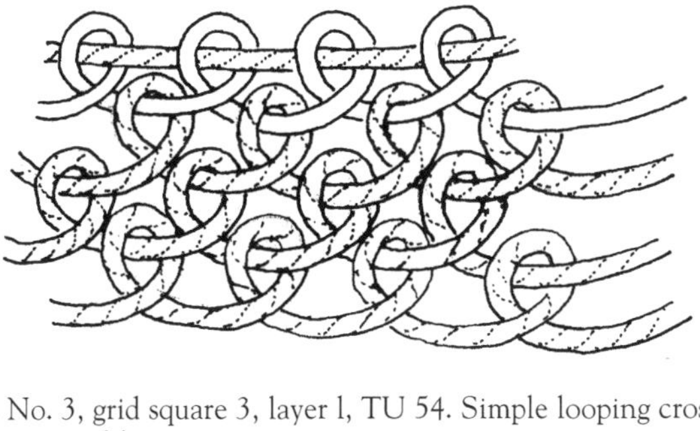

Figure 8.7 Fabric No. 3, grid square 3, layer 1, TU 54. Simple looping crossed right-over-left. Compact fabric.

Fabrics excavated from site Chiu Chiu 200, located near the confluence of the Loa and Salado rivers, provide an interesting set of correlations and contrasts with the TU 54 material. María Antinonia Benavente (1981; 1982) reported the excavations at Chiu Chiu 200, a settlement complex of habitation structures.[9]

The fabrics were located in midden in the walling of the house structures. The excavator Benavente stated that the midden constitutes a single stratigraphic unit (Benavente 1982: 81).[10] She does not discuss the possibility of disturbance or the presence of intrusive material from later periods. The twined fabrics reported from this site have been misidentified (Benavente 1981: 79; 1982: 77, plate 1). These putative twined items are not textiles; instead they are lumps of camelid fleece sewn together by means of long running stitches, using yarns spun and plied from camelid fibre. In addition, I saw no evidence for the following fabric structures: cross-knit loops, simple looping with two single elements (one of the structures that occurs at TU 54) or interlinking, all of which have been listed by Benavente (1981: 79; 1982: 77). Perhaps it would be wise to suspend judgement on the existence of these three structures at Chiu Chiu 200 for the time being.

However, the Chiu Chiu 200 material contains a greater variety of fabric structures than TU 54. True weaving exists at the former site, as well as narrow obliquely interlaced bands using one set of elements (plate 8.7). In addition, looped fabrics are represented. The two looped structures for which I am able to attest are simple looping (single element) crossed right-over-left, and loop-and-twist (one twist, single element) crossed right-over-left (plate 8.8). Like the TU 54 fabrics, the looped stitches from these fabrics cross right-over-left. There is also evidence for red dye in at least two Chiu Chiu 200 fabrics, one of simple looping, camelid fibre yarn, the other a loop-and-twist fabric of vegetal fibre yarn. The presence of dyed vegetal fibre yarn is of interest, since the dyed yarns at TU 54 are only of camelid fibre. I did not observe any other dyed colours. Another type of patterning, not present at TU 54, occurs in another Chiu Chiu 200 fabric. Two different natural colours of yarn were observed to create areas of colour with a stepped diagonal division, but until the fabric is softened and unrolled, it is impossible to ascertain the nature of the design.

The fabric structures I identified at the sites of TU 54 and Chiu Chiu 200 are as follows:

TU 54	CHIU CHIU 200
Simple looping (single element) crossed right-over-left	Simple looping (single element) crossed right-over-left
Simple looping (two single elements) crossed right-over-left (interlocked stitches)	Loop-and-twist (single element) crossed right-over-left
	Plain oblique interlacing (one set of elements, some doubled)
	Weaving (1/1 interlacing)

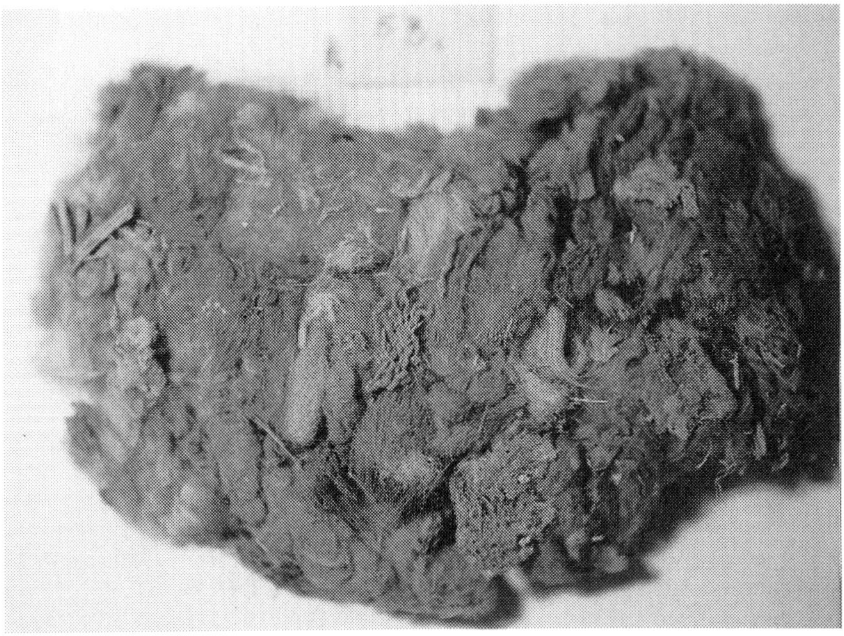

Plate 8.7 Wad consisting of various pieces of camelid fleece, and a short length of oblique, interlaced braid made with cotton yarns. From site Chiu Chiu 200. By courtesy of M. Antonia Benavente.

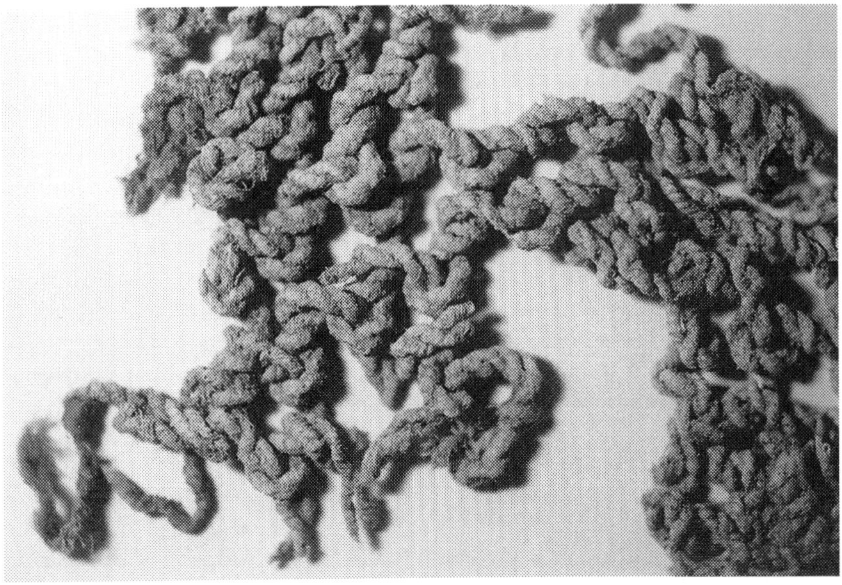

Plate 8.8 Loop-and-twist fabric of vegetal fibre from site Chiu Chiu 200. Note the short length of mid-brown camelid fibre spun on to the loose end of vegetal fibre yarn. By courtesy of M. Antonia Benavente.

In addition, the stitching of hides is present at both sites, and the stitching together of lumps of camelid fleece at Chiu Chiu 200.

The study of Chiu Chiu 200 fabrics suggests that yarns of vegetal fibre were selected to form open-textured loop-and-twist fabrics, and yarns of camelid fibre for closely worked simple looped fabrics. The Chiu Chiu 200 simple looped fabrics are very tightly worked, lacking the foundation element used in four out of the five TU 54 fabrics. In addition, they have the appearance of having been worked spirally, unlike the horizontal lie of the stitches in the TU 54 fabrics.

At both sites, two-ply camelid fibre yarns (Z spun and S plied) were normally employed to make the simple looped fabrics with a compact texture. In contrast, at Chiu Chiu 200, vegetal yarns were employed in open-worked loop-and-twist fabrics. These vegetal yarns are either Z spun and S plied, or S spun and Z plied. Thus the construction of the vegetal fibre yarns is less consistent than that of the camelid fibre yarns used in the looped fabrics. Site TU 54 displays a far greater commitment to the use of camelid fibre than does Chiu Chiu 200.

Certain patterns may be observed in the association between fibre and structure in the Chiu Chiu 200 fabrics. It appears that the more flexible fibre was selected for closely worked, compact fabrics and more rigid fibre for loose-mesh, open-work fabrics. There are hints in the literature that similar patterns of association were observed elsewhere. Bird, for example, mentions that the 'brown refuse' pottery-bearing layers at Punta Pichalo yielded three simple looped bags and thirty-two loop-and-twist bags. He does not provide full details, but at this coastal site plant fibre is said to be the most commonly represented material in the yarns, and of the bags he says, 'practically all those with loose mesh are made of plant fibre cord' (Bird 1946: 257). At the Argentinian site of Inca Cueva cave 7, simple looping is reported to be loosely or tightly worked, and the compact fabrics are made from camelid fibre yarns (Aguerre *et al.* 1973: 205, 207).[11]

In the north of Peru, in a geographical area where camelid fibre was not exploited at an early age, patterns in fibre content and fabric structure also may be observed. For example, Grieder contrasts the use of non-cotton hard fibres in fabrics made by double interconnected looping and simple linking, with that of cotton, which was employed in two-colour simple looped fabrics (Grieder 1986: 19–20). Not only was a particular fibre associated with specific fabric structures, but the fibres are also associated with a marked contrast in yarn construction. The bast fibres were Z spun, single ply, whereas the looped fabrics are of S-spun and Z-plied cotton yarn. Bird (in Bird *et al.* 1985: 107) reported spinning patterns similar to La Galgada for the site of Huaca Prieta, and they are the reverse of the normal patterns in the yarns used in fabrics at TU 54.[12]

Basketry

Three basketry fragments are known from TU 54. One of the items is a recognizable object, the base of a coiled basket. The other two are small and fragmentary. One of these pieces is interesting as it is made of interlaced warp and weft. The presence of this piece suggests that the people of TU 54 knew of the principles of true weaving, even though textiles have yet to be excavated from this site.

Evidently the making of basketry was not practised to any great extent at TU 54, as suggested by the paucity of such remains and the fact that two of the items were repaired before they were discarded in the midden deposits. Item No. 4, a rod wrapped with strips of vegetal stitches, was broken and the pieces held together with a fairly coarse yarn of camelid fibre. Basketry No. 2 is the flat base of a coiled basket. It was repaired by sewing long, straight stitches of cut leather to hold the coils together (plate 8.9; Dransart 1992c: fig. 148). These repairs suggest that camelid fibre yarns and leather were more easily available than the raw material used to make the basket in the first place. Chiu Chiu 200 has also yielded evidence for coiled basketry repaired in a similar manner with long, straight stitches of cut leather strips (Benavente 1982: plate 1, '*cestería c*').

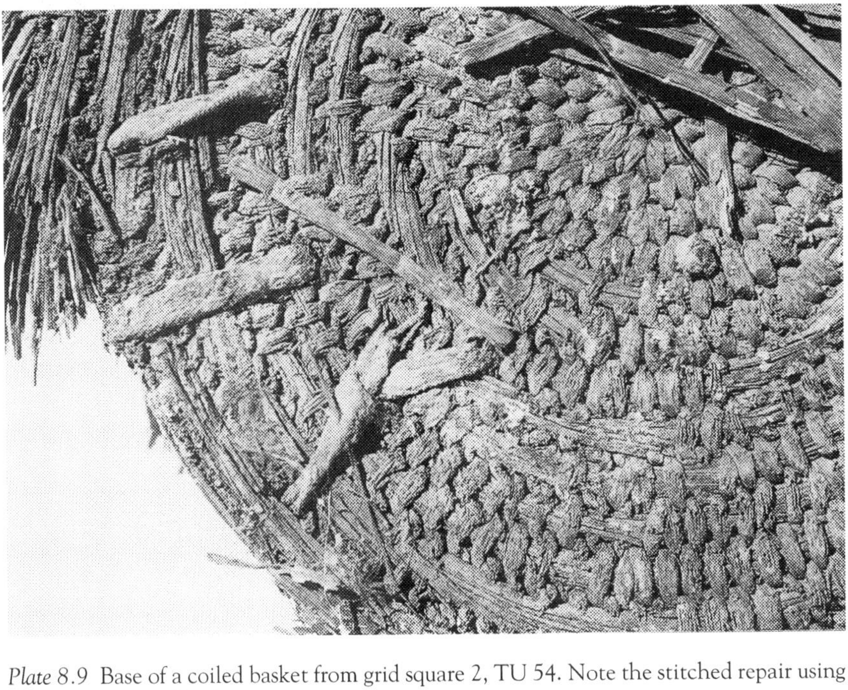

Plate 8.9 Base of a coiled basket from grid square 2, TU 54. Note the stitched repair using strips of leather.

In the north of Chile and at Inca Cueva cave 7, simple looping was a much-used fabric-making technique. It was virtually always accompanied by coiled basketry. Indeed, coiled basketry may be seen as an analogous structure to looping with two single elements (figure 8.8). The preference for looped fabric structures and coiled basketry in the South-Central Andes contrasts with the situation in the Central Andes, where twining was preferred over looping and the basketry types are different from the coiled structures preferred further south (Bird *et al.* 1985: 92–8).

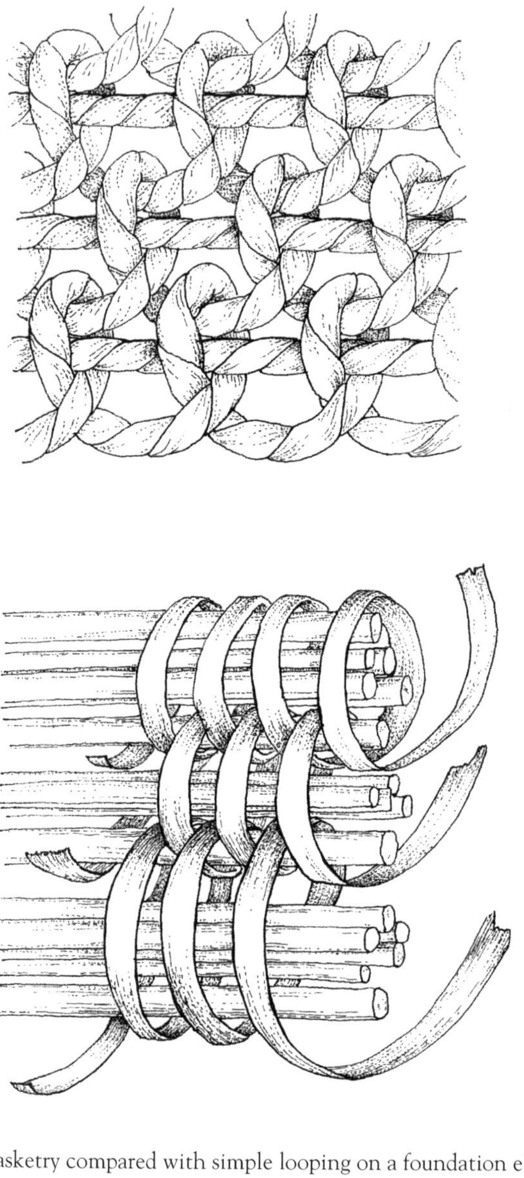

Figure 8.8 Coiled basketry compared with simple looping on a foundation element.

Items of hide and animal skin

Some of the yarns listed in the inventory were pierced through and/or tied to rodent skins, one yarn was stitched through a piece of rodent skin and, in another instance, a piece of camelid hide was stitched. In addition to these items, there are pieces of skin, hide and animal tails, which were evidently cut, and the remains of pelts, which were dyed red (plate 8.10). Item No. 538 in the yarn inventory consists of a piece of camelid hide stitched using two different yarns (figure 8.9). The presence of stitched hide in the TU 54 deposits raises the possibility that the people wore at least some items of hide clothing. Garments made from animal hides are rarely mentioned in archaeological literature on the area, although bodies in coastal graves are frequently mentioned as having been wrapped in animal or bird skins. Standen and Núñez (1984: 141) report the use of tanned camelid skin in twisted strips for pubic coverings and as bracelets, bags and shrouds at the earlier site of Tiliviche 2, but pelican skins were also used for shrouds and seal skins for sandals. In any case, the piece from TU 54 did not have the fleece removed. Although the evidence is slight, it is possible the people wore stitched garments of untanned hide. Yarn inventory No. 606 is a yarn stitched through a piece of rodent hide, and the furs of rodents might have been used to make garments or coverings. Both these items came from the same layer, P, of grid square 3 and, in both cases, the stitching was evidently quickly executed.

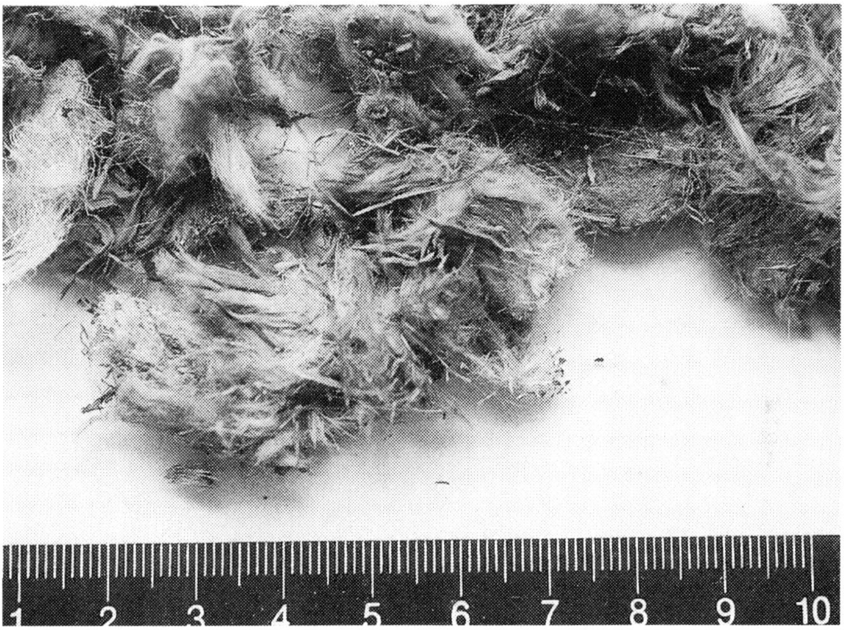

Plate 8.10 Unidentified animal pelt, of which the fur was painted or dyed red. Grid square B3, layer 18, TU 54.

Tulan 55

Material from grid squares 1, 2 and some of the grid square 3 layers was available for study. Although Popper (1977: 46) mentions a large number of yarns, of which eight were made from plant fibres (not cotton), only eighteen yarns are listed here in the inventory and all are of camelid fibres. The post-Hispanic pit in grid square

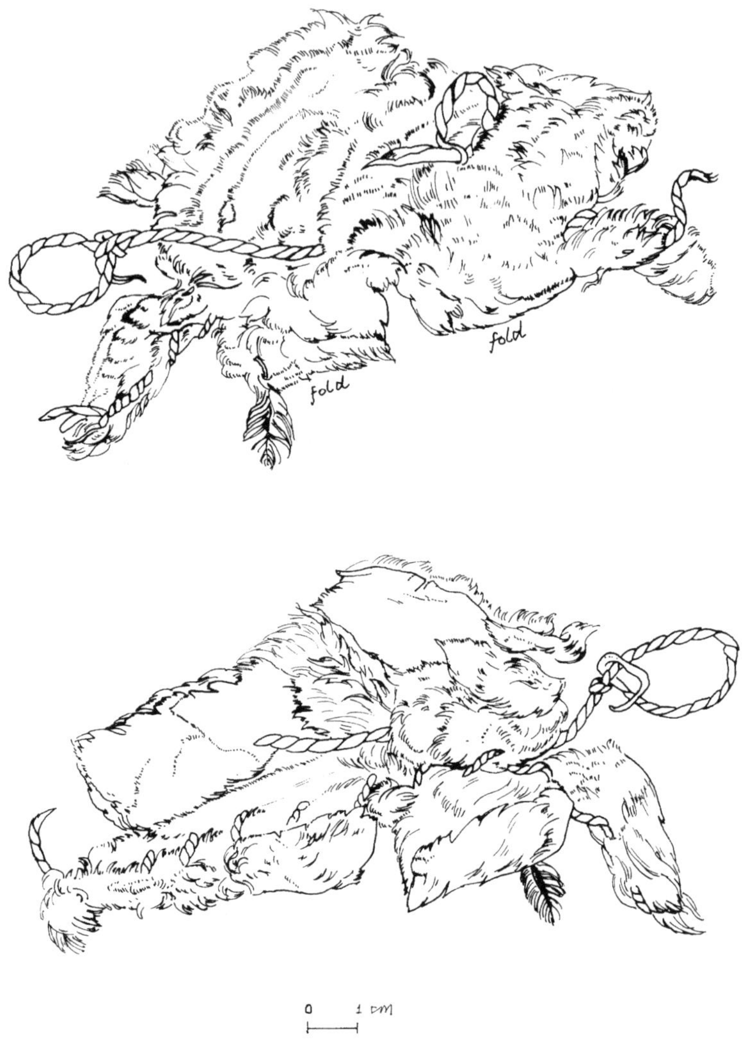

Figure 8.9 Fragment of camelid hide stitched with camelid fibre yarns (No. 538a and 538(b), grid square 3, layer n, TU 54.

2 yielded a strip of silk material (fabric No. 7 in the Appendix). Other notable items are a dark brown/black yarn (No. 742) and the first, and as yet only, occurrence in the yarn inventory of a yarn spun from grey fleece (No. 743). As previously discussed in Chapter 5, archaeological fabrics using grey fleece are generally rare. Even though a greater range of fleece colour from domesticated camelids was evidently available, one of the yarns (No. 746) was perhaps spun from the fleece of a wild camelid to judge from the dimorphism between the fine, woolly fibres and the coarse, hairy fibres.

A small bundle of what looks like sinew and hide was wrapped with a Z2S yarn in an object from layer IX-o of grid square 3. To date, this is the sole occurrence of sinew from all the sites that have yielded yarn remains in the sample of sites studied here. Clement Meighan reported only five sinew yarns out of a total sample of 892 yarns excavated from the residential site of Guatacondo on the River Loa (Meighan 1980: 112).

Tulan 82

Many of the fifty-two yarns from site TU 82 listed in the yarn inventory in the Appendix are fine and regular in diameter. Nearly all are of camelid fibre, with three yarns of human hair. Most of the camelid fibre yarns were evidently intended for weaving into cloth, and the fabric inventory includes a fragment of textile in which both warp and weft were dyed a blackish-blue. The warp and weft counts are about twice as dense as those of the textile found at TU 67. A closer examination of the TU 82 textile shows an underlying light brown natural colour in both systems.

At least two dyed colours were used in the yarns at TU 82: red and blue. One yarn contains both red and blue (yarn No. 803). In many cases, the blue is faded and, as with the TU 54 dyed yarns, it can be seen that light natural colours were preferentially selected for dyeing. Pieces of unspun fleece are also present in the deposits, ranging in colour from white to mid-brown, but one small sample (layer II) is black, with fine fibres. In addition, there is a single unplied yarn from layer III spun from black fleece (No. 788).

Layers II and III contained twelve sections of what looks like the decomposed remains of a knotted net. These pieces are very light brown to mid-brown in colour, and have an S2Z construction with a tight ply angle. The diameter is 1 mm or less. These knotted fragments may have come from an open-worked knotted net with fixed knots. If this is the case, the net would have been unusual due to the fineness of the yarn, and also to the spinning and plying of the yarn in the opposite direction compared with the constituent yarns reported for all of the Tulan fabrics included in this inventory. I do not know of any other knotted nets reported in the literature from northern Chile.

Coarser yarns are also present in the TU 82 sample and they tend to be softer and even to mimic the structure of natural staples of fleece (for example, No. 753). Yarn No. 803, resembles those used in the so-called *mantas apelfadas*, literally meaning, 'shaggy shawls', but perhaps 'fleecy shawls' would be more appropriate. These garments are found wrapped round corpses in cemeteries of the north of Chile, and they occur frequently from the Alto Ramírez period onwards (Focacci and Erices 1972–3: 50). In the San Pedro de Atacama cemeteries, fleecy tunics are found with the same shaggy appearance, mimicking staples of fleece. The use of such garments that imitate the hides of camelids is a technical achievement, since they replace the hides themselves, yet are much more durable and flexible than animal skins, as well as being lighter to wear and warm.

Spinning and fabric-making implements

Among the yarns and material discussed in this chapter are yarns in various stages of construction. The great wealth of material from TU 54 alone gives some indication of the large-scale production from one site. It is therefore somewhat surprising that few artefacts have been identified as implements suitable for spinning yarn or making fabric.

Objects that might have served as spindle whorls are very scarce. The column excavated at TU 52 that produced organic remains including nine yarns also contained a charred and distorted piece of wood, pierced with a hole (layer IIIb; figure 8.10). At TU 54, layer h in grid square 3 produced a worked, perforated, flat piece of pumice weighing 0.7 g (figure 8.11). Finally, a ground and perforated sherd of black burnished pottery from site TU 57 is a third candidate for the status of spindle whorl (figure 8.12). These items might have served as whorls, but there is no certainty about their function. The situation occurring in the Tulan sites is in contrast to Pirincay in the southern highlands of Ecuador, where bone and ceramic artefacts were identified as spinning and weaving implements (Bruhns 1988–9). It is thought that camelids were introduced into the Pirincay area late in the first millennium BC, a major change accompanied by a sudden increase in the numbers of spindle whorls. Of 116 whorls recovered in the archaeological deposits, 106 are said to have come from contexts associated with camelids (ibid.: 76).

Figure 8.10 Pierced wooden object, TU 52.

Figure 8.11 Perforated and shaped pumice, TU 54.

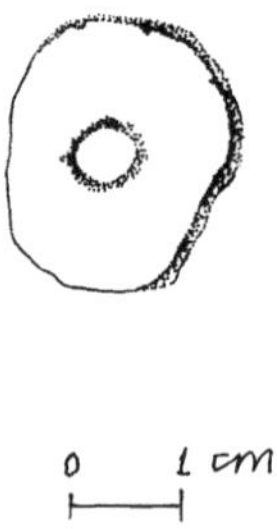

Figure 8.12 Ground and perforated pottery sherd, TU 57.

Of course, it is possible to spin fibre without the aid of a whorl.[13] A straight stick is used to produce *mismiña* in Isluga and elsewhere in the north of Chile, or fibre may be rolled on the spinner's thigh to make thread. If a spinner wishes to spin or ply with a spindle, he or she may improvise a whorl using some object which is close to hand, but which might not survive in the archaeological record. In 1982, I observed a woman spinning using a stick shaft weighted with two small potatoes in Andahuaylas, south of Ayacucho in the Peruvian highlands. However, little is known about the use of spindles at the early sites in the San Pedro de Atacama area. Wooden whorls are found in cemeteries of the later San Pedro de Atacama culture. They are generally rectangular and are flat or rounded in cross-section, often decorated with notches or knobs (Costa 1988: 112; Llagostera and Costa 1984: 53, illustration 59).[14]

Cactus spines were probably used as needles and piercing tools. Layer m, grid square 3, at site TU 54 contained a fire-blackened cactus spine, 80 mm long, with the tip cut off and the end tapered. Such tools were perhaps used for piercing pelts and hides. A possible cactus spine needle was found in the post-Hispanic pit in the cave TU 55; it is 78 mm long, and the tip was cut off and a new, shaped point cut, pierced with a hole. Nowadays, long cactus spines with the tips burned off are used for knitting in the San Pedro de Atacama area (Lindberg 1967: 5).

It is possible that worked bone was used to make implements, but as yet there has been no detailed study of the bone artefacts from the Tulan sites, nor from PU 1. So far, no weaving implements have been securely identified from the sites considered in this study. Lindberg uses ethnographic analogy to argue that pieces of marine bivalve shell (*ostión*) were used to beat the weft against the fell of the textile during the weaving process. This action produces wear on the outer, undulating edge of the shell, and pieces of shell with similar wear patterns are said by Lindberg (1967: 5) to occur in archaeological sites of the area, including tombs of the Quitor cemeteries. It is of interest that three small pieces of bivalve shell have been found in the TU 54 deposits (layers l, f and r of grid square 3). The piece in layer l measures 26 mm wide by 18 mm, and its edge is worn smooth. However, it is not known whether the weaving of cloth took place at this site and so the presence of this worn piece of shell remains enigmatic.

Yarns and fabrics excavated from some of the Tulan sites provide evidence for the exploitation of camelid fibre over a period of time from the Late Archaic to the Formative period. Camelid fleece was the only fibre identified in yarns from sites dated to the Late Archaic period in the Tulan Quebrada (TU 52 and the lower layers of sites TU 67 and TU 55).[15] These yarns would have been spun from the fleece of wild camelids, whether vicuña or guanaco. Sample sizes from early periods are small, but the findings are in keeping with other inland sites, such as Inca Cueva cave 4 yarns, where camelid was predominant over vegetal fibre in the yarns.

The quantity of remains from TU 54 demonstrates the importance of camelid fibre as a raw material to the people who used the site early in the Formative period. These yarns vary from fine and carefully made to coarse yarns, which have the appearance of being quickly spun, perhaps to hold some objects together, after which they were seemingly quickly discarded. Among this mass of material, a small group of yarns was discussed in detail above on the basis of their anomalous, asymmetric structure or for their special effects. This group of varied yarns demonstrates that spinners were fully in control of their medium, and the yarns bear witness to an aesthetic response on the part of the spinners to the raw material they were using. In contrast, yarns chosen for making into fabric conform to a different aesthetic: they have a regular Z2S construction, a medium-ply angle and a diameter of 2 mm. They are mostly of light brown natural colours, but some other colours are introduced (a lighter natural colour, or yarn dyed red).

Spinners were evidently keen to maximize the number of natural colours available in their fleece: white was reserved as such, and light brown colours were selected for dyeing. In the Late Archaic and Early Formative samples, the colour range largely resembled that available in the fleeces of vicuña and guanaco. However, one dark brown yarn and a tiny piece of black fleece at TU 54 suggest that colour variation was already underway in the coats of domesticated camelid stock. Such a proposition is supported by the presence of a textile made from black camelid warp and weft at TU 85. However, it would seem that many of the yarns were not intended for making into fabric, and useful balls of yarn and unfinished yarns were discarded in the TU 54 midden deposits. The apparent wastefulness of camelid fibre yarn is remarkable, since the people evidently spent a lot of time spinning but did not always convert the yarn into an end product.

In the analysis presented above, the TU 54 fabrics were compared with fabrics reported from other sites such as Conanoxa W(a), Tiliviche 2, Inca Cueva cave 7 and Camarones 15, all sites belonging to the Late Archaic period. Seemingly, the corpus of knowledge and experience that enabled the people of TU 54 to make yarn and cloth was not very different from that possessed by Archaic period hunters and gatherers. One exception is that the people of TU 54 and Chiu Chiu 200 probably did not use weft twining. Lumbreras (1974: 44) considered weft twining to be a diagnostic element and a temporal marker in the Central Andes, since its use in Peru did not continue beyond the end of the Archaic period. Perhaps his observation is valid for the north of Chile as well.[16] Looped fabric structures, on the other hand, continued to be used over a longer period of time, as seen clearly in the Alto Ramírez sites, where it was accompanied by weaving. Textiles and looped fabrics occur at Chiu Chiu 200, a site contemporary with TU 54 but, to date, true weaving has not been identified at the latter site. At the moment, it is not clear when textiles were first woven in the Tulan Quebrada.

In time, yarn was used less wastefully than at TU 54, as can be seen in the heavily darned and repaired clothing in the later San Pedro de Atacama cemeteries. Site TU 82 has produced evidence for fine, regular yarns, which were probably intended for weaving into cloth for garments such as mantles and tunics. In addition, there are fragments of a tentatively identified knotted net, made from yarn spun and plied in the opposite to normal direction (S2Z). This site also yielded small pieces of yarn with a fleecy appearance used perhaps to make fleecy shawls or tunics, which became current in the San Pedro de Atacama culture. These fleecy garments would have created the impression that the wearer was wrapped in camelid hide. They represent a technical achievement, since the textiles are more durable and flexible than camelid skins, but they presumably also had an aesthetic value in societies ideologically and economically bound to their herds of camelids.

Coda

This discussion of the yarns and fabrics of Tulan societies has focused particularly on camelid fibre yarn and how it was applied to the human body. Yarns were wrapped in swathes round people's heads and they were used to string beads to adorn the human frame. They were converted into fringe-like pubic coverings. Meighan also mentions the existence of what he calls 'string skirts' at Guatacondo (Meighan 1980: 117, plate 13a). To a minor extent, yarns were also used in looped fabrics, which in later times and generally throughout the South-Central Andes, constituted caps, cloths for wrapping human heads, tunics and bags. Much Late Archaic and Early Formative camelid fibre yarn was spun for purposes other than looping or weaving fabrics and textiles.

In planning this book, I decided not to include a discussion on the topic of the *quipu*, the Andean knot record. Although Wari *quipus* are known (for an example see Morris and Von Hagen [1993: 116, fig. 106]), the device is associated with the Inka period in most people's minds. I felt that a simple reference to an artefact that consists of a primary cord from which a series of pendant yarns are hung and into which knots are tied would be insufficient given the complexity of current debates on the *quipu*. However, an anonymous reviewer commented on the manuscript of this book saying 'it would surely be pertinent to mention *quipus*, even if there is no record of them in the area under discussion'.

Here, then, is a brief review of some literature that the interested reader may consult. Radicati di Primeglio (1984) has drawn attention to the existence of *quilcas*, a small type of *quipu* consisting of pendant and subsidiary yarns. The uppermost sections of the pendant yarns were tightly wrapped with coloured yarns, forming tubular blocks of colour, which Radicati di Primeglio calls 'cartouches' following the adoption of the term by Radamés A. Altieri (1941). Radicati di Primeglio (1984: 20–6) compared these tubular 'cartouches' with similar devices on wigs dating from the Wari period. These wigs consist of looped caps from which are hung plaits of human hair, the ends being tightly wound with dyed camelid fibre yarns. He treated *quilcas* and *quipus* as 'two modalities of a same art of writing' (ibid.: 27). The implication in his article is that both modalities were contemporary. However the *quilcas* he describes were perhaps seven hundred years older than Inka *quipus*, dating from Wari times, like the wound ends of the wigs with which he made the perceptive comparison. Conklin (1990) investigated Wari *quipus*, examining the wound pendant yarns combined with simple knots. He found evidence for base-five counting.

Leland Locke (1923) explored the numerical character of Inka *quipus* and the use of the decimal system in the knots. More recent studies emphasizing the numerical aspects of Inka record keeping and the structural make-up of the *quipu* include Radicati di Primeglio (1979) and Ascher and Ascher (1981). In an anthropological study, Marcia Ascher (1991) has investigated how mathematical ideas were embedded in Inka culture. Gary Urton's 1997 study of the ontology of numbers among the Quechua owes a debt to Jadran Mimica's book on the character and ideology of numbering among the Iqwaye of Papua New Guinea (Mimica 1992). Urton uses the *quipu* as evidence expressing the result of arithmetical calculations in the context of Quechua culture and ideology (Urton 1997: 139, 179–81).

The work of John V. Murra (1975d) and John H. Rowe (1979: 4; 1985: 197–8) has shown that historical records, events and information about holy places were stored in *quipu* form, or that colonial records retained the structure of having been recorded in such a form. Denise Arnold *et al.* (2002: 365–70) have observed *quipu* structures in contemporary Andean practices such as the making of libations. In colonial times, European priests compared rosaries with *quipus*, and both rosaries and *quipus* served to keep account of either good deeds or proper religious observance (Dransart 1998: 139–40).[17] However, Martti Pärssinen comments that we do not know 'how much information it was possible to transmit autonomously without additional oral texts' (Pärssinen 1992: 32). He has examined how the basic variables of colour, order and number were used to encode verbal data in the pendant yarns of the *quipu*, trying out his method on tribute and storage records and historical events (ibid.: 34–46). Admitting that his interpretation is not necessarily free from errors, Pärssinen comments (ibid.: 47) that his method shows how it is possible to encode personal and place names on *quipus* through the use of phonetic and non-phonetic systems.

Laura Laurencich Minelli has studied a 'literary *quipu*' dating from the Colonial period. It is provided with knots as well as ideograms consisting of woven or plaited devices hung from the pendant yarns. She explains that in order to read this *quipu*, one proceeded from left to right using key words represented by the ideograms or syllables, consisting of ideograms combined with knots, which indicated which syllable to extract from the key word (Laurencich Minelli 1996; 2000).

Denise Arnold *et al.* (2000: 347–50) provide an ethnographically informed examination of the embodied quality of the terminology used to name *quipu* parts, in the manner of reading *quipus* from one hand to another, and the body dynamics that are involved in moving from right to left and from head to foot.[18] This work is an important reminder that the component yarns of *quipus* are the product both of the necessary bodily skills and a corpus of culturally relevant knowledge.

The studies cited here demonstrate how an outwardly simple system of recording information through the use of pendant and subsidiary yarns was a product of a particular cultural milieu. As scholars continue their researches into the relationships between the recording of numerical and verbal information in ever greater depth, it might be possible to situate the development of the *quipu* as a product of societies that had already specialized in weaving. In European languages the terms 'text' and 'textiles' are commonly known to share a root in the Latin *texere*, 'to weave' and scholars have explored, often through semiotic analysis, the 'webs of significance' that people spin in the course of their lives.[19] The relationship between numbering systems and weaving has been given less attention, yet the importance of arithmetical skills for weavers is paramount. In contrast, the Late Archaic and Early Formative societies of the Tulan Quebrada did not rely on weaving to any great extent. Their spinning and looping skills were perhaps more conducive to the conveying of understanding of other orders of knowledge, of a more kinetic character to do with rotation, direction, spiralling movements and the like, stemming from their appreciation of the raw materials they had at their disposal and the domains that provided them (Frame 1986: 56; Dransart 1995).

9 Conclusions

Earth, water, fleece and fabric

In this book, I have explored some of the historical particularities in the relationships negotiated between herders and their herd animals in the Andes. The method used was to juxtapose different aspects of a herding way of life in order to bring into focus certain trajectories through time. I fully recognize that the quality of the information differs in the two main historically located junctures studied here: the Purifica and Tulan Quebradas between five thousand and one thousand five hundred years ago, and Isluga in the 1980s and 1990s. Obviously, the ethnographic material from Isluga is much more fine-grained than a study based on the material culture of archaeological societies allows. Nevertheless, I have taken a long-term view of the relationships established and maintained between herd animals and human beings in this study of the exploitation of a renewable resource: fleece. The social relations that characterize the societies studied here involve the assumption of responsibility by human owners over their herd animals. In conventional terms, these social relations arise from a pastoral way of life. Pastoral societies began to co-exist with hunting and gathering societies in the Late Archaic period (over four thousand years ago) in arid and highland areas of the South-Central Andes.

My particular focus has been to study the spinning of camelid fleece as a fundamentally important product in Andean societies through time. I incorporated the study of yarns and fabrics excavated from sites in the Tulan Quebrada into a broader cultural context. A dynamic perspective was adopted, examining the consequences of domesticating camelids on the system of production and manufacture. A long-term view was undertaken in this study of the exploitation of camelid fibre, a usage that continues up to the present and is attested here by the ethnographic data from Isluga.

Camelid fleece is not a secondary product. In Eurasia, the wool of sheep has been regarded as a 'new use' or a 'secondary product' that was only exploited later in the history of domesticated sheep (Bökönyi 1969: 222; Sherratt 1981: 261–3). In contrast, the fleece of vicuña and guanaco was spun as soon as human groups begun to occupy sites in the South-Central Andes. In the Andes, a fabric-making tradition developed that relied heavily on camelid fibre. The domestication of the South American camelids was not a prerequisite for that development.

The community of people who built their houses at site TU 54 and who herded their llamas in the Tulan Quebrada had ample supplies of camelid fleece at their disposal. Among the large assemblage of yarns from the midden deposits, there is little evidence for heavy-duty ropes for strapping firewood to the backs of their llamas. Tim Holden noted the very slight character of the evidence for cultivated crops at TU 54 (1990: 301; 1991). The people would not have produced cultivated crops for transportation. They probably did not use the potential of labour service that llamas can supply as beasts of burden.[1]

Although both the TU 54 community and the people of Isluga either lived or live by a herding way of life, the character of their pastoralism is very different. With this in mind, I wish to return to some questions I posed in Chapter 1. Did the herding of camelids in the Atacama three thousand years ago resemble Chukchi herding practices in which the human owners respect the 'wild' behaviour and appearance of their herd animals, while using those animals as a source of raw materials, or had they already forged strategies comparable with those used in Isluga, recognizing their own animals as named individuals? Were the herd animals a focus of elaborate ritual? The strong colour marking that exists in contemporary herds of llamas would not have been so pronounced in the TU 54 herds. It is perhaps unlikely that comparable systems of classification devised by Aymara and Quechua herders, as discussed in Chapter 3, would have been used three thousand years ago in the Tulan Quebrada. A large proportion of the animals had light brown fleece colours with some white (probably under the belly). Many more of the herd closely resembled guanacos than is the case in contemporary herds. The TU 54 herds were surrounded by demographic reservoirs of wild animals, and there might have been more interactions between the two types of camelid population than occurs at present.

Yet yarn No. 254 from TU 54 is of a darker brown colour and black fleece was available in small quantities, too, at TU 85. According to Gilmore (1950: 449), melanism does not occur in guanaco populations. It is possible that at least some of the llamas in TU 54 herds were strikingly different from the rest of the animals. There are indications that changes in fleece colours were occurring that would have important consequences for the availability of fleece colours. These changes would have encouraged the spinners to introduce changes into their classification of fleece as a material for technical activity.

It is possible, too, that the herds were the focus of elaborate ritual. The TU 54 people developed a tradition started in the Late Archaic period for incising boulders with drawings of camelids. At PU 1 and Kalina panel 101-UR-I, people incised rock with the schematic outlines of camelids in profile. At TU 54, the drawings include profile heads of camelids.[2] The presence of shell pendants with ornithomorphic ornament provides a tantalising glimpse of the symbolic expression of the value of birds, which, to judge by the presence of the parts of snares discarded in the midden, the TU 54 people hunted. By the time the great panel of rock art was incised on to the rock face above the Tulan spring, herders had developed complex classifications of birds and camelids. The panel possibly dates from the late pre-Hispanic period, before or around the arrival of the Inkas in the area in the late fourteenth to early fifteenth century. In it, large camelids are arranged above a section that incorporates small camelids and aquatic birds. There is one large feline among the large camelids, drawn in profile and confronting a fissure in the rock face. The association between birds, felines and camelids was and still is widespread throughout Andean herding communities and, as discussed in Chapters 3 and 4, it has acquired specific meanings that are expressed in the beliefs about the origins of camelids and in ritual in contemporary Isluga.

With the adoption of a herding way of life from the Late Archaic period onwards in the South-Central Andes, people became owners of animals and they were able to ensure access to their own supplies of food and raw materials. However, there is a tension between caring for generation after generation of herd animals at the same time as using them as sources of food and raw materials. Herders do their utmost to cater for the well-being of their herds, yet they also have to sacrifice individual animals in order to obtain food. The situation is different with fleece, which can be harvested from camelids while they are still alive. Clothing is arguably of equal importance to food in order to sustain human life in the South-Central Andes, especially at high altitude, where climatic variability is extreme. Most historians of dress and costume consider clothing to go beyond warmth and protection. They see dress as expressive of a realm of cultural values as well as being implicated in the shaping of those very values (Barnes and Eicher 1992). If herders may obtain, say, four times the amount of fleece from their own animals as from hunted camelids, it is pertinent to ask: what difference does owning one's animals have on yarn and fabric production?

Initially, the extra fleece available to herders seems to have made little difference to the spinning of yarn and the making of fabrics. In the millennium that elapsed between the beginnings of herding at PU 1 and the occupation of TU 54, fibre technology seems to have relied on many of the techniques that are characteristic of the Archaic period. Certainly, spinners were very active at TU 54, and the amount of perfectly usable yarn that found its way into the midden is a source of surprise for contemporary spinners. In Isluga, spinners carefully wrap balls of yarn in cloth, or they hang large balls from plaited ropes strung across the end of a house. All usable yarn is carefully hoarded until required.

A 'typical' TU 54 yarn is two-ply (Z2S), has a medium ply angle, and it measures approximately 2 mm in diameter. In appearance, it corresponds most closely to the yarn that is spun for the weft of a textile in Isluga. However, my own work in Isluga reported in Chapter 5 and MacKenzie's work (1991) among the Telefol in Papua New Guinea show that the spinning of yarn is intimately interrelated with the fabric-making processes for which it is intended. A 'typical' yarn in Isluga is two-ply (Z2S), with a very tight ply angle and a very smooth profile, so that the threads will not cling together during the weaving of a warp-faced textile. As explained in Chapter 5, the increasing use of factory-spun acrylic yarn is changing the quality of the cloth woven in Isluga. In contrast, TU 54 yarn requirements for making looped fabrics were different. The smooth profile was not critical, and the yarns have a medium rather than a tight ply angle. MacKenzie observed that a workable length of yarn is important for looping fabrics: if it is too long, it tends to tangle, and if it is too short, it is used too quickly (1991: 210). She suggested that looping was developed before the practice of spinning large amounts of thread became commonplace. In northern Chile, fabric-making techniques emerged with the development of looping using yarn spun from camelid fibre (especially at inland sites) or from vegetal fibre (especially at coastal sites).

Most studies of fabric-making techniques and weaving assume that the spinning of yarn is undertaken as the first step in order to transform it into fabrics or textiles. A fascinating outcome of the study of the TU 54 yarns is the realization that the spinners did not always intend their yarn to be transformed into cloth. Finely spun camelid fibre yarn was important for threading beads of various types. Many of the coarser yarns, which would have been more quickly spun, were intended to hold bundles of items, around which they were knotted. They served a temporary purpose and were quickly discarded in the midden.

Another use to which yarns were put was to wrap round the head in the form of 'turbans'. This distinctive form of headgear is more commonly associated with Late Archaic period sites on the coast of the far north of Chile (Agüero Piwonka 1995). During the Formative period, the 'turbans' became more elaborate and more full. The discovery of a 'turban' wrapped round the head of a child at TU 85 shows that the custom was widespread. Large quantities of yarn would have been required for wrapping in swathes round the head. Spinning for such a purpose perhaps encouraged the production of theoretically endless balls of yarn.[3] The easily available supplies of fleece harvested from herds of domesticated animals would have facilitated such a production.

The earliest true textiles in the north of Chile take the form of narrow woven bands. In the Formative period, weaving became more common and wider cloth was produced. It shared the distinctive Andean characteristic of being warped from a continuous thread that was then interlaced with the weft to produce a web with four selvedges. In Asian and European treadle looms, weavers cut the warp ends and thread them through heddles before tying them to the breast beam of the loom. After weaving has been completed, the warp ends are cut to release the textile from the loom. Traditionally in the Andes, the weaver does not cut the warp, but enters the last few weft threads with a needle. The finished textile has two side selvedges, formed from the turns of the weft and, at the top and bottom of the web, selvedges formed from the turns of the warp.

It is possible that the spinning of theoretically endless balls of yarn for 'turbans' then lent itself to the adoption of weaving loom widths in the Tulan Quebrada. By the time site TU 82 was occupied in the fourth century AD, it appears that fleece was used less 'wastefully' than at TU 54. Spinning and weaving technology would have been integrated with herding practices, and it would have responded to the needs and ideology of a pastoral way of life. Evidently, many of the TU 82 yarns were intended for weaving into loom products. Fragments of textile from the site of TU 82 may be compared with better preserved items from tombs of the San Pedro de Atacama Culture. One of these fragments (fabric No. 10 in the Appendix) resembles the warp-faced plain weave in tunics worn by the dead in cemeteries surrounding San Pedro de Atacama. It has a count of twenty-two warp and six weft threads per 10 mm. Although it has a greater warp and weft density than fabric No. 1, from the lowest layer at TU 67, it is, in fact, of an average quality.[4]

Yarn No. 803 at TU 82 was compared with the warp of the fleecy shawls and tunics found in graves in northern Chile from the Alto Ramírez period onward. Most of these heavy mantles were woven in natural-coloured yarns, and they have long staples of unspun fleece incorporated into a plain-weave textile. There are exceptions: tomb 3 at Coyo Oriente contained such a fabric with red-dyed, unspun fleece applied to an undyed 'barber pole' warp, plied from two strands of different natural colours. Some of the staples of fleece used in these mantles are very long. In one case, a body from the cemetery of Quitor 2 (No. 1,905) was wrapped in a very coarsely woven fabric (a balanced plain weave with approximately five warp and five weft threads per 50 mm) from which unspun fibre staples protrude, up to 300 mm long in the white fleece, and up to 180 mm long in the light brown fleece.

Staples as long as these imply that the spinners had alpaca fleece at their disposal, although, theoretically, it is possible for llama hair to attain such lengths (Gilmore 1950: 439). They suggest that the animals were kept unshorn for several years to obtain the necessary fleece growth. Although the shawls are coarsely woven, they are well made and they constitute distinctive items of clothing (tunics made in a similar manner also occur in graves in northern Chile). It is probable that the length of the staples hanging from the weave added to the prestige of the garment, and the long staples might even have represented what in Isluga are considered to be prized *saxsali* animals. The idea of a camelid with fleece hanging down to the ground is an impossible ideal for most contemporary herders, who say that alpacas lose their fleece on shrubs and bushes. Consequently their owners choose to shear alpacas regularly so that the fleece does not become over-long and tangled. If one were serious about growing especially long fleece, it would be necessary to keep the alpaca in a pen.[5] It is not known whether numerous herds of alpacas were kept locally in the past; nowadays, only llamas are herded in the valleys to the east of the Salar de Atacama. Fine Tiwanaku tapestry garments have been identified in Coyo Oriental tombs by Amy Oakland (1986), who also suggests that locally made provincial copies were woven. It is likely that textiles imported from Tiahuanaco would have been made from alpaca fibre. However, the possibility that even fine textiles were made from llama fleece in the San Pedro de Atacama area cannot be discounted. Wheeler, *et al.* (1995) identified a fine-fleeced llama among the mummified camelids from Moquegua whose fleece they examined.

In the course of this work, I have sometimes de-emphasized the difference between the fleeces of llamas and alpacas because of the difficulty in distinguishing between them.[6] Lange Topic *et al.* (1987) published a commentary in response to an article published by Shimada and Shimada in 1985, criticizing the Shimadas for de-emphasizing the differences between llamas and alpacas. Lange Topic *et al.* (1987: 832) insisted that alpacas are kept primarily as producers of fibre. In reply, Shimada and Shimada pointed out that Wheeler's work in Junín indicates that alpacas were domesticated by 6000 BP, a period for which there is no evidence that alpacas were bred specially for fibre production, and they stressed the importance of the alpaca as a source of meat (Shimada and Shimada 1987: 836–7).

In the context of this book, I wish to stress the importance of the llama as a producer of fibre. The published literature tends uncritically to adopt the generalizing propositions that the alpaca was and is primarily a fibre-bearing animal and that the llama was and is a beast of burden. However, my experience in Isluga and San Pedro de Atacama has shown me the importance of the llama as a producer of fibre. We should keep in mind that the economic and cultural significance of the South American camelids is a complex affair, depending on the local conditions in which these animals were and are used.

Heather Lechtman emphasized that the activity of practising techniques (whatever the medium) is an important factor in understanding the character of technologies as systems generated by a given society, and that they are inseparable from the cultural systems in which they are manifested (Lechtman 1981: 18). The material evidence excavated from the Tulan Quebrada demonstrates how the use of certain fabric-making techniques is interrelated with the qualities of fleece and the yarn spun from it. In its reliance on a particular medium (camelid fibre), it also helps shed light on how people interacted with the camelid populations. The study of the Tulan material also shows how a need for reliable supplies of fleece became an important factor in the process of *maintaining* herds of camelids and a herding way of life, but not of *establishing* such a way of life.

The aesthetics of spinning in the Andes

From the analysis of the yarns in Chapter 8, it appears that the spinners were sensitive to the fleece colours available in their raw material. In the TU 54 material, white was not abundant, and it was never observed to have been dyed. The yarns selected for dyeing red were spun from the plentiful supplies of fleece in the light brown to mid-brown range of the colour spectrum. In contemporary Isluga, there is greater range of fleece colours, and weavers recognize as many as eighteen different natural colours. They use these colours in culturally specific roles in the garments, blankets, carrying cloths and sacks that they weave. It is likely that the Tulan societies in the past also used their fleece colours in a culturally specific manner to serve different purposes.

However, no evidence as yet has been detected suggesting that fleeces were blended to achieve certain effects. Lila O'Neale reported that in Paracas Cavernas fabrics, grey appears as a colour blend produced by mixing what she says were 'Brown and cream wool fibers prior to spinning' (O'Neale 1942: 178). However, this may be a misunderstanding, since grey fleece, as grown by the animal itself, is a mixture of pigmented and white fibres. Grey is, in fact, a relatively rare colour in surviving archaeological textiles. It occurs in but one late yarn in the material from the TU 55 cave site. Unspun camelid fibre excavated from the late pre-Hispanic cemetery site of Chacance near María Elena on the River Loa also includes grey fleece (plate 9.1). The natural fleece colours from the earlier Tulan sites discussed here fall within a more restricted white to mid-brown range in which black and dark brown appear as exotic, or at least highly unusual, occurrences. Although the spinners seem to have been sensitive to the variations available in

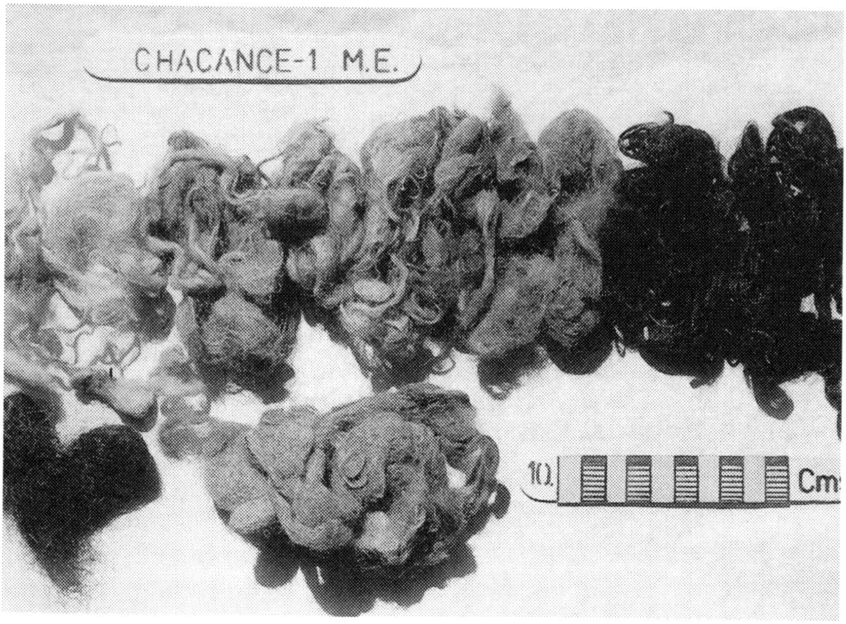

Plate 9.1 Different coloured camelid fleeces from the cemetery site of Chacance, near María Elena, said to have been found in a woven bag. Museum inventory No. 810228. By courtesy of Museo Escolar, María Elena, Chile.

their raw material, they did not attempt to extend the range by blending fibres from different fleeces.

It is worth considering the TU 54 assemblage of yarns in the light of A.H. Gayton's comments on the character of Andean yarns:

> Yarns whether of cotton or wool were spun to an extraordinary degree of perfection. The aesthetic aim seems to have been just that: uniform diameter, tension and color perfection in all grades including those of exquisite fineness.
> (Gayton 1967: 287)

Gayton associated 'hasty, slipshod spinning' with only the poorest quality of garments. Otherwise, she considered Andean spinning to be characterized by a uniformity that prevented spinners from exploring other possibilities, such as combining differently coloured strands in one yarn, nodular colour flecks, or intentional irregularities of diameter.[7] The TU 54 yarns used in looped fabrics and the yarns used in the TU 85 'turban' are characterized by their uniformity of diameter, tension and colour.

Equally, fabric-making techniques in the Andes have been regarded as the result of a 'spare technology', which made possible the most sophisticated of textiles (Stone-Miller 1994: 20). Contemporary warp-faced textiles are woven on a loom set up with a heddle rod and a shed rod as devices to separate the threads of the warp into two sheds, in order to produce a plain-weave textile, with an over-one-under-one order of interlacing. Weavers produce pattern stripes, selecting and picking up threads with the fingers in order to introduce areas of complementary warp-weave with two good faces, or two 'right' sides (Rowe 1977). This method of weaving depends on the manual dexterity of the weaver, which is highly praised in Isluga. Young girls spend hours playing five stones, consequently developing their manual dexterity. To pick up both stones and warp threads is *pallaña* in Aymara. The textiles they go on to produce as teenagers have closely spaced warp threads that are clearly visible in the finished product. Isluga spinners and weavers do not like fuzzy threads, which obscure the structure both of the yarn and of the fabric. When the structure of the yarn and weave are clearly visible, women describe the result as *q'ara*, meaning 'bare'. This is high praise indeed.

Earth, land, water and fleece in the Andes

The great panel of rock art above the spring at Tulan (TU 60) indicates that people using the saline waters of the spring and the pastures shared some of the concerns of contemporary herders of camelids in the Andes. The artists recognized the importance of water in the production of fleece. This association may be documented from at least the early seventeenth century, in the Huarochirí narrative of a celestial llama who was said to drink water from all sources and came down to earth in an abundance of fleece described as a deluge (Taylor 1987: Chapter 29).

There is ethnographic evidence with respect to spinning that links this activity with water. Spinning is described as a ritual activity by Ana Macías de Arias (n.d.) among the Quechua people of Calchaqui, Iruya and the *puna* of north-west Argentina, for whom the Pachamama is said to be the divinity who presides over spinning at sunset, when the time/earth mother begins to rest, and to whom invocations are made to ensure that the animals give birth and produce abundant fleece. Spinners also ask of the Pachamama that the spindle should not grab their hand and prevent them from spinning.[8] Moreover, Ana Macías de Arias says that a special decorated votive whorl is used to begin the spinning of the first fleece harvested in a year. Such whorls may be of a white stone and decorated with a lightning-serpent symbol.[9] She does not explain the full significance of the connection between lightning and spinning, but there is a clear association between water and the spinning of fleece. Teasing of the fleece is done after the animals are shorn on the banks of the Iruya river. In the case of vicuña fleece, she says that it must be spun on the banks of a river, or that the spinner should constantly sprinkle water on the ground, after first performing a rite to the spring (*puquio*) (Macías de Arias n.d.). Her account of spinning in north-west Argentina emphasizes a conceptual association between water and fleece, which we also saw in the context of Isluga herding and spinning practices (Chapters 3 and 4). In Chapter 5 the double meaning of *jawi* as 'fleece' and 'wet' or 'water' emerged from a discussion of rivers and their celestial counterpart, the Milky Way. Thus, richly expressed symbolism combining water and fleece occurs widely in the Andes.

In Isluga, one of the important concerns for people is to ensure that they have adequate supplies of fleece for their needs. Herds are valued as sources of consumable produce, and clearly the self-regenerating powers of fibre mean that owners of fleece-bearing animals attach great value to such an important raw material.

Chapters 3 and 4 examined the parallel character of human and animal lineages. In Isluga society, rights to possess animals as property are transmitted from parents to children, and ownership is claimed for the progeny of these animals in the uterine line of the camelids. The production of fibre and fabric and the reproduction of herding labour and social relations between people are combined in an economic system which relies on the successful reproduction of the herds themselves. Herding is based on the social principles of divided access to live animals and common access to pastures (Ingold 1980).

The ethnography of Isluga as a herding society provides material for re-envisaging the process of 'domestication' when people assume responsibility over another animal species. Isluga herders tame each generation of camelid into appropriate llama or alpaca behaviour. In my analysis, human and camelid lineages both participate in Isluga cultural life, but they are considered to be distinct, and herders do not view their animals as surrogate children. Just as the personhood of human beings is respected, people recognize the llama-hood or alpaca-hood of their camelids. The processes of incorporating herd animals into Isluga cultural life makes possible the production of fleece that plays such an important role in the material culture of the community.

When a llama, alpaca or sheep is sacrificed, its blood is libated to the ground, and litanies are cited of place names of the pasture grounds that the animal used to frequent.[10] In Isluga, the pasture grounds and water sources form part of the nourishing terrains that sustain both herd animals and human herders. The landforms are considered to be imbued with living qualities. Elsewhere in the Andes, a woman in Huaquirca, Peru, told Peter Gose: 'For us, the earth is a living thing' (Gose 1994: 3). Certain hills in Isluga are seen as 'herders', and a particularly beneficent spirit of the hills is known as a 'provider', or *aviador* (Martínez 1976: 281). Yet Isluga people also recognize that the terrains they inhabit must be nourished to restore the vitality of the earth and her resources. Camelids are providers of life-sustaining food and fleece, but the ability of camelids to transform pasture into fleece is jeopardized if human herders do not play their part in sustaining the vitality of the land. The vital landforms of Isluga terrain are gendered beings, as they are female or male, but they are also ambiguous, since they are clearly not simply modelled on human beings. Marietta Ortega Perrier recognized the reluctance of the people 'to eliminate any ambiguity in their explanations' of these entities (Ortega 1998: 116). Isluga people respect the particular animalhood of their herd animals, and they do not expect their life cycles and behaviour to mirror exactly those of human beings. In a similar vein, the animate landforms and water sources are respectfully attributed their own qualities. They share some of these qualities, such as being gendered, with human beings and animals.

In an analysis of a ritual in which water is joined, mixed and redistributed in San Pedro de Condo, Bolivia, Lynn Sikkink referred to the 'commons of land and people' (Sikkink 1997: 183). In the context of this book, I would rephrase her statement to incorporate land, people and herd animals. The interaction between the individual person and his or her environment is a complex process in Isluga. The landscape has been, and still is in process of being, sculpted by the activities of the Wirjin Tayka, the *uywiri*, people and animals. People, herd animals and environments are mutually dependent. In Isluga, the interdependence between humans (*jaqi*), herded animals (*uywa*) and the land (the Wirjin Tayka and the *uywiri*) is a vital process that has brought into being the Isluga landscape and the Isluga herding way of life of its inhabitants.

Appendix

Inventory of yarns, fabrics and basketry from sites TU 52, TU 67, TU 54, TU 55, TU 85, TU 58 and TU 82 in the Tulan Quebrada

All items are listed by site, grid square and stratigraphic layer. The yarns are listed first, in tabular form, followed by a listing of the fabrics and, finally, the basketry. All yarns are given separate inventory numbers even though, in some cases, they may represent fragments of the same yarn.

Yarn inventory

Characteristics:

Length: approximate length in mm
Diameter: approximate diameter in mm
Strand twist angle: angle of twist or spin direction: S, Z or I
Number of strands: unplied (one strand) or plied (two or more strands)
Tension of twist: T (tight), M (medium), L (loose)
Fibre: C (camelid), V (vegetal), H (human hair), B (bast), A (cotton)
Colour of fibre

Also included in the yarn inventory are specimens whose identification is doubtful. They might be fragments of fleece that resemble yarns, or they might be small remnants of yarns, the spin of which has slackened. Specimens with an insecure identification are indicated by (?).

Fabric inventory

Characteristics:

Greatest measurements in mm
Analysis of yarns used and structure of fabric
Warp and weft counts (where applicable)
All yarns are camelid fibre, unless otherwise stated

Yarns

REGION	SITE	GRID SQUARE	LAYER	ITEM NUMBER	PART	LENGTH	DIAMETER	STRAND TWIST ANGLE	NUMBER OF STRANDS	PLY TWIST ANGLE	NUMBER OF PLIES	RE-PLY TWIST ANGLE	NUMBER OF RE-PLIES	RE-RE-PLY TWIST ANGLE	NUMBER OF RE-RE-PLIES	ATYPICAL YARN	TENSION OF TWIST	FIBRE	NUMBER OF WHITE STRANDS	NUMBER OF VERY LIGHT BROWN STRANDS	NUMBER OF LIGHT BROWN STRANDS	NUMBER OF BEIGE STRANDS	NUMBER OF YELLOW-BROWN STRANDS	NUMBER OF MID-BROWN STRANDS	NUMBER OF DARK BROWN STRANDS	NUMBER OF GREY STRANDS	NUMBER OF DYED-RED STRANDS	NUMBER OF REEF KNOTS	NUMBER OF GRANNY KNOTS	NUMBER OF OVERHAND KNOTS	NUMBER OF SLIP KNOTS	NUMBER OF FIGURE-OF-EIGHT KNOTS	NUMBER OF LARKSHEAD KNOTS	NUMBER OF HALF-HITCHES	NUMBER OF UNIDENTIFIED KNOTS	THREE-ELEMENT PLAIT
TU	52	0	c	1		152	4	S	2	Z	1						M	C					2													
TU	52	0	e	2		37	2	Z	2	S	1						M	C					2													
TU	52	0	i	3		78	1	Z	1								T	C									1									
TU	52	0	i	4		51	1	Z	1								T	C									1									
TU	52	0	j	5		103	2	Z	2	S	1						M	C					2													
TU	52	0	j	6		71	2	Z	2	S	1						M	C					2													
TU	52	0	j	7		58	2	Z	2	S	1						M	C					2													
TU	52	0	j	8		22	2	Z	2	S	1						M	C						1												
TU	52	0	j	9		28	2	Z	1								M	C					1													
TU	67	4	a	10		61	1	S	2	Z	1						M	C					2													
TU	67	4	b	11		44	1	Z	2	S	1						M	C					2													
TU	67	4	b	12		12	1	Z	2	S	1						M	C					2													
TU	67	4	b	13		31	1	Z	4	S	1						T	C				4														
TU	67	4	b	14		44	1	Z	2	S	1						M	C					2													
TU	67	4	b	15		24	1	Z	2	S	1						M	C					2													
TU	67	4	b	16		13	1	Z	2	S	1						M	C					2													
TU	67	4	b	17		13	1	Z	2	S	1						M	C					2													
TU	67	4	b	18		12	1	Z	2	S	1						M	C					2													
TU	67	4	b	19		20	1	Z	2	S	1						L	C					2													
TU	67	4	b	20		20	1	Z	2	S	1						L	C					2													
TU	67	4	b	21		32	1	Z	2	S	1						M	C						2			2									
TU	67	4	b	22		45	2	I	1									C		1																
TU	67	4	c	23		35	1	Z	2	S	1						M	C					2													
TU	67	4	c	24		33	1	Z	2	S	1						M	C					2													
TU	67	4	c	25		14	2	Z	2	S	1						M	C				2														
TU	67	4	c	26		18	2	Z	1								M	C		1							1									
TU	67	4	c	27		41	2	Z	2	S	1						L	C						2												
TU	67	4	c	28		14	4	Z	2	S	1						M	C						2												
TU	67	4	c	29		11	4	Z	2	S	1						M	C						2												
TU	67	4	c	30		10	4	Z	2	S	1						M	C						2												
TU	67	4	c	31		33	1	Z	1								T	C				1														
TU	67	4	c	32		23	3	Z	1								M	C						1												
TU	67	4	c	33			3	I	1									V																	1	
TU	67	4	d	34		24	1	S	1								L	C					1													
TU	67	4	d	35		23	1	S	1								L	C			1															
TU	67	4	d	36		20		S	1								M	C		1																
TU	67	4	d	37		16	1	S	1								M	C					1													
TU	67	4	d	38		15	2	Z	1								T	C						1												
TU	67	4	d	39		23	1	S	1								M	C					1													
TU	67	4	d	40		16	1	S	2	Z	1						T	C		2							1									
TU	67	4	d	41		17	1	S	2	Z	1						M	C					2													
TU	67	4	d	42		17	3	Z	2	S	1						T	C				2														
TU	67	4	d	43		43	3	Z	2	S	1						M	C						2												
TU	67	4	d	44		42		S	2	S							M	C		2																
TU	67	4	d	45		55	2	Z	2	S	1						T	C						2											1	
TU	67	4	d	46		20		I	1									C			1															
TU	67	4	d	47		15		I	1									C		1																
TU	67	4	d	48		5		I	1									C									1									
TU	67	4	e	49		24	1	Z	2	S	1						M	C					1													
TU	67	4	e	50		28	1	Z	2	S	1							C				2														
TU	67	4	e	51		15	4	Z	2	S	1						M	C	2																	
TU	67	4	e	52		41	2	Z	1								M	C					1													
TU	67	4	e	53		25	1	S	1								T	C					1													
TU	67	4	e	54		46	1	S	1								T	C			1															
TU	67	4	e	55		26	2	Z	2	S	1						M	C			2															
TU	67	4	e	56		20	2	Z	1								M	C			1															
TU	67	4	e	57		11	2	S	1								T	C					1													
TU	67	4	e	58		10	1	Z	1								M	C					1													
TU	67	4	e	59		12	1	Z	1								M	C				1														
TU	67	4	e	60		10	2	S	1								T	C			1															
TU	67	4	e	61		16	2	S	1									C					1													
TU	67	4	e	62		4	2	S	1									C			1															
TU	67	4	e	63		13		I	1									C																		
TU	67	4	e	64		7		I	1									C																		
TU	54	1	a	65		145	5	S	2	Z	1						T	C						2												
TU	54	1	a	66		75	2	Z	3	S	1						M	C						3												
TU	54	1	a	67		60	2	Z	2	S	1						M	C						2												
TU	54	1	a	68		140	4	Z	2	S	1						T	C						2												
TU	54	1	a	69		180	1	S	2	S	1						T	C						2												
TU	54	1	a	70		30		Z	2	S	1							C																	1	

Yarns

REGION	SITE	GRID SQUARE	LAYER	ITEM NUMBER	PART	LENGTH	DIAMETER	STRAND TWIST ANGLE	NUMBER OF STRANDS	PLY TWIST ANGLE	NUMBER OF PLIES	RE-PLY TWIST ANGLE	NUMBER OF RE-PLIES	RE-RE-PLY TWIST ANGLE	NUMBER OF RE-RE-PLIES	ATYPICAL YARN	TENSION OF TWIST	FIBRE	NUMBER OF WHITE STRANDS	NUMBER OF VERY LIGHT BROWN STRANDS	NUMBER OF LIGHT BROWN STRANDS	NUMBER OF BEIGE STRANDS	NUMBER OF YELLOW-BROWN STRANDS	NUMBER OF MID-BROWN STRANDS	NUMBER OF DARK BROWN STRANDS	NUMBER OF GREY STRANDS	NUMBER OF DYED-RED STRANDS	NUMBER OF REEF KNOTS	NUMBER OF GRANNY KNOTS	NUMBER OF OVERHAND KNOTS	NUMBER OF SLIP KNOTS	NUMBER OF FIGURE-OF-EIGHT KNOTS	NUMBER OF LARKSHEAD KNOTS	NUMBER OF HALF-HITCHES	NUMBER OF UNIDENTIFIED KNOTS	THREE-ELEMENT PLAIT
TU	54	1	b	71		355	3	Z	2	S	1						T	C					2													
TU	54	1	b	72		50	2	Z	2	S	1						L	C						2												
TU	54	1	b	73		85	2	Z	3	S	1						M	H							3											
TU	54	1	c	74		120	3	Z	2	S	1						M	C						2												
TU	54	1	c	75		240		Z	3	S	1							C					3													
TU	54	1	c	76			2	Z	2	S	1						M	C					2													
TU	54	1	c	77		150		Z	4	S	2	Z	1					H							4											
TU	54	1	c	78		285	4	Z	2	S	1						M	C						2												
TU	54	1	c	79			2	Z	1									C			1															
TU	54	1	c	80		60	2	Z	4	S	1						M	C						4												
TU	54	1	e	81		100	1	Z	2	S	1						T	C						2						1						
TU	54	1	e	82			2	Z	1									C						1												
TU	54	2	d	83		193	2	Z	2	S	1						T	C					2													
TU	54	2	d	84		80	2	Z	2	S	1						M	C				2														
TU	54	2	e	85		340	1	S	1								T	C			1															
TU	54	2	e	86		212	1	Z	2	S	1						T	C				2														
TU	54	2	e	87		160	3	Z	1								M	C			1															
TU	54	2	f	88		60		Z	2	S	1							C				2														
TU	54	2	f	89		170	4	Z	3	S	1						T	C						3												
TU	54	2	f	90		186	2	Z	2	S	1						M	C			1			1												
TU	54	2	f	91		130	1	Z	4	S	2	Z	1				M	C					4													
TU	54	2	f	92		225	1	Z	4	S	2	Z	1				M	C					4													
TU	54	2	f	93		180	1	Z	4	S	2	Z	1				T	C					2													
TU	54	2	f	94		70	2	S	2	Z	1						M	C				2														
TU	54	2	f	95		60	2	Z	3	S	1						M	C			1			2												
TU	54	2	f	96				Z	2	S	1							C	1			1														
TU	54	2	f	97				Z	2	S	1							C	1			1														
TU	54	2	f	98		70	3	Z	4	S	1							C	1					3												
TU	54	2	g	99		70	2	Z	2	S	1							C	1																	
TU	54	2	g	100		35	3	Z	2	S	1						M	C					2													
TU	54	2	g	101		25	3	Z	2	S	1						M	C					2													
TU	54	2	g	102		30	3	Z	2	S							T	C						2												
TU	54	2	g	103		45	3	Z	2	S	1						T	C						2												
TU	54	2	g	104		50	2	Z	2	S	1						T	C					2												1	
TU	54	2	g	105		133	2	Z	2	S	1						M	C						2												
TU	54	2	h	106		80	2	Z	1								T	C	1																	
TU	54	2	h	107		260	3	Z	2	S	1						T	C						2												
TU	54	2	h	108		100	2	Z	2	S							M	C					2													
TU	54	2	h	109		45	2	Z	2	S	1						M	C					2													
TU	54	2	h	110		45	1	S	2	Z	1						T	C					2													
TU	54	2	h	111		60		Z	2	S	1							C					2													
TU	54	2	h	112		70		Z	2	S	1							C					2													
TU	54	2	h	113		75	3	Z	4	S	1						M	C					4													
TU	54	2	h	114		40		?	1									C						1			1								1	
TU	54	2	i	115			5	Z	10	S	5	Z	1				T	C						10			10			2						
TU	54	2	i	116		56	3	Z	4	S	1						M	C						4												
TU	54	2	i	117		100	2	Z	4	S	1						T	C						4			4									
TU	54	2	i	118		40		Z	2	S	1							C					2													
TU	54	2	j	119		90	2	Z	4	S	2	Z	1				L	H							4										1	
TU	54	2	j	120		60	1	Z	2	S	1						M	C						2												
TU	54	2	j	121		40		S	1									C																		
TU	54	2	k	122		70	3	Z	4	S	1						M	C						4			4									
TU	54	2	k	123		55	3	S	4	Z	2	S	1				M	C									2									
TU	54	2	k	124		45	3	S	4	Z	2	S	1				M	C									4								1	
TU	54	2	k	125		160	3	Z	4	S	2	Z	1					C			4															
TU	54	2	k	126		310	3	Z	4	S	1						T	C				4														
TU	54	2	k	127		40	3	Z	4	S	1						T	C				4														
TU	54	2	k	128		95	1	Z	2	S	1						M	C			1	1														
TU	54	2	k	129		40	3	Z	4	S	1						M	C						4												
TU	54	2	l	130		115	5	Z	2	S	1						L	C				2														
TU	54	2	l	131			2	Z	2	S	1						M	C					2													
TU	54	2	l	132			4	Z	4	S	2	Z	1				T	C			4															
TU	54	2	l	133		150	2	Z	1								T	C				1														
TU	54	2	l	134		70	2	Z	1								T	C				1														
TU	54	2	l	135		90	2	Z	1								T	C				1														
TU	54	2	l.	136					1									C																		
TU	54	2	l	137					2		1							C																		
TU	54	2	m	138		160	4	Z	2	S	1						T	C				2														
TU	54	2	m	139		58	1	S	2	Z	1						M	C			2						2								1	
TU	54	2	m	140	a		2	Z	1								M	C				1													1	

Yarns

REGION	SITE	GRID SQUARE	LAYER	ITEM NUMBER	PART	LENGTH	DIAMETER	STRAND TWIST ANGLE	NUMBER OF STRANDS	PLY TWIST ANGLE	NUMBER OF PLIES	RE-PLY TWIST ANGLE	NUMBER OF RE-PLIES	RE-RE-PLY TWIST ANGLE	NUMBER OF RE-RE-PLIES	ATYPICAL YARN	TENSION OF TWIST	FIBRE	NUMBER OF WHITE STRANDS	NUMBER OF VERY LIGHT BROWN STRANDS	NUMBER OF LIGHT BROWN STRANDS	NUMBER OF BEIGE STRANDS	NUMBER OF YELLOW-BROWN STRANDS	NUMBER OF MID-BROWN STRANDS	NUMBER OF DARK BROWN STRANDS	NUMBER OF GREY STRANDS	NUMBER OF DYED-RED STRANDS	NUMBER OF REEF KNOTS	NUMBER OF GRANNY KNOTS	NUMBER OF OVERHAND KNOTS	NUMBER OF SLIP KNOTS	NUMBER OF FIGURE-OF-EIGHT KNOTS	NUMBER OF LARKSHEAD KNOTS	NUMBER OF HALF-HITCHES	NUMBER OF UNIDENTIFIED KNOTS	THREE-ELEMENT PLAIT
TU	54	2	m	140	b		2	Z	2	S	1						M	C				2														
TU	54	2	m	141		140	2	Z	1								M	C				1														
TU	54	2	m	142		63	1	Z	1								T	C					1													
TU	54	2	m	143		40	1	Z	1								T	C					1													
TU	54	2	m	144		430	2	Z	2	S	1						M	C				2														
TU	54	2	m	145		165	2	Z	2	S	1						T	C					2													
TU	54	2	m	146		100	1	Z	2	S	1						M	C				2								1					1	
TU	54	2	m	147		110	5	Z	2	S	1						L	C			2															
TU	54	2	m	148		60	6	Z	2	S	1						L	C			1	1														
TU	54	2	m	149		75	2	Z	2	S	1						M	C						2												
TU	54	2	m	150		16	3	Z	2	S	1						L	C						2												
TU	54	2	n	151		345	2	Z	2	S	1						M	C				2														
TU	54	2	n	152		150	2	Z	1								T	C				1														
TU	54	2	n	153			4	Z	4	S	2	Z	1				M	C			2	2													1	
TU	54	2	o	154		60	1	S	2	Z	1						M	C	2																	
TU	54	2	o	155		180	2	Z	2	S	1						L	C				2														
TU	54	2	o	156			3	Z	2	S	1						M	C				2													1	
TU	54	2	o	157		165	3	Z	2	S	1						M	C				2														
TU	54	2	o	158		190	3	Z	2	S	1						M	C				2														
TU	54	2	o	159		140	2	Z	1								T	C						1												
TU	54	2	o	160		225	2	Z	1								T	C						1												
TU	54	2	o	161		130	3	Z	2	S	1						M	C						2												
TU	54	2	o	162		65	1	Z	2	S	1						L	CB			1			1												
TU	54	2	o	163		150	2	Z	2	S	1						M	C			2															
TU	54	2	o	164		150	2	S	2	Z	1						M	C			2						2									
TU	54	2	o	165		155	1	Z	2	S	1						M	C			2						2									
TU	54	2	o	166		125	3	Z	2	S	1						M	C						2												
TU	54	2	o	167		415	2	Z	2	S	1						M	C			2															
TU	54	2	o	168		115	2	Z	2	S	1						M	C			2															
TU	54	2	o	169		50	5	Z	4	S	2	Z	1					C					4													
TU	54	2	o	170		60	3	Z	2	S	1						M	C						2												
TU	54	2	o	171		125	2										M	C			2															
TU	54	2	o	172		315	3	Z	2	S	1						T	C					2													
TU	54	2	o	173		100	3	Z	4	S	1						M	C			4														1	
TU	54	2	o	174		95		Z	2	S	1							C					2													
TU	54	2	o	175		105		Z	2	S	1							C																	
TU	54	2	o	176		45	2	Z	2	S	1						M	C			2															
TU	54	2	o	177		35	2	Z	2	S	1						M	C			2															
TU	54	2	o	178		140		Z	2	S	1							C																	
TU	54	2	o	179		100		Z	1									C																	
TU	54	2	o	180		70	1	Z	2	S	1						L	C		2																
TU	54	2	o	181		65	2	Z	1								T	C						1												
TU	54	2	o	182				Z	2	S	1							C																	
TU	54	2	p	183		80	1	S	2	Z	1						M	C			2						2									
TU	54	2	p	184		50	2										1		CB																	
TU	54	2	p	185		110	3	Z	6		1	S	1			2	T	C						6			4									
TU	54	2	p	186		65	3		5				1	S	1	3	T	CA	2					2			1									
TU	54	2	p	187		300	2	Z	2	S	1						L	B						2												
TU	54	2	p	188		106	2	Z	2	S	1							CV						2												
TU	54	2	p	189		130	1	Z	2	S	1						L	C					2													
TU	54	2	p	190		255	1	S	2	Z	1						T	C			2															
TU	54	2	p	191			2	S	3	Z	1						T	C			3						3								1	
TU	54	2	p	192			2	Z	2	S	1						M	C					2							1						
TU	54	2	p	193		155	2	Z	2	S	1						M	C						2												
TU	54	2	p	194		250	3	Z	2	S	1						M	C					2													
TU	54	2	p	195		130	3	Z	2	S	1							C					2													
TU	54	2	p	196		85		Z	2	S	1							C					2													
TU	54	2	q	197		115	4	Z	4	S	2	Z	1				T	C			4															
TU	54	2	q	198		390	3	Z	2	S	1						M	C						2												
TU	54	2	q	199		155	2	Z	2	S	1						M	C			2															
TU	54	2	q	200		280	2	Z	1								M	C					1													
TU	54	2	q	201		110	2	Z	1								T	C	1																	
TU	54	2	r	202		75	1	Z	2	S	1						T	C						2			2									
TU	54	2	r	203		200	4	Z	3	S	1						M	C						2												
TU	54	2	r	204		60	2	Z	2	S	1						M	C						2				1								
TU	54	2	r	205		65	2	Z	2	S	1						M	C						2												
TU	54	2	r	206		63	2	Z	2	S	1						M	C						2												
TU	54	2	r	207		170		Z	2	S	1							C					2													
TU	54	2	r	208		143	3	Z	2	S	1						M	C						2												
TU	54	2	r	209		55		Z	2	S	1							C					2							1						

Yarns

REGION	SITE	GRID SQUARE	LAYER	ITEM NUMBER	PART	LENGTH	DIAMETER	STRAND TWIST ANGLE	NUMBER OF STRANDS	PLY TWIST ANGLE	NUMBER OF PLIES	RE-PLY TWIST ANGLE	NUMBER OF RE-PLIES	RE-RE-PLY TWIST ANGLE	NUMBER OF RE-RE-PLIES	ATYPICAL YARN	TENSION OF TWIST	FIBRE	NUMBER OF WHITE STRANDS	NUMBER OF VERY LIGHT BROWN STRANDS	NUMBER OF LIGHT BROWN STRANDS	NUMBER OF BEIGE STRANDS	NUMBER OF YELLOW-BROWN STRANDS	NUMBER OF MID-BROWN STRANDS	NUMBER OF DARK BROWN STRANDS	NUMBER OF GREY STRANDS	NUMBER OF DYED-RED STRANDS	NUMBER OF REEF KNOTS	NUMBER OF GRANNY KNOTS	NUMBER OF OVERHAND KNOTS	NUMBER OF SLIP KNOTS	NUMBER OF FIGURE-OF-EIGHT KNOTS	NUMBER OF LARKSHEAD KNOTS	NUMBER OF HALF-HITCHES	NUMBER OF UNIDENTIFIED KNOTS	THREE-ELEMENT PLAIT
TU	54	2	r	210		70	3	Z	4	S	1						L	CB					3	1												
TU	54	2	r	211		50		Z	2	S	1							C					2													
TU	54	2	t	212		21	1	Z	1								T	C	1																	
TU	54	3	a	213		60	3	Z	2	S	1						M	C						2				1								
TU	54	3	a	214		43	4	Z	2	S	1						M	C					2													
TU	54	3	a	215		76	2	Z	4	S	2	Z	1				L	C					4													
TU	54	3	c	216		43	3	Z	1								M	C			1						1									
TU	54	3	d	217			2	S	2	Z	1						L	C	2																	
TU	54	3	e	218		102	1	Z	2	S	1						M	C									2									
TU	54	3	e	219		144	1	Z	3	S	1						T	C						3												
TU	54	3	e	220		95	2	S	2	Z	1						M	C						2												
TU	54	3	e	221		29	2	Z	1								M	C					1													
TU	54	3	e	222		70	1	Z	2	S	1						M	C			2															
TU	54	3	f	223		101	3	Z	2	S	1						M	C						2			2									
TU	54	3	f	224		122	2	Z	2	S	1						M	C				2														
TU	54	3	f	225		200	2	Z	2	S	1						M	C				2														
TU	54	3	f	226	a	40	1	Z	2	S	1						M	C			1	1													1	
TU	54	3	f	226	b	160	4	S	2	Z	1						L	C				2														
TU	54	3	f	227		92	2	Z	1								M	C			1			1												
TU	54	3	f	228		81	2	Z	3	S	1						M	C			2			1												
TU	54	3	f	229		56	5	Z	1								T	C			1															
TU	54	3	g	230		89	5	Z	4	S	1						M	C									4									
TU	54	3	g	231		154	3	Z	2	S	1						M	C			2						2									
TU	54	3	g	232		90	4	Z	4	S	2	Z	1				M	C						4												
TU	54	3	g	233		90	2	Z	2	S	1						M	C					2													
TU	54	3	g	234		144	1	Z	2	S	1						M	C					2													
TU	54	3	g	235		62	2	Z	2	S	1						L	B				2														
TU	54	3	g	236		68	5	Z	1								T	C			1															
TU	54	3	g	237		70	2	Z	2	S	1						M	C			2															
TU	54	3	g	238			3	I	1									V																		
TU	54	3	g	239		29	1	Z	1								L	C			1						1									
TU	54	3	g	240		42	5	I	1									C		1																
TU	54	3	g	241		100	2	I	2	Z	1						M	V					2							1	1					
TU	54	3	g	242		177	2	Z	2	S	1						L	B			2															
TU	54	3	h	243		40	3	Z	2	S	1						T	C			2						1									
TU	54	3	h	244		57	2	Z	2	S	1						T	C						2												
TU	54	3	h	245		75	2	Z	2	S	1						M	C						2												
TU	54	3	h	246		137	1	Z	1								T	C			1															
TU	54	3	h	247	a		1	Z	1								T	C					1													
TU	54	3	h	247	b		2	Z	2	Z	1						L	C					2													
TU	54	3	h	248		63	3	Z	2	S	1						M	C					2													
TU	54	3	h	249		43	2	Z	2	I	1						L	C						2												
TU	54	3	h	250		33	2	Z	1								M	C						1			1									
TU	54	3	j	251		50	3	Z	1								T	C						1			1									
TU	54	3	j	252		45	1	Z	1								T	C						1			1									
TU	54	3	j	253		25	3	Z	1								T	C						1			1									
TU	54	3	j	254		90	2	S	2	Z	1						M	C							2					2						
TU	54	3	j	255		53	2	Z	3	S	1					4	T	C	1		2															
TU	54	3	j	256		123	1	S	2	Z	1						M	C						2												
TU	54	3	j	257		128	1	S	2	Z	1						M	C						2												
TU	54	3	j	258		73	2	Z	2	S	1						M	C			1	1														
TU	54	3	j	259		175	3	Z	2	S	1						M	C						2												
TU	54	3	j	260			2	Z	1								T	C						1												
TU	54	3	j	261		160	2	S	1								T	C				1														
TU	54	33	j	262		80	2	S	1								T	C				1														
TU	54	3	j	263		58	2	S	1								T	C				1														
TU	54	3	j	264		50	1	S	2	Z	1						T	CB			1															
TU	54	3	j	265		90	3	Z	3	S	1						M	C					2	1												
TU	54	3	j	266		40	1	Z	1								M	C	1																	
TU	54	3	j	267	a		4	Z	3	S	1						M	C					3						1							
TU	54	3	j	267	b		3	Z	4	S	1						M	C					4													
TU	54	3	j	268		130	1	S	2	Z	1						M	C	2																	
TU	54	3	j	269		37	3	Z	3	S	1						L	C			3															
TU	54	3	j	270		65	2	Z	2	S	1						L	C			1			1												
TU	54	3	j	271		100	3	Z	3	S	1						L	C					3													
TU	54	3	j	272		62	3	Z	2	S							M	C						2												
TU	54	3	j	273		66	2	Z	2	S	1						M	C						2												
TU	54	3	j	274		73	2	Z	1								T	C					1													
TU	54	3	j	275		63	2	S	1								M	C			1															
TU	54	3	j	276		80	2	Z	1								M	C					1													

Yarns

REGION	SITE	GRID SQUARE	LAYER	ITEM NUMBER	PART	LENGTH	DIAMETER	STRAND TWIST ANGLE	NUMBER OF STRANDS	PLY TWIST ANGLE	NUMBER OF PLIES	RE-PLY TWIST ANGLE	NUMBER OF RE-PLIES	RE-RE-PLY TWIST ANGLE	NUMBER OF RE-RE-PLIES	ATYPICAL YARN	TENSION OF TWIST	FIBRE	NUMBER OF WHITE STRANDS	NUMBER OF VERY LIGHT BROWN STRANDS	NUMBER OF LIGHT BROWN STRANDS	NUMBER OF BEIGE STRANDS	NUMBER OF YELLOW-BROWN STRANDS	NUMBER OF MID-BROWN STRANDS	NUMBER OF DARK BROWN STRANDS	NUMBER OF GREY STRANDS	NUMBER OF DYED-RED STRANDS	NUMBER OF REEF KNOTS	NUMBER OF GRANNY KNOTS	NUMBER OF OVERHAND KNOTS	NUMBER OF SLIP KNOTS	NUMBER OF FIGURE-OF-EIGHT KNOTS	NUMBER OF LARKSHEAD KNOTS	NUMBER OF HALF-HITCHES	NUMBER OF UNIDENTIFIED KNOTS	THREE-ELEMENT PLAIT
TU	54	3	j	277		27	2	Z	1								L	C					1													
TU	54	3	j	278		35	2	Z	1								L	C					1													
TU	54	3	j	279		47	2	Z	1									C				1														
TU	54	3	j	280		50	3	Z	2	S							L	C						2												
TU	54	3	j	281		40	2	S	1								T	C				1							1							
TU	54	3	j	282		51	1	S	1								T	C			1															
TU	54	3	j	283		93	3	Z	3	S	1						L	C					3													
TU	54	3	j	284		65	3	Z	2	S	1						L	C																		
TU	54	3	j	285		76	3	Z	1									C					1													
TU	54	3	j	286		50	1	Z	1								T	C				1														
TU	54	3	j	287		17	1	Z	2	S	1						M	C									2									
TU	54	3	j	288		154	3	Z	4	S	1						M	C	1	1		2														
TU	54	3	j	289		55	1	Z	2	S	1						M	C			2															
TU	54	3	j	290		50	1	Z	2	S	1						L	C			2															
TU	54	3	j	291		55	3	Z	3	S	1						M	C						3												
TU	54	3	j	292		90	3	Z	1								T	C						1												
TU	54	3	j	293		7	1	S	2	Z	1						M	C									2									
TU	54	3	k	294		19	1	Z	2	S	1						M	C									2									
TU	54	3	k	295		220	3	Z	2	S	1						M	C			2															
TU	54	3	k	296		110	3	Z	2	S	1						L	C			2															
TU	54	3	k	297		40	3	Z	2	S	1						L	C		1	1															
TU	54	3	k	298		120	3	Z	2	S							L	C			2															
TU	54	3	k	298		58	3	Z	2	S	1						M	C			2															
TU	54	3	k	300		238	2	Z	2	S	1						L	C			2															
TU	54	3	k	301		65	2	Z	2	S	1						M	C	2																	
TU	54	3	k	302		99	3	Z	2	S	1						L	C			2															
TU	54	3	k	303		75	2	Z	2	S	1						M	C			2															
TU	54	3	k	304		62	2	Z	2	S	1						M	C		1	1															
TU	54	3	k	305		134	4	Z	3	S	1						M	C			2	1														
TU	54	3	k	306		195	3	Z	3	S	1						M	C				3														
TU	54	3	k	307		40	3	Z	2	S	1						M	C						2												
TU	54	3	k	308		44	3	Z	2	S	1						M	C						2												
TU	54	3	k	309		50	3	Z	2	S	1						M	C						2												
TU	54	3	k	310		21	2	Z	1								T	C						1												
TU	54	3	k	311		86	2	Z	5	S	1						T	C						5												
TU	54	3	k	312		42	3	Z	5	S	1							C						5												
TU	54	3	k	313		65	2	Z	3	S	1						T	C						3												
TU	54	3	k	314		74	2	Z	3	S	1						T	C						3												
TU	54	3	k	315		120	2	Z	2	S	1						M	C						2												
TU	54	3	k	316		80		Z	3	S	1						T	C						3												
TU	54	3	k	317		48	1	Z	2	S	1						M	C			2															
TU	54	3	k	318		26	1	Z	1								M	C	1																	
TU	54	3	k	319		78	3	Z	2	S	1						M	C						2												
TU	54	3	k	320		40	1	S	3	Z	1						M	B						3												
TU	54	3	k	321		120	3	Z	2	S	1						T	C			2															
TU	54	3	k	322		100	3	Z	2	S	1						T	C			2															
TU	54	3	k	323		103	2	Z	2	S	1						T	C					2												1	
TU	54	3	k	324	a		2	Z	4	S	1						M	C			4							1								
TU	54	3	k	324	b		2	Z	3	S	1						M	C			3															
TU	54	3	k	325		203	4	Z	2	S							M	C			1	1							1							
TU	54	3	k	326		51	6	Z	2	S	1						M	C			2															
TU	54	3	k	327		110	1	Z	1								T	C				1														
TU	54	3	k	328		62	1	Z	1								T	C				1														
TU	54	3	k	329		64	3	Z	2	S	1						M	C						2												
TU	54	3	k	330		50	3	Z	3	S	1						T	C					3													
TU	54	3	k	331		101	3	Z	1								T	C						1												
TU	54	3	k	332		84	3	Z	1								T	C						1												
TU	54	3	k	333		93	1	Z	2	S	1						M	C				2														
TU	54	3	k	334			4	Z	2	S	1						M	C				2						1								
TU	54	3	k	335			2	Z	2	S	1						T	C						2						1						
TU	54	3	k	336		69	3	Z	4	S	1						M	C					4													
TU	54	3	k	337	a		1	Z	2	S	1						L	C						2											1	
TU	54	3	k	337	b		5	Z	2	S	1						L	C						2												
TU	54	3	k	338		40	2	Z	1								M	C	1																	
TU	54	3	k	339			2	Z	2	S	1						M	C					2												1	
TU	54	3	k	340		63	3	Z	2	S	1						T	C						2												
TU	54	3	k	341		40	1	Z	1								M	C																		
TU	54	3	k	342		350	2	Z	2	S	1						T	C						2												
TU	54	3	k	343		33	1	Z	1								T	C						1					1							
TU	54	3	l	344		110	1	S	1								T	C			1						1									

Yarns

REGION	SITE	GRID SQUARE	LAYER	ITEM NUMBER	PART	LENGTH	DIAMETER	STRAND TWIST ANGLE	NUMBER OF STRANDS	PLY TWIST ANGLE	NUMBER OF PLIES	RE-PLY TWIST ANGLE	NUMBER OF RE-PLIES	RE-RE-PLY TWIST ANGLE	NUMBER OF RE-RE-PLIES	ATYPICAL YARN	TENSION OF TWIST	FIBRE	NUMBER OF WHITE STRANDS	NUMBER OF VERY LIGHT BROWN STRANDS	NUMBER OF LIGHT BROWN STRANDS	NUMBER OF BEIGE STRANDS	NUMBER OF YELLOW-BROWN STRANDS	NUMBER OF MID-BROWN STRANDS	NUMBER OF DARK BROWN STRANDS	NUMBER OF GREY STRANDS	NUMBER OF DYED-RED STRANDS	NUMBER OF REEF KNOTS	NUMBER OF GRANNY KNOTS	NUMBER OF OVERHAND KNOTS	NUMBER OF SLIP KNOTS	NUMBER OF FIGURE-OF-EIGHT KNOTS	NUMBER OF LARKSHEAD KNOTS	NUMBER OF HALF-HITCHES	NUMBER OF UNIDENTIFIED KNOTS	THREE-ELEMENT PLAIT
TU	54	3	1	345		365	3	Z	1								T	C			1															
TU	54	3	1	346		215	3	Z	1								T	C						1												
TU	54	3	1	347		90	3	Z	1								M	C				1														
TU	54	3	1	348		50	4	I	1									C	1																	
TU	54	3	1	349		212	2	Z	1								M	C			1															
TU	54	3	1	350		178	3	Z	1								M	C					1													
TU	54	3	1	351	a		2	Z	2	S	1						M	C	2											1						
TU	54	3	1	351	b		1	Z	2	S	1						M	C						2												
TU	54	3	1	352		116	2	Z	2	S	1						T	C		2																
TU	54	3	1	353		80	1	Z	2	S	1						M	C			2															
TU	54	3	1	354		97	2	Z	2	S	1						M	C	1		1															
TU	54	3	1	355		33	1	Z	2	S	1						M	C				2													1	
TU	54	3	1	356		58	3	Z	2	S	1						M	C			1	1								1						
TU	54	3	1	357			1	Z	2	S	1						M	C						2											1	
TU	54	3	1	358		105	3	Z	2	S	1						L	C			2															
TU	54	3	1	359		232	2	Z	2	S	1						M	C			2								1							
TU	54	3	1	360		152	2	Z	2	S	1						M	C			2															
TU	54	3	1	361		50	2	Z	2	S	1						M	C						2												
TU	54	3	1	362		34	2	Z	2	S	1						M	C						2												
TU	54	3	1	363		82	3	Z	2	S	1						T	C						2												
TU	54	3	1	364			4	Z	2	S	1						M	C			2													2		
TU	54	3	1	365		230	3	Z	3	S	1						M	C				3														
TU	54	3	1	366		90	4	Z	3	S	1						M	C			3															
TU	54	3	1	367		142	2	Z	2	S							L	C			2															
TU	54	3	1	368		136	1	Z	2	S	1						L	C				2														
TU	54	3	1	369		63	1	Z	1								T	C				1														
TU	54	3	1	370		225	2	Z	2	S	1						M	C						2												
TU	54	3	1	371		70	3	Z	2	S	1						T	C			1			1												
TU	54	3	1	372		113	1	Z	2	S	1						M	C	1								1									
TU	54	3	1	373		83	3	Z	2	S	1						M	C						2												
TU	54	3	1	374		76	2	Z	2	S	1						T	C			2															
TU	54	3	1	375		63	3	Z	2	S	1						M	C			2															
TU	54	3	1	376		84	2	Z	2	S	1						M	C				2														
TU	54	3	1	377		100	2	Z	2	S	1						L	C			2															
TU	54	3	1	378		176	2	Z	2	S	1						M	C			2															
TU	54	3	1	379		67	3	Z	4	S	1						T	C		4																
TU	54	3	1	340		45	5	Z	2	I								C			2															
TU	54	3	1	381		139	3	Z	2	S	1						M	C						2												
TU	54	3	1	382		85	2	Z	2	S	1						M	C						2												
TU	54	3	1	383		113	5	Z	2	S							M	C				2														
TU	54	3	1	384		155	3	Z	2	S	1						M	C	2																	
TU	54	3	1	385		142	2	Z	2	S	1						M	C						2												
TU	54	3	1	386		205	3	Z	2	S	1						M	C					2													
TU	54	3	1	387	a	37	2	Z	2	S	1						M	C			2														1	
TU	54	3	1	387	b	38	2	Z	2	S	1						M	C			2														1	
TU	54	3	1	388		105	4	Z	6	S	2	Z	1			5	M	C				2		2												
TU	54	3	1	389	a		3	Z	4	S	2	Z	1				M	C					4					1								
TU	54	3	1	389	b		4	Z	4	S	2	Z	1				M	C					4					2								
TU	54	3	1	389	c		2	Z	2	S	1						M	C					2						2							
TU	54	3	1	390	a		3	Z	4	S	2	Z	1				M	C					4							1					1	
TU	54	3	1	390	b		2	Z	2	S	1						M	C					2													
TU	54	3	1	391			2	Z	2	S	1						M	C			2														2	
TU	54	3	1	392		50	6	Z	4	S	2	Z	1				T	C			4															
TU	54	3	1	393			3	Z	6	S	3	Z	1				M	CB				5		1									1			
TU	54	3	1	394		65	1	Z	2	Z	1						T	C				2														
TU	54	3	1	395		150	3	Z	2	S	1						T	C		1	1															
TU	54	3	1	396	a	175	3	Z	2	S	1						M	C			2								1							
TU	54	3	1	396	b	70	4	Z	4	S	1						M	C			4															
TU	54	3	1	397		76	2	Z	1								T	C																		
TU	54	3	1	398		27	6	Z	2	S	1							C			2															
TU	54	3	1	399		58	3	Z	2	S	1							C	2																	
TU	54	3	1	400		100	3	Z	2	S	1							C	1		1															
TU	54	3	1	401		146	2	Z	1									C	1																	
TU	54	3	1	402		26	1	Z	1								T	C				1														
TU	54	3	1	403		31	2	Z	2	S	1							C	2																	
TU	54	3	1	404		35	2	I	1									C			1															
TU	54	3	1	405		42	1	Z	1									C			1															
TU	54	3	1	406		45	3	Z	2	I	1							C																		
TU	54	3	1	407		10		Z	1									C	1																	
TU	54	3	1	408		20	5	Z	2	S	1						L	C	2																	

Yarns

REGION	SITE	GRID SQUARE	LAYER	ITEM NUMBER	PART	LENGTH	DIAMETER	STRAND TWIST ANGLE	NUMBER OF STRANDS	PLY TWIST ANGLE	NUMBER OF PLIES	RE-PLY TWIST ANGLE	NUMBER OF RE-PLIES	RE-RE-PLY TWIST ANGLE	NUMBER OF RE-RE-PLIES	ATYPICAL YARN	TENSION OF TWIST	FIBRE	NUMBER OF WHITE STRANDS	NUMBER OF VERY LIGHT BROWN STRANDS	NUMBER OF LIGHT BROWN STRANDS	NUMBER OF BEIGE STRANDS	NUMBER OF YELLOW-BROWN STRANDS	NUMBER OF MID-BROWN STRANDS	NUMBER OF DARK BROWN STRANDS	NUMBER OF GREY STRANDS	NUMBER OF DYED-RED STRANDS	NUMBER OF REEF KNOTS	NUMBER OF GRANNY KNOTS	NUMBER OF OVERHAND KNOTS	NUMBER OF SLIP KNOTS	NUMBER OF FIGURE-OF-EIGHT KNOTS	NUMBER OF LARKSHEAD KNOTS	NUMBER OF HALF-HITCHES	NUMBER OF UNIDENTIFIED KNOTS	THREE-ELEMENT PLAIT
TU	54	3	1	409		38	2	Z	2	S	1						M	C			2															
TU	54	3	1	410			1	I	1									V																1		
TU	54	3	1	411		26	2	Z	2	S	1						M	C			2						2									
TU	54	3	1	412		45	2	Z	1									C						1												
TU	54	3	1	413		46	2	Z	1									C			1															
TU	54	3	1	414		60	3	Z	1									C			1															
TU	54	3	1	415		31	3	Z	1									C			1															
TU	54	3	1	416		19	5	Z	2	S	1						M	C			2															
TU	54	3	1	417		26	6	Z	2	S	1						M	C						2												
TU	54	3	1	418		19	2	Z	2	S	1						M	C			2															
TU	54	3	1	419		17	2	Z	2	S	1						M	C	2																	
TU	54	3	1	420		41	7	I	1									C			1															
TU	54	3	1	421			2	Z	2	S	1						M	C		2																
TU	54	3	1	422		460	2	Z	2	S	1						M	C				2														
TU	54	3	1	423		190	2	Z	2	S	1						M	C				2														
TU	54	3	1	424		100	3	Z	2	S	1						L	C			2															
TU	54	3	1	425		145	3	Z	1								L	C				1														
TU	54	3	1	426		93	2	Z	2	S	1						T	C	2																	
TU	54	3	1	427		84	4	Z	2	S	1						T	C	2																	
TU	54	3	1	428		420	2	Z	7	S	1							C			7															
TU	54	3	1	429		155	3	Z	7	S	1						M	C						7												
TU	54	3	1	430		265	4	Z	2	S	1						L	C			1			1												
TU	54	3	1	431		50	4	Z	2	S	1						L	C			1			1												
TU	54	3	1	432		206	3	Z	2	S	1						M	C			2			2												
TU	54	3	1	433		68	3	Z	3	S	1						M	C			3															
TU	54	3	1	434		75	4	Z	4	S	2	Z	1				L	CB				2		2												
TU	54	3	1	435		70	2	Z	2	S	1							CB			1			1												
TU	54	3	1	436			2	Z	2	S	1						M	C						2			2			2						
TU	54	3	1	437		240	4	Z	4	S	1						M	C			4															
TU	54	3	1	438		84	3	Z	2	S	1						M	C			2															
TU	54	3	1	439		75	3	Z	2	S	1						M	C			2															
TU	54	3	1	440		40	2	Z	2	S	1						M	C			2															
TU	54	3	1	441		439	3	Z	2	S	1						M	C			2															
TU	54	3	1	442		42	1	S	1								T	C			1						1								1	
TU	54	3	1	443		65	2	Z	1								M	C						1												
TU	54	3	1	444		96	2	Z	2	S	1						L	C			2															
TU	54	3	1	445		30	2	Z	1									C						1												
TU	54	3	1	446		39	2	Z	2	S	1						M	C						2												
TU	54	3	1	447		54	2	Z	1									C			1															
TU	54	3	1	448		19	2	Z	2	S	1						L	C						2												
TU	54	3	1	449		22	2	Z	3	S	1						M	C			3															
TU	54	3	1	450		15	1	Z	1									C	1																	
TU	54	3	1	451		36	3	Z	2	S	1						M	C			2									1						
TU	54	3	1	452		75	5	I	1									C	1																	
TU	54	3	1	453		70	13	I	1									C		1																
TU	54	3	1	454		22	1	?	1									C									1									
TU	54	3	m	455		50	2	Z	2	S	1						L	C									2									
TU	54	3	m	456	a		5	Z	8	S	1						T	C		8							8				1					
TU	54	3	m	456	b		3	Z	2	S	1						M	C		1	1															
TU	54	3	m	457		302	3	Z	2	S	1						M	C				2														
TU	54	3	m	458		163	3	Z	2	S	1						M	C			2															
TU	54	3	m	459		223	3	Z	2	S	1						L	C	2																	
TU	54	3	m	460	a			Z	2	S	1						M	C												2						
TU	54	3	m	460	b			Z	2	I								C																		
TU	54	3	m	461		165	2	Z	2	S	1						M	C						2												
TU	54	3	m	462	a	110	5	Z	2	S	1						M	C				2									1					
TU	54	3	m	462	b	240	3	Z	2	S	1						T	C						2												
TU	54	3	m	463		196	3	Z	2	S	1						M	C			2															
TU	54	3	m	464		55	3	Z	2	S	1						M	C			2															
TU	54	3	m	465		130	3	Z	2	S	1						M	C		1	1															
TU	54	3	m	466		212	2	Z	2	S	1						M	C			2															
TU	54	3	m	467		53	3	Z	2	S	1						M	C			2															
TU	54	3	m	468		158	2	Z	2	S	1						T	C			2															
TU	54	3	m	469		118	3	Z	2	S	1						M	C			2															
TU	54	3	m	470		220	2	Z	2	S	1						M	C			2															
TU	54	3	m	471		168	2	Z	2	S	1						M	C						2												
TU	54	3	m	472		195	2	Z	2	S	1						T	C						2												
TU	54	3	m	473		154	3	Z	2	S	1						M	C				2														
TU	54	3	m	474		109	3	Z	2	S	1						M	C				2														
TU	54	3	m	475		237	3	Z	2	S	1						M	C			2															

Yarns

REGION	SITE	GRID SQUARE	LAYER	ITEM NUMBER	PART	LENGTH	DIAMETER	STRAND TWIST ANGLE	NUMBER OF STRANDS	PLY TWIST ANGLE	NUMBER OF PLIES	RE-PLY TWIST ANGLE	NUMBER OF RE-PLIES	RE-RE-PLY TWIST ANGLE	NUMBER OF RE-RE-PLIES	ATYPICAL YARN	TENSION OF TWIST	FIBRE	NUMBER OF WHITE STRANDS	NUMBER OF VERY LIGHT BROWN STRANDS	NUMBER OF LIGHT BROWN STRANDS	NUMBER OF BEIGE STRANDS	NUMBER OF YELLOW-BROWN STRANDS	NUMBER OF MID-BROWN STRANDS	NUMBER OF DARK BROWN STRANDS	NUMBER OF GREY STRANDS	NUMBER OF DYED-RED STRANDS	NUMBER OF REEF KNOTS	NUMBER OF GRANNY KNOTS	NUMBER OF OVERHAND KNOTS	NUMBER OF SLIP KNOTS	NUMBER OF FIGURE-OF-EIGHT KNOTS	NUMBER OF LARKSHEAD KNOTS	NUMBER OF HALF-HITCHES	NUMBER OF UNIDENTIFIED KNOTS	THREE-ELEMENT PLAIT
TU	54	3	m	476		68	1	Z	2	S	1						M	C						2											1	
TU	54	3	m	477		63	1	Z	2	S	1						M	C						2												
TU	54	3	m	478			3	Z	2	S	1						T	C						2					1						1	
TU	54	3	m	479			2	Z	2	S	1						M	C	2										1							
TU	54	3	m	480		173	1	Z	1								T	C						1			1									
TU	54	3	m	481		43	2	Z	2	S	1						L	C			2															
TU	54	3	m	482		20	5	Z	2	I								C			2															
TU	54	3	m	483		64	2	Z	2	S	1						M	C						2												
TU	54	3	m	484		52	3	Z	2	S	1						M	C						2												
TU	54	3	m	485		185	2	Z	1									C			1															
TU	54	3	m	486		35	6	Z	1								L	C	1																	
TU	54	3	m	487		122	2	Z	2	S	1						M	C	2																	
TU	54	3	m	488		253	2	Z	2	S	1						M	C			2															
TU	54	3	m	489		172	3	Z	2	S	1						M	C			2															
TU	54	3	m	490		206	3	Z	2	S	1						M	C						2												
TU	54	3	m	491		100	3	Z	2	S	1						M	C						2												
TU	54	3	m	492		75	3	Z	2	S	1						M	C						2												
TU	54	3	m	493		33	2	Z	1									C	1																	
TU	54	3	m	494	a		4	Z	2	S	1						T	C					2													
TU	54	3	m	494	b			Z	1								T	C					1				1									
TU	54	3	m	495		69	2	Z	1									C						2												
TU	54	3	m	496		61	1	Z	1									C						1												
TU	54	3	m	497		47	6	I	1									C			1															
TU	54	3	n	498	a		2	Z	2	S	1						M	C						2												
TU	54	3	n	498	b		2	Z	2	S	1						M	C				2														
TU	54	3	n	499		192	2	Z	2	S	1						T	C		2																
TU	54	3	n	500		185	2	S	3	Z	1						M	C						3			3									
TU	54	3	n	501		67	2	Z	4	S	2	Z	1				T	C					4													
TU	54	3	n	502		160	2	Z	4	S	2	Z	1				M	C				4														
TU	54	3	n	503		170	2	Z	4	S	2	Z	1				M	C			4															
TU	54	3	n	504		195	5	Z	2	S	1						T	C						2												
TU	54	3	n	505		69	2	Z	2	S	1						T	C			2															
TU	54	3	n	506		160	4	Z	3	S	1						L	CB	1	1				1												
TU	54	3	n	507		455	3	Z	2	S	1						T	C		1	1															
TU	54	3	n	508	a	250	3	Z	2	S	1						M	C						2												
TU	54	3	n	508	b	115	1	Z	2	S	1							C									2									
TU	54	3	n	509			3	Z	2	S	1						M	C					2					1								
TU	54	3	n	510		54	2	Z	2	S	1						M	C			2															
TU	54	3	n	511		63	2	Z	2	S	1						M	C			2															
TU	54	3	n	512		421	3	Z	2	S	1						L	C					2													
TU	54	3	n	513		60	4	Z	2	S	1						M	C					2													
TU	54	3	n	514		108	2	Z	1								T	C					1													
TU	54	3	n	515		36	2	Z	2	S	1						M	C		2																
TU	54	3	n	516		75	3	Z	2	S	1						L	C						2												
TU	54	3	n	517		68	2	Z	2	S	1						L	C						2												
TU	54	3	n	518		200	1	Z	1								T	C						1												
TU	54	3	n	519		160	2	Z	1								T	C						1												
TU	54	3	n	520		45	1	Z	1								T	C	1																	
TU	54	3	n	521		60	3	Z	2	S	1						M	C						2												
TU	54	3	n	522		40	2	Z	2	S	1						M	C			2															
TU	54	3	n	523	a	59	2	Z	2	S	1						M	C					2												1	
TU	54	3	n	523	b	54	2	Z	2	S	1						M	C					2												1	
TU	54	3	n	524		160	3	Z	4	S	2	Z	1				T	C			4											1				
TU	54	3	n	525	a		2	Z	2	S	1						L	C						2						1						
TU	54	3	n	525	b		1	Z	1								T	C						1												
TU	54	3	n	526		144	5	Z	4	S	2	Z	1				L	C						4												
TU	54	3	n	527		60	3	Z	4	S	1						L	C			4															
TU	54	3	n	528		125	5	Z	4	S	1						L	C			4															
TU	54	3	n	529		80	5	Z	4	S	1						L	C			4															
TU	54	3	n	530		156	4	Z	2	S	1						M	C			2															
TU	54	3	n	531		60	4	Z	2	S	1						M	C			2															
TU	54	3	n	532		79	3	Z	2	S	1						M	C			2															
TU	54	3	n	533		72	4	Z	2	S	1						L	C			2															
TU	54	3	n	534		143	3	Z	2	S	1						T	C			2															
TU	54	3	n	535		77	3	Z	1								M	C		1																
TU	54	3	n	536		190	7	Z	2	S	1						L	C		2									1							
TU	54	3	n	537		295	3	Z	2	S	1						M	C			2															
TU	54	3	n	538	a	166	2	Z	2	S	1						M	C			2														2	
TU	54	3	n	538	b		2	Z	2	S	1						M	C	1		1															
TU	54	3	n	539		118	4	Z	2	S	1						M	C						2												

Yarns

REGION	SITE	GRID SQUARE	LAYER	ITEM NUMBER	PART	LENGTH	DIAMETER	STRAND TWIST ANGLE	NUMBER OF STRANDS	PLY TWIST ANGLE	NUMBER OF PLIES	RE-PLY TWIST ANGLE	NUMBER OF RE-PLIES	RE-RE-PLY TWIST ANGLE	NUMBER OF RE-RE-PLIES	ATYPICAL YARN	TENSION OF TWIST	FIBRE	NUMBER OF WHITE STRANDS	NUMBER OF VERY LIGHT BROWN STRANDS	NUMBER OF LIGHT BROWN STRANDS	NUMBER OF BEIGE STRANDS	NUMBER OF YELLOW-BROWN STRANDS	NUMBER OF MID-BROWN STRANDS	NUMBER OF DARK BROWN STRANDS	NUMBER OF GREY STRANDS	NUMBER OF DYED-RED STRANDS	NUMBER OF REEF KNOTS	NUMBER OF GRANNY KNOTS	NUMBER OF OVERHAND KNOTS	NUMBER OF SLIP KNOTS	NUMBER OF FIGURE-OF-EIGHT KNOTS	NUMBER OF LARKSHEAD KNOTS	NUMBER OF HALF-HITCHES	NUMBER OF UNIDENTIFIED KNOTS	THREE-ELEMENT PLAIT
TU	54	3	n	540		19	3	Z	1								L	C							1											
TU	54	3	o	541		22	4	Z	1									C									1								1	
TU	54	3	o	542		80	1	S	1								L	C									1									
TU	54	3	o	543		324	2	S	2	Z	1						T	C	2								2									
TU	54	3	o	544		260	1	S	2	Z	1						L	H							2						1					
TU	54	3	o	545		78	1	Z	2	S	1						T	C			2															
TU	54	3	o	546		171	2	Z	2	S	1						L	C						2												
TU	54	3	o	547		324	3	Z	4	S	2	Z	1				M	C			4									1						
TU	54	3	o	548		160	4	Z	4	S	2	Z	1				M	C			4															
TU	54	3	o	549		330	5	Z	8	Z		S	2	S	1	6	T	C			8															
TU	54	3	o	550		95	2	Z	2	S	1						M	C			2															
TU	54	3	o	551		70	2	Z	2	S	1						M	C					2												
TU	54	3	o	552		49	1	Z	2	S	1						M	C	2																	
TU	54	3	o	553		87	1	Z	2	S	1						L	C			2						2									
TU	54	3	o	554		90	3	Z	2	S	1						M	C					2												
TU	54	3	o	555	a	197	1	S	2	Z	1						T	A	2												1					
TU	54	3	o	555	b	64	1	S	2	Z	1						T	A	2																	
TU	54	3	o	556			3	Z	2	S	1						T	C					2						1		1					
TU	54	3	o	557			3	Z	4	S	1						M	C			4														1	
TU	54	3	o	558		75	2	Z	1								M	C			1															
TU	54	3	o	559		22	2	Z	1								M	C			1															
TU	54	3	o	560		49	4	Z	2	S	1						M	C						2												
TU	54	3	o	561	a	275	4	I	1									C			1															
TU	54	3	o	561	b	135	4	I	1									C						1												
TU	54	3	p	562		52	1	S	2	Z	1						M	C									2									
TU	54	3	p	563		65	1	Z	2	S	1						T	C						2			2									
TU	54	3	p	564	a		2	S	4	Z	1						M	C			4						4									3	
TU	54	3	p	564	b		3	Z	2	S	1						M	C		2																
TU	54	3	p	564	c		3	Z	2	S	1						T	C						2												
TU	54	3	p	564	d		3	Z	2	S	1						M	C		2																
TU	54	3	p	565		74	3	Z	2	S	1						M	C		2																
TU	54	3	p	566		394	2	Z	2	S	1						M	C			2						2									
TU	54	3	p	567		363	3	Z	6	S	3	Z	1				M	C					6												
TU	54	3	p	568		425	4	Z	8	S	1						T	C					8												
TU	54	3	p	569		342	4	Z	6	S	1						T	C					6												
TU	54	3	p	570		240	3	Z	4	S	1						T	C					4												
TU	54	3	p	571		115	4	Z	6	S	1						T	C					6												
TU	54	3	p	572		393	5	Z	4	S	2	Z	1				M	C				4							1						
TU	54	3	p	573		218	4	Z	2	S	1						M	CB						2												
TU	54	3	p	574		288	3	Z	2	S	1						T	C		2																
TU	54	3	p	575		278	2	Z	2	S	1						M	C		1	1															
TU	54	3	p	576		165	3	Z	2	S	1						M	C			2															
TU	54	3	p	577		215	2	Z	2	S	1						M	C		1	1															
TU	54	3	p	578		293	2	Z	2	S	1						M	C			2															
TU	54	3	p	579		100	3	Z	2	S	1						M	C			2															
TU	54	3	p	580		101	3	Z	2	S	1						M	C					2												
TU	54	3	p	581		90	3	Z	2	S	1						M	C					2												
TU	54	3	p	582		85	3	Z	2	S	1						M	C						2												
TU	54	3	p	583		140	5	Z	1								M	C					1												
TU	54	3	p	584		50	5	Z	2	S	1						M	C			2															
TU	54	3	p	585		130	2	Z	1								M	C			1															
TU	54	3	p	586		147	3	Z	2	S	1						M	C			2															
TU	54	3	p	587		200	3	Z	3	S	1						M	C					3												
TU	54	3	p	588		176	2	Z	4	S	1						T	C						4						3						
TU	54	3	p	589		95	2	Z	2	S	1						M	C						2						1						
TU	54	3	p	590	a	134	3	Z	4	S	1						M	C						4						1						
TU	54	3	p	590	b		1	Z	2	S	1						M	C						2												
TU	54	3	p	591	a		4	Z	4	S	1						L	C					4							1			1			
TU	54	3	p	591	b		2	Z	2	S	1						L	C					2												
TU	54	3	p	592		377	2	Z	2	S	1						M	C				2								1						
TU	54	3	p	593		98	3	Z	2	S	1						T	C					2												
TU	54	3	p	594		312	2	Z	2	S	1						M	C			2															
TU	54	3	p	595		65	2	Z	2	S	1						M	C			2															
TU	54	3	p	596		60	2	Z	2	S	1						M	C			2															
TU	54	3	p	597		50	2	Z	2	S	1						M	C			2															
TU	54	3	p	598	a			Z	2	S	1							C		2														1		
TU	54	3	p	598	b		2	Z	2	S	1						T	C		2																
TU	54	3	p	599		144	3	Z	1								T	C			1								1	3						
TU	54	3	3	600			2	S	2	Z	1						M	C			2															
TU	54	3	p	601		70	3	Z	2	S	1						M	C					2							1						

Yarns

REGION	SITE	GRID SQUARE	LAYER	ITEM NUMBER	PART	LENGTH	DIAMETER	STRAND TWIST ANGLE	NUMBER OF STRANDS	PLY TWIST ANGLE	NUMBER OF PLIES	RE-PLY TWIST ANGLE	NUMBER OF RE-PLIES	RE-RE-PLY TWIST ANGLE	NUMBER OF RE-RE-PLIES	ATYPICAL YARN	TENSION OF TWIST	FIBRE	NUMBER OF WHITE STRANDS	NUMBER OF VERY LIGHT BROWN STRANDS	NUMBER OF LIGHT BROWN STRANDS	NUMBER OF BEIGE STRANDS	NUMBER OF YELLOW-BROWN STRANDS	NUMBER OF MID-BROWN STRANDS	NUMBER OF DARK BROWN STRANDS	NUMBER OF GREY STRANDS	NUMBER OF DYED-RED STRANDS	NUMBER OF REEF KNOTS	NUMBER OF GRANNY KNOTS	NUMBER OF OVERHAND KNOTS	NUMBER OF SLIP KNOTS	NUMBER OF FIGURE-OF-EIGHT KNOTS	NUMBER OF LARKSHEAD KNOTS	NUMBER OF HALF-HITCHES	NUMBER OF UNIDENTIFIED KNOTS	THREE-ELEMENT PLAIT
TU	54	3	p	602		295	4	Z	1									C			1															
TU	54	3	p	603		47	3	Z	2	S	1						M	C	1		1															
TU	54	3	p	604			5	Z	3	S	1						M	C						3												
TU	54	3	p	605	a	128	3	S	2	Z	1						M	C				2														
TU	54	3	p	605	b	22	2	Z	2	S	1						M	C			2															
TU	54	3	p	606		120	3	Z	2	S	1						M	C				2								2						
TU	54	3	p	607		132	3	Z	2	S	1						M	C					2													
TU	54	3	p	608		51	2	Z	2	S	1						M	C					2													
TU	54	3	p	609		82	2	Z	1									C			1															
TU	54	3	p	610		61		Z	1									C			1															
TU	54	3	p	611		21	3	Z	2	S	1						M	C					2													
TU	54	3	p	612		33	3	Z	1									C		1																
TU	54	3	p	613		50	2	Z	1									C			1															
TU	54	3	p	614		21	3	?	1									C			1															
TU	54	3	p	615		55	4	Z	1									C						1												
TU	54	3	q	616		25	3	S	6	Z	3	S	1				T	C		6							6									
TU	54	3	q	617	a		2	S	4	Z	2	S	1				M	C									4			1						
TU	54	3	q	618	b		1	S	2	Z	1						M	C									2									
TU	54	3	q	618		476	5	Z	8	S	4	Z	2	S	1		M	C					8							1						
TU	54	3	q	619		96	1	Z	2	S	1						M	C						2												
TU	54	3	q	620		200	3	Z	4	S	1						M	C						4												
TU	54	3	q	621		106	1	Z	2	S	1						M	C						2												
TU	54	3	q	622		200	4	Z	2	S	1						M	C						2												
TU	54	3	q	623		100	2	Z	2	S	1						M	C			2															
TU	54	3	q	624		110	4	Z	2	S	1						M	C				2							1							
TU	54	3	q	625		95	3	Z	2	S	1						M	C				2														
TU	54	3	q	626		40	3	Z	2	S	1						M	C				2														
TU	54	3	q	627		121	3	Z	2	S	1						M	C						2												
TU	54	3	q	628		298	3	Z	2	S	1						L	C					2													
TU	54	3	q	629		146	5	Z	2	S	1						M	C			2															
TU	54	3	q	630		178	3	Z	2	S	1						M	C			2															
TU	54	3	q	631		260	2	Z	2	S	1						M	C			2															
TU	54	3	q	632		116	2	Z	2	S	1						M	C			2															
TU	54	3	q	633		77	2	Z	2	S	1						M	C						2												
TU	54	3	q	634		107	2	Z	2	S	1						M	C						2												
TU	54	3	q	635		100	3	Z	2	S	1						L	C			2															
TU	54	3	q	636		156	3	Z	1								T	C			1															
TU	54	3	q	637		65	2	Z	4	S	1						M	C					4													
TU	54	3	q	638		340	5	Z	2	S	1						M	C					2						1							
TU	54	3	q	639		90	3	Z	1								T	C					1													
TU	54	3	q	640		73	3	Z	2	S	1						M	C						2												
TU	54	3	q	641		170	8	I	2	Z	1						M	V			2															
TU	54	3	q	642			3	I	1									V																		
TU	54	3	q	643			3	Z	2	S	1						L	C						2						1				3		
TU	54	3	q	644	a		3	Z	2	S	1						M	C						2					3							
TU	54	3	q	644	b		3	Z	2	S	1						M	C		1	1															
TU	54	3	q	645		25	6	Z	1									C						1												
TU	54	3	r	646		133	1	S	3	Z	1						M	C						3			3									
TU	54	3	r	647		239	4	Z	4	S	1						M	C					4													
TU	54	3	r	648		51	4	Z	4	S	1						M	C					4													
TU	54	3	r	649			2	Z	1								T	C						1										2	1	
TU	54	3	r	650		37	2	Z	1								T	C				1														
TU	54	3	r	651		130		I	1									H							1					1						
TU	54	3	r	652		90	12	I	3									V						3						1						Y
TU	54	3	r	653		140	16	I	3									V						3						1						Y
TU	54	3	r	654		100	7	I	3									V							3					1						Y
TU	54	3	s	655		713	4	Z	3	S	1						M	C					3													
TU	54	3	s	656		51	2	Z	1								T	C					1													
TU	54	3	s	657		160	2	Z	1								T	C			1															
TU	54	3	s	658		144	2	Z	1								T	C				1														
TU	54	3	s	659			2	Z	2	S	1						M	C					2											1		
TU	54	3	s	660		11	2	Z	2	S	1						M	C					2													
TU	54	3	s	661		21	2	Z	2	S	1						M	C					2													
TU	54	4	f	662		94	4	Z	8	S	4	Z	2	Z	1		M	C						8				1								
TU	54	4	f	663		117	2	S	2	S	1						T	C					2													
TU	54	4	j	664	a		1	Z	2	S	1						M	C			2							1								
TU	54	4	j	664	b		2	S	3	Z	1						T	C									3									
TU	54	4	j	665		57	1	Z	2	S							M	C			2															
TU	54	4	j	666		257	3	Z	2	S	1						M	C				2														
TU	54	4	j	667		41	3	Z	2	S	1						M	C		2																

Yarns

REGION	SITE	GRID SQUARE	LAYER	ITEM NUMBER	PART	LENGTH	DIAMETER	STRAND TWIST ANGLE	NUMBER OF STRANDS	PLY TWIST ANGLE	NUMBER OF PLIES	RE-PLY TWIST ANGLE	NUMBER OF RE-PLIES	RE-RE-PLY TWIST ANGLE	NUMBER OF RE-RE-PLIES	ATYPICAL YARN	TENSION OF TWIST	FIBRE	NUMBER OF WHITE STRANDS	NUMBER OF VERY LIGHT BROWN STRANDS	NUMBER OF LIGHT BROWN STRANDS	NUMBER OF BEIGE STRANDS	NUMBER OF YELLOW-BROWN STRANDS	NUMBER OF MID-BROWN STRANDS	NUMBER OF DARK BROWN STRANDS	NUMBER OF GREY STRANDS	NUMBER OF DYED-RED STRANDS	NUMBER OF REEF KNOTS	NUMBER OF GRANNY KNOTS	NUMBER OF OVERHAND KNOTS	NUMBER OF SLIP KNOTS	NUMBER OF FIGURE-OF-EIGHT KNOTS	NUMBER OF LARKSHEAD KNOTS	NUMBER OF HALF-HITCHES	NUMBER OF UNIDENTIFIED KNOTS	THREE-ELEMENT PLAIT
TU	54	4	j	668		224	4	Z	2	S	1						M	C		1				1												
TU	54	4	j	669		93	2	Z	2	S	1						M	CB			2															
TU	54	4	j	670		72	3	Z	2	S	1						L	C				2														
TU	54	4	j	671	a	79	3	Z	2	S	1						T	C						2					1							
TU	54	4	j	671	b		2	Z	1								T	C									1									
TU	54	4	l	672		115	3	Z	2	S	1						L	C			2															
TU	54	4	l	673		52	4	Z	2	S	1						M	C			2															
TU	54	4	l	674		75	2	Z	2	S	1						M	C			2															
TU	54	4	l	675	a	176	3	Z	4	S	1						L	C						3	1				2							
TU	54	4	l	675	b	51	3	Z	3	S	1						L	C						2	1											
TU	54	4	l	676		49	2	Z	2	S	1						M	C						2												
TU	54	4	l	677		25	1	Z	1								T	C			1															
TU	54	4	l	678		17	2	Z	1								M	C		1										1						
TU	54	4	l	679		36	2	Z	1								T	C			1															
TU	54	4	l	680		22	2	Z	2	S	1						M	C			2															
TU	54	4	l	681		34	1	Z	1								T	C			1															
TU	54	4	l	682		50	1	Z	2	S	1						L	C			2						2									
TU	54	4	l	683		49	1	Z	1								T	C									1									
TU	54	4	l	684			2	Z	2	S	1						M	C									2		1					2		
TU	54	4	l	685		60	3	Z	2	S	1						M	C						2												
TU	54	4	l	686		39	4	Z	4	S	1						M	C						4												
TU	54	4	l	687		60	4	Z	2	S	1						M	C			2															
TU	54	4	l	688		166	3	Z	2	S	1						M	C		1	1															
TU	54	4	l	689		37	2	Z	2	S	1						M	C		1	1															
TU	54	4	l	690		96	2	Z	2	S	1						M	C		1																
TU	54	4	l	691		83	2	Z	2	S	1						M	C		1																
TU	54	4	l	692		55	2	Z	1								T	C		1																
TU	54	4	l	693		23	2	Z	2	S	1						M	C		2																
TU	54	4	l	694		25	2	Z	1								T	C		1																
TU	54	4	l	695		15	1	Z	1								T	C		1																
TU	54	4	l	696		67	2	Z	1								T	C		1																
TU	54	4	l	697			4	Z	10	S	1						M	C			9						1			1						
TU	54	4	l	698		12	3	Z	2	S	1						M	C						2												
TU	54	4	l	699		162	4	Z	1									C						1												
TU	54	4	l	700	a	135	3	Z	2	S	1						M	C			2								2							
TU	54	4	l	700	b	50	2	Z	2	S	1						M	C			2															
TU	54	4	l	701		250	3	Z	4	S	1						M	C	4										1							
TU	54	4	l	702		110	3	Z	2	S	1						M	C			2								1							
TU	54	4	l	703		110	2	I	2	Z	1						L	V						1	1						1					
TU	54	4	l	704		70	4	I	2	Z	1						M	V			2															
TU	54	4	l	705		160	4	S	2	Z	1						L	B						2												
TU	54	4	l	706		90	6	Z	1								M	C			1															
TU	54	4	m	707		110	2	Z	2	S	1						M	C						2												
TU	54	4	m	708		66	2	Z	1								M	C			1															
TU	54	4	m	709		52	2	Z	1								T	C			1															
TU	54	4	m	710		60	2	Z	1								T	C		1																
TU	54	4	m	711		62	2	Z	2	S	1						M	C			2															
TU	54	4	m	712		42	1	Z	1								T	C			1									1						
TU	54	4	m	713		32	1	Z	2	S	1						T	C			2									1						
TU	54	4	m	714		10	2	Z	1								L	C			1															
TU	54	4	m	715		40	2	Z	2	S	1						L	C						2						1						
TU	54	4	m	716		59	2	Z	1								T	C		1																
TU	54	4	m	717		31	3	Z	2	S	1						L	C			2															
TU	54	4	m	718		34	2	Z	2	S	1						M	C						2												
TU	54	4	m	719		23	2	Z	1								T	C						1												
TU	54	4	m	720		13	1	Z	1									C			1															
TU	54	4	m	721		20	2	Z	1								T	C			1															
TU	54	4	m	722			2	Z	2	S	1							C						2						1						
TU	54	4	m	723		20	2	Z	2	Z	1						M	C						2												
TU	54	4	m	724		31	2	Z	1								T	C		1																
TU	54	4	n	725		37	3	Z	3	S	1						M	C		3																
TU	54	4	n	726		101	3	Z	3	S	1							C		1	2													1		
TU	54	5	a	727		200	6	Z	8	S	1						T	C				8														
TU	54	5	a	728		70	2	Z	2	S	1						M	CB			1				1											
TU	54	5	a	729		69	5	Z	4	S	1						M	CB				3		1												
TU	54	6	a	730		160	2	Z	2	S	1						M	C			2															
TU	55	1	4	731		370	3	Z	4	S	1						M	C			4															
TU	55	1	4	732		15	1	Z	2	S	1						M	C		2																
TU	55	2	6	733		29	1	Z	2	S	1						T	C						2												
TU	55	2	6	734		94	3	Z	4	S	2	Z	1				M	C		2				2												

Yarns

REGION	SITE	GRID SQUARE	LAYER	ITEM NUMBER	PART	LENGTH	DIAMETER	STRAND TWIST ANGLE	NUMBER OF STRANDS	PLY TWIST ANGLE	NUMBER OF PLIES	RE-PLY TWIST ANGLE	NUMBER OF RE-PLIES	RE-RE-PLY TWIST ANGLE	NUMBER OF RE-RE-PLIES	ATYPICAL YARN	TENSION OF TWIST	FIBRE	NUMBER OF WHITE STRANDS	NUMBER OF VERY LIGHT BROWN STRANDS	NUMBER OF LIGHT BROWN STRANDS	NUMBER OF BEIGE STRANDS	NUMBER OF YELLOW-BROWN STRANDS	NUMBER OF MID-BROWN STRANDS	NUMBER OF DARK BROWN STRANDS	NUMBER OF GREY STRANDS	NUMBER OF DYED-RED STRANDS	NUMBER OF REEF KNOTS	NUMBER OF GRANNY KNOTS	NUMBER OF OVERHAND KNOTS	NUMBER OF SLIP KNOTS	NUMBER OF FIGURE-OF-EIGHT KNOTS	NUMBER OF LARKSHEAD KNOTS	NUMBER OF HALF-HITCHES	NUMBER OF UNIDENTIFIED KNOTS	THREE-ELEMENT PLAIT
TU	55	2	6	735			3	Z	3	S	1							C	1					2											1	
TU	55	2	8	736		53	1	Z	2	S	1						M	C			2						2									
TU	55	2	8	737		107	2	Z	2	S	1						T	C			2															
TU	55	2	8	738		170	2	Z	2	S	1						M	C				2														
TU	55	2	8	739		68	1	Z	2	S	1						T	C		2																
TU	55	2	8	740		60	3	Z	2	S	1						L	C			1				1											
TU	55	2	8	741		58	4	Z	2	S	1						T	C			1			1												
TU	55	2	P	742		15	1	S	2	Z	1						M	C							2											
TU	55	2	P	743		70	2	Z	1								M	C								1										
TU	55	2	P	744			3	S	2	S	1						L	C			1	1														
TU	55	2	P	745		90	3	Z	2	S	1						T	C	1		1															
TU	55	2	P	746		28	2	Z	2	S	1						T	C			2															
TU	55	2	4	747		150	2	S	2	Z	1						T	C	2																	
TU	55	2	9	748			2	Z	2	S	1						M	C			2															
TU	85	2	2	749		24	2	Z	1								T	C				1														
TU	85	2	2	750		10	1	Z	1									C			1															
TU	85	2	3	751		21	3	Z	3	I	1						M	C				3														
TU	85	7		752	a		1	Z	2	S	1						T	C					2													
TU	85	7		752	b		1	Z	2	S	1						T	C									2									
TU	82	1	1	753		105	9	Z	3	S	1	Z	1				L	C			1			1	1											
TU	82	1	1	754		47	1	Z	2	S	1						M	C						2												
TU	82	1	1	755		60	4	Z	3	S	1						T	C		1	2															
TU	82	1	1	756		50	1	Z	1								T	C			1															
TU	82	1	1	757		48	3	Z	2	I	1							C			1			1												
TU	82	1	2	758		85	1	S	6	Z	3	S					T	C			6															
TU	82	1	2	759		52	2	Z	3	S	1						M	C			3							3								
TU	82	1	3	760		21	1	Z	1								T	C		1																
TU	82	1	3	761		23	1	S	2	Z	1						M	C		1				1												1
TU	82	1	3	762		10	2	S	1								M	C						1			1									
TU	82	1	3	763		23		Z	2	S	1						L	C		1	1															
TU	82	1	3	764		12	1	Z	1								M	C			1															
TU	82	1	3	765		18	1	S	2	Z	1						T	C		2																1
TU	82	1	3	766		14	1	S	2	Z	1						T	C			2															1
TU	82	1	3	767		52	1	S	6	Z	3	S	1				T	C		6																
TU	82	1	3	768		53	1	S	2	Z	1						T	C			2										1					1
TU	82	1	3	769		52	1	S	2	Z	1						M	C			2															2
TU	82	1	3	770		34	1	S	2	Z	1						T	C			2															1
TU	82	1	3	771		36	1	S	2	Z	1						T	C			2															1
TU	82	1	3	772		17	1	S	2	Z	1						T	C			2															1
TU	82	1	3	773		88	1	S	2	Z	1						T	C			2															3
TU	82	1	3	774		113	1	S	2	Z	1						T	C			2															4
TU	82	1	3	775		38	1	Z	2	S	1						T	C		2								2								
TU	82	1	3	776		46	1	Z	2	S	1						M	C			2															
TU	82	1	3	777		39	1	Z	2	S	1						M	C										2								
TU	82	1	3	778		50	1	Z	2	S	1						M	C			2															
TU	82	1	3	779		53	3	Z	2	S	1						M	C		2											1					
TU	82	1	3	780		17	1	S	2	Z	1						T	C			2															1
TU	82	1	3	781			1	S	2	Z	1						T	C			2															1
TU	82	1	3	782		58	3	Z	16	S	4	Z	1				M	C		16								16								
TU	82	1	3	783		72	2	Z	2	S	1						M	C						2												
TU	82	1	3	784		61	3	Z	5	S	1						T	C			5															
TU	82	1	3	785		68	1	Z	2	S	1						L	H							2											
TU	82	1	3	786		84	2	Z	2	S	1						L	H							2											
TU	82	1	3	787		60	2	Z	2	S	1						L	C			2															
TU	82	1	3	788		78	1	Z	1								M	C							1											
TU	82	1	3	789		19	3	Z	2	S	1						T	C						2			2									
TU	82	1	3	790		36	1	Z	1									C			1															
TU	82	1	3	791		24	1	Z	1									C			1															
TU	82	1	3	792		19	2	S	1								T	C		1							1									
TU	82	1	3	793		40	3	Z	1								T	C		1																
TU	82	1	3	794		69	1	Z	4	S	1						T	C		1	3															
TU	82	1	3	795		24	2	Z	3	S	1						T	C		2	1															
TU	82	1	3	796		17	1	Z	1									C			1															
TU	82	1	3	797		34	2	Z	1								T	C		1																
TU	82	1	3	798		16	2	Z	3	S	1						T	C		2	1															
TU	82	1	3	799		53	1	Z	1								T	C			1															
TU	82	1	3	800		54	2	Z	2	S	1						M	C		2																
TU	82	1	3	801		50	1	Z	2	S	1						M	H							2											
TU	82	1	4	802		35	3	S	1								T	C			1															
TU	82	1	4	803		65		Z	3	S	2					7	T	C						2			2	1								

Yarns

REGION	SITE	GRID SQUARE	LAYER	ITEM NUMBER	PART	LENGTH	DIAMETER	STRAND TWIST ANGLE	NUMBER OF STRANDS	PLY TWIST ANGLE	NUMBER OF PLIES	RE-PLY TWIST ANGLE	NUMBER OF RE-PLIES	RE-RE-PLY TWIST ANGLE	NUMBER OF RE-RE-PLIES	ATYPICAL YARN	TENSION OF TWIST	FIBRE	NUMBER OF WHITE STRANDS	NUMBER OF VERY LIGHT BROWN STRANDS	NUMBER OF LIGHT BROWN STRANDS	NUMBER OF BEIGE STRANDS	NUMBER OF YELLOW-BROWN STRANDS	NUMBER OF MID-BROWN STRANDS	NUMBER OF DARK BROWN STRANDS	NUMBER OF GREY STRANDS	NUMBER OF DYED-RED STRANDS	NUMBER OF REEF KNOTS	NUMBER OF GRANNY KNOTS	NUMBER OF OVERHAND KNOTS	NUMBER OF SLIP KNOTS	NUMBER OF FIGURE-OF-EIGHT KNOTS	NUMBER OF LARKSHEAD KNOTS	NUMBER OF HALF-HITCHES	NUMBER OF UNIDENTIFIED KNOTS	THREE-ELEMENT PLAIT
TU	82	1	4	804		60	1	Z	4	S	1						T	C		2	2															

Figure A.1 Yarn No. 421, grid square 3, layer l, TU 54

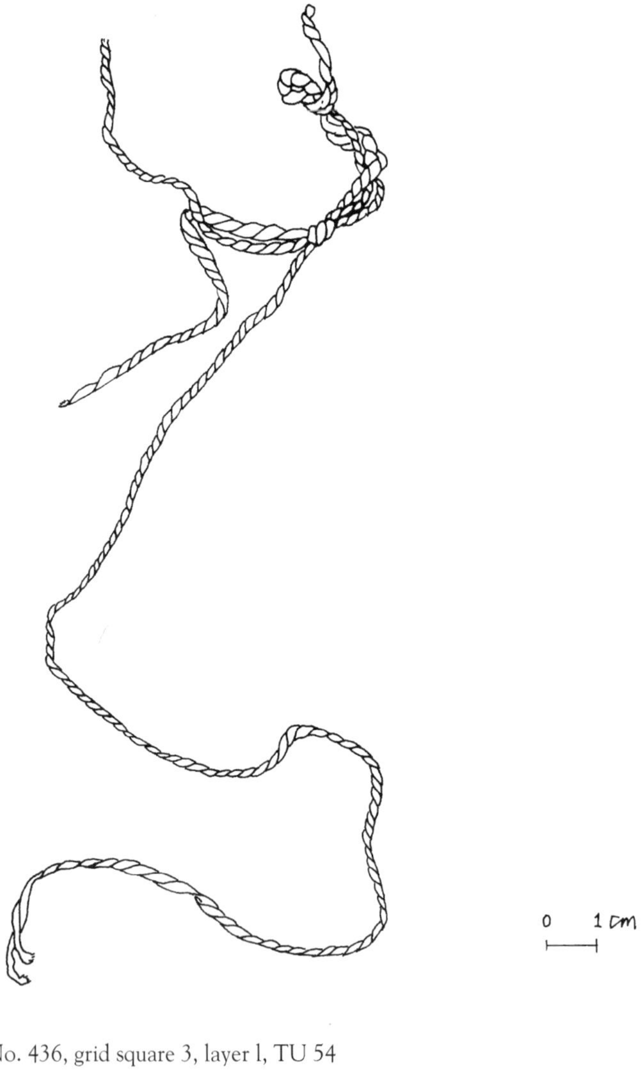

Figure A.2 Yarn No. 436, grid square 3, layer l, TU 54

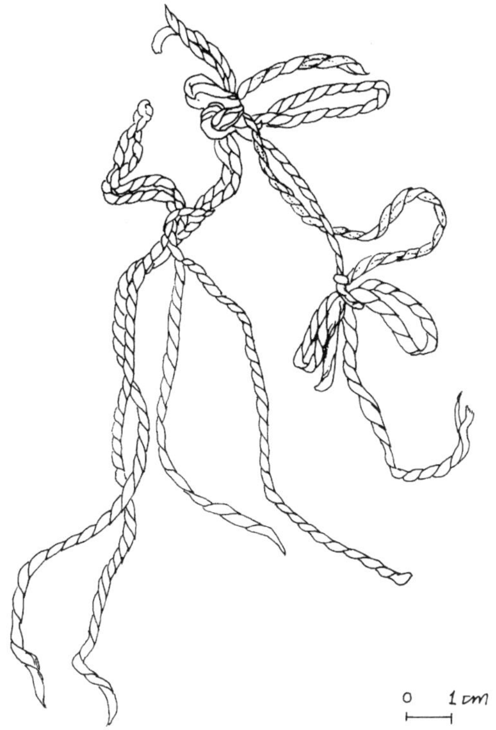

Figure A.3 Yarn No. 564, grid square 3, layer p, TU 54

Fabrics

TU 67

Grid square 4, layer G: fabric No. 1

Fragment of weaving, 8 mm (warp direction) by 11 mm (weft direction). No selvedges are present. This perhaps represents a fragment of a warp-faced textile. Plain weave, 1/1. Warp: Z2S, M ply. Light yellow-brown. 0.5 mm diameter Weft: Z spun, unplied. 1 mm diameter 10.5 warp per 10 mm, three weft per 5 mm.

TU 54

Grid square 2, layer 65–70 cm (layer L): fabric No. 2

Fragment of looped fabric, 40 mm by 5 mm. Simple looping on a foundation element, crossed right-over-left (interlocked stitches). Compact fabric. There are approximately four and a half loops in 20 mm widthways. At the bottom of the fabric there is a fringe formed of yarns caught round the lag between the loops of the bottom row of looping; the remains of six such yarns exist, the longest length extant is 30 mm. Yarn: Z2S, M ply; yellowish brown. The foundation element is similar, but it is a single, unplied yarn. (See figure 8.4 on p. 220.)

Grid square 3, layer L: fabric No. 3

Fragment of looped fabric, 74 mm by 5 mm. Simple looping crossed right-over-left. Compact fabric. There are approximately 5 loops in 20 mm widthways. Yarns: Z2S, M ply. The uppermost row is of a light beige colour, 1 mm diameter; it is looped over a mid-brown foundation yarn. The next three rows are of mid-brown colour, 2 mm diameter. (See figure 8.7 on p. 221.)

Grid square 3, layer P: fabric No. 4

Fragment of looped fabric, 52 mm by 20 mm. Simple looping on a foundation element crossed right-over-left (interlocked stitches). Compact fabric. There are approximately four loops in 20 mm widthways. Yarns: Z2S, M ply; 2 mm diameter. The foundation and looping yarns are of a light brown colour, except for two rows and the uppermost fragmentary row of looping, which were dyed. This colour is now an orange-pink. (See figure 8.6 on p. 220.)

Grid square B3, layer 15: fabric No. 5

Fragment of looped fabric approximately 23 mm by 40 mm. Simple looping on a foundation element; compact fabric. Yarn: Z2S, M–T ply; 1–2 mm diameter; light brown. This fragment is dirty and in need of conservation for the full details to be observed.

Grid square B3, layer 17: fabric No. 6

Fragment of looped fabric, 230 mm by 19 mm, plus three small pieces (22 mm by 16 mm; 13 mm by 13 mm; 10 mm by 6 mm). Simple looping on a foundation element crossed right-over-left (interlocked stitches). Compact fabric. There are a maximum of five rows extant. There are approximately eight loops in 20 mm widthways and 3 rows in 10 mm. Yarn: Z2S, M ply, 1.5 mm diameter. Yellowish brown; the yarns of the smaller pieces are somewhat lighter in colour than those of the larger fragment. (See plate 8.6 on p. 220.)

TU 55

Grid square 2, pit: fabric No. 7

A strip of dark pink silk material, 50 mm by 5 mm in rep weave (1/1 interlacing).

TU 85

Grid square 7, human burial 3: fabric No. 8

The body of a child had the remains of the hair on the cranium, which had been covered with a turban. It is in a very fragile state of preservation. Immediately on top of the hair was placed a textile. No selvedges remain, so it is not possible to distinguish warp from weft. One element is used singly, and the other is used double. Both elements are Z3S, but the doubled element is coarser in diameter having a density of three elements in 10 mm. The yarns listed in the yarn inventory as No. 752) were wrapped round this textile. (See plate 8.2 on p. 207.)

Grid square 11, layer VII: fabric No. 9

Textile, folded, measuring about 55 mm by 25 mm as folded. This fragment is in a fragile condition. No selvedges were observed. One set of elements are so closely spaced that the second set of elements can scarcely be seen, but both warp and weft in this plain weave textile (1/1 interlacing) are of black camelid fibre. The closely spaced set of elements uses yarn that is Z2S, with a T ply angle and a diameter of about 1.5 mm. There are eleven such yarns in 10 mm. The hidden element appears to have a Z2S structure and there are three such yarns in 10 mm.

TU 82

Fabric No. 10

Fragment of weaving, 42 mm by 21 mm. The fragment is in a brittle condition and no selvedges are present. Since the ply angle of the longer set of elements is tighter, this is regarded as the warp direction, and this weaving would then represent a piece of a warp-faced textile. Plain weave, 1/1. Warp: Z2S, T ply angle; beige or light brown, dyed a blackish-blue; less than 1 mm diameter. Weft: Z2S, L ply angle, beige or light brown, dyed a blackish-blue; less than 1 mm diameter. There are twenty-two warp per 10 mm, and six weft per 10 mm.

Basketry

TU 67

Grid square 4, layer G: basketry No. 1

Fragment of coiled basketry, 24 mm long and 65 mm wide. Vegetal fibre. A foundation element, approximately 3 mm diameter, is wrapped with dark brown cane 1 mm wide forming simple stitches (eleven are extant). The stitches are simple and non-interlocking, three of them are split by the remains of the stitch which would have passed round the coil immediately above.

TU 54

Grid square 2, layer o or p: basketry No. 2

The flat base of a coiled basket, with a diameter of approximately 140 mm. There is a bundle foundation forming a spiral construction, which is wrapped with narrow strips of cane in simple and non-interlocking stitches. Each stitch pierces through the foundation of the previous spiral. Work proceeded from the centre and, in places, there are two stitches on top of each other on the same spot, to allow the circle to become bigger. The right stitch is on top of the left where this occurs, implying that the stitching was worked from left to right. The basket was worn and mended in antiquity; there are nine long, straight stitches holding the coils together, sewn with strips of leather. (See plate 8.9 on p. 225.)

Grid square 3, layer P: basketry No. 3

A small piece of woven cane, 62 mm by 17 mm. Four strips of cane are interlaced 1/1 by a slightly narrower strip of cane; seven passes of this cane exist. Both elements are a dark brown colour.

Grid square 3, layer P: basketry No. 4

A fragmentary basket construction, consisting of rod foundation, 59 mm long, with wrapping stitches first worked with the stitches leaning from lower left to upper right, then a second row was worked with the wrapping stitches leaning from lower right to upper left. The result is a series of X shapes. The wrapping stitches are of narrow, dark brown cane. This section of basketry was broken and mended in antiquity, using a yarn of camelid fibre, Z3S, T ply; mid-brown; 4.5–6 mm diameter. (This yarn is listed and described as No. 601 in the yarn inventory.)

TU 58

Tomb 4: basketry No. 5

Various fragments in a very fragile condition of an object which might have served as a handle. The pieces measure 21 mm, 20 mm, 10 mm, and 8 mm long, with another half-width fragment measuring 9 mm long. There are two rigid warp elements, of cane. The weft interlaces 1/1, and is closely packed to form a weft-faced plain weave, with the weft lying at an oblique angle to the warp. The weft yarn is of camelid fibre, darkened to a yellow-blackish colour, and is probably Z2S; T-ply angle; 1 mm diameter.

Glossary

Altiplano (Spanish) High-altitude, treeless grasslands between approximately 3,500 and 4,500 m asl. See *puna*.

Amaya (Aymara) Popularly seems to refer 'the soul' or 'spirit' of a human being, the equivalent of the Spanish *alma*. Dictionary meanings list the following: (1) Dead (adj.); (2) a corpse; (3) *ch'ullpa* (q.v.); (4) species of medicinal plant (botanical) (Büttner and Condori Cruz 1984: 8; Ayala Loayza 1988: 68; Deza Galindo 1989: 31). In Isluga, people use the term *amaya* to refer to a certain quality of light when the setting sun turns everything yellow, the colour of death.

AMS (archaeological term) Accelerator Mass Spectrometry. To find the amount of radiocarbon in a sample, scientists either measure its radioactivity (the conventional *beta* counting method) or they count the radiocarbon atoms using AMS. This second method can be used on very small samples. In this book, AMS was used for dating yarns from sites TU 67 and TU 54.

Arax pacha (Aymara) The upper world in Aymara cosmology.

Araxsaya (Aymara) The upper moiety in Isluga.

Assemblage (archaeological term) A collection of artefacts found together.

Avestruz (Spanish) Rhea, a small Andean ostrich. In Aymara, it is called *suri*.

Awayu (Aymaraized Spanish) Term for a carrying cloth or shawl. The Quechua equivalent is *llijlla*.

Bofedal (Spanish) Marshy bottomlands consisting of areas of succulent vegetation in the *altiplano* (q.v.). They occur naturally along the banks of rivers and are fed by melted snow water, but they are often irrigated by herders to extend the area of moist vegetation. See *champial*, *uq"u* and *potrero*.

Cacique (Spanish) 'chief' derived from a Carib word which gained currency in the Andes during the colonial period.

Chaku (Quechua) Method of hunting herd animals by driving them into an enclosure of some kind. The Inkas often used this method.

Ch'alla (Aymara) A religious rite, the blessing of objects by sprinkling with alcohol or water (Büttner and Condori Cruz 1984: 40). In Isluga, the *ch'alla* rite is observed on 1 August. The word *ch'alla* is also used to refer to confetti, which is liberally sprinkled over people at festivals and birthday celebrations, as well as white flour, which is sprinkled over people at Carnival.

Champial (Hispanicized Aymara) In Isluga, used as a synonym for *bofedal* (from the Aymara word *ch'ampa*, 'turf', with a Spanish ending).

Chicha (Spanish term derived from the Taíno of the Antilles) Fermented maize drink. In Isluga it is made, accompanied by various ritual observances, for the *wayñu* ceremony (q.v.).

Ch'ullpa (Aymara) (1) Name of a homestead near Enquelga; (2) pre-Inkaic funerary monument (Büttner and Condori Cruz 1984: 32); (3) future time, when the *bofedal* (q.v.) in Enquelga will turn yellow (dried up) and *ch'ullpa* will arrive with a rush of wind.

Coca (Hispanicized term for Aymara *kuka*) Two species of a shrub (Latin terms: *Erythroxylon coca* and *E. novogranatense*), the leaves of which are chewed with a lime-based catalyst to act as a mild stimulant. In Isluga, ash made from a certain type of cactus from the *precordillera* (q.v.) is favoured as the base of the catalyst. *Coca* is chewed in order to relieve hunger and fatigue. It also has important ritual and medicinal uses, and people who know how to 'read' the leaves use them in divination. The use of *coca* by the Quechua and Aymara peoples should not be confused with the processing of *coca* in order to obtain cocaine alkaloid.

Comadre (Spanish) Co-mother. This and *compadre* (q.v.) are ritual kinship terms usually employed between the parents and godparents of a child. In Isluga, such relationships are established at the baptism of a baby, at a child's first haircutting and at marriage. However, the Día Comadre and the Día Compadre referred to in Chapter 4 in the discussion of the *wayñu* ceremony (q.v.) are ritual names for the Thursdays and Saturdays before Carnival (which coincides with Ash Wednesday) on which different families celebrate their individual *wayñu* festival. Thus this usage of the words *comadre/compadre* differs from that of the rite designated Jueves de Compadre on the eve of the second Thursday before Carnival and that called Jueves de Comadre on the eve of the Thursday preceding Carnival, reported for the Aymara of Chucuito by Tschopik (1951: 280–1).

Compadre (Spanish) Literally, co-father or co-parent. See *comadre*.

Complejo cultural (Spanish archaeological term) 'Cultural complex': a group of related industries in an area.

Corrales para sembrar (Spanish) Large corrals with dry-stone walls used for both containing animals and for planting crops (potatoes and *quinua*). After the harvest of the crops, they are left fallow for at least a year, during which they serve as corrals where llamas and alpacas are kept overnight. *Corrales para sembrar* constitute heritable property in Isluga and they are owned by individuals.

Costa (Spanish) Literally, the 'coast'. In Isluga it is used to describe the valleys of the *precordillera* to where the people take their herds during the windy season. The act of going to the *costa*, and the period of time spent there, are called the *costeo*.

Estancia (Spanish) (1) In Isluga, this word refers to one of the main settlements, e.g. Enquelga or Arawilla; (2) in the valleys to the east of the Salar de

Atacama it refers to a homestead with access to land; (3) in Peru, it often implies landed property. An *Empresa Asociativa* is a large estate run by a consortium.

Etic An 'etic' view is an outsider's version, the converse being 'emic', an insider's perspective.

Industry (archaeological term) A set of archaeological artefacts which re-occurs at various sites (Mostny 1988: 37). The proportions of the constituent artefacts are not necessarily the same at each site, due to functional variation.

Islug marka (Aymara) The name given to the ritual centre of the Isluga people. It has the appearance of a small town, with a large church at its centre. However, people only congregate there for festivals. Traditionally, the whole community gathers together for All Souls (1 November), the feast of St Thomas (21 December) and Carnival (a movable festival which begins the Saturday before Ash Wednesday).

Jallu pacha (Aymara) The rainy season, corresponding to summer (November to March) in the South-Central Andes.

Jañachu (Aymara) Male llama, stud animal (Büttner and Condori Cruz 1984: 73).

Jurk"u (Aymara) This is the Isluga pronunciation of *urk"u*, the traditional dress worn by the women.

Juturi (Aymara) In Isluga, a deep hollow considered to come from the inner world, a point of creation, from where llamas, alpacas and sheep may emerge. See Chapter 3.

Kancha (Spanish) A rectangular corral of dry-stone walling for containing llamas and alpacas in Isluga. Smaller corrals, circular or sub-rectangular in shape, for sheep are called *uwij uyu*.

K'isa (Aymara) In Isluga, a coloured stripe in a textile consisting of narrow bands of colour. Nowadays, these usually consist of tonal gradations of one or two different hues. The word *k'isa* means, literally, 'wrinkled fruit' (Büttner and Condori Cruz 1984: 108).

Lomas (Spanish) *Loma* is a 'slope', or it may refer to a low hill or rising ground in the middle of a plain. However, the term *lomas* (vegetation) is often used to refer to the vegetational cover supported by winter fog moisture on the coast in the desert of southern Peru. It is also sometimes used to refer to desert scrub vegetation further inland which grows on the foothills of the western slopes of the Andes (Craig 1985: 28–9).

Mallku (Aymara) Term of respect for a condor or for a (human) *cacique*. In this book it is used to designate a male *uywiri* (q.v.) or *juturi* (q.v.), the sacred sites in the landscape discussed in Chapter 3. It can also have the meaning of 'lord'.

Mamaqullu (Aymara) Female llamas (literally, 'hill females').

Manq"a pacha (Aymara) The inner world (underworld) in Aymara cosmology.

Manq"asaya (Aymara) The lower (literally, inner) moiety in Isluga.

Misa (Aymaraized Spanish) Term used in the Andes for a ritual table, borrowed from the Spanish *mesa*, 'table'.

Miscanti (archaeological term) A stone tool industry defined by Niemeyer and Schiappacasse (1968), based on the use of basalt and lutite (a laminated slate) and characterized by pentagonal points.

Pachamama (Aymara and Quechua) In Aymara, *pacha* means time or earth, and *mama* means woman, mistress, a term of address for all adult women. The Pachamama is the virgin earth/time mother of the Andes. See *Wirjin Tayka.*

Páramo (Spanish) Alpine plain with grassy vegetation, particularly in Ecuador and northern Peru. On the eastern slopes of the Andes the *páramo* is perpetually covered by low clouds.

Pirka (Aymara) Dry stone-walling.

Potrero (Spanish) (1) In Isluga, a synonym for *bofedal* (q.v.); (2) in San Pedro de Atacama, an irrigated and cultivated plot of land.

Precordillera (Spanish) The valleys to the west of the *altiplano* or the Cordillera de los Andes (the Andean chain of mountains), leading to the coastal desert in the north of Chile.

Puna (Quechua) The term for high-altitude, treeless grasslands between approximately 3,500 and 4,500 m asl. See *altiplano.*

Puripica (archaeological term) A lithic industry characterized by the use of flakes rather than blades, and a greater proportion of knives rather than projectile points. The most abundantly used material is basalt, but silicified rock, quartz and volcanic glass were also exploited. Puripica stone tools tend to be much smaller in size than Tulan tools (Núñez and Santoro 1988: 48–52).

Q'ara (Aymara) 'Bare' breed (or variety) of llama. It can be distinguished by the sparse amount of fleece on its neck, particularly under the throat. See *t'awrani.*

Quebrada (Spanish) Valley along which flows an intermittent river or stream.

Quinua (Spanish) Latin: *Chenopodium quinoa*, a grain crop of the Andes.

Quipu (Spanish spelling. In modern Quechua: *khipu* and Aymara: *kipu*) A characteristically Andean form of recording information by tying series of knots into yarns that hang from a main cord. In an early seventeenth-century dictionary, *Qquipu* is listed as a 'knot, or account by knots' and the verb *Qquipuni* 'to count by knots' (Gonzalez Holguín 1952 [1608]: 309). The Spanish noun *cuenta* that Gonzalez Holguín uses for *qquipu* bears the meanings of 'account, count or reckoning' and 'account, narrative'. It is thought that *quipus* recorded numerical and verbal information (see Chapter 8).

Qullu (Aymara) 'Hill'.

Rep or **repp** (English textile term derived from French *reps*) A plain or tabby woven textile in which the compactness of the warp and weft differ: one of these elements is so densely packed that it conceals the other, giving the textile a characteristic ribbed appearance. In a rep, only the weft may be visible; alternatively, only the warp may be seen. Both types occur in Andean textiles (Emery 1966: 77, 86–7; d'Harcourt 1974: 19–20).

Roving (spinning term) The name given to prepared fleece or fibre when it is teased out then pulled into a long length prior to spinning. Synonym: rolag.

Salar (Spanish) Salt-covered basin floor.

Saxsali (Aymara) In Isluga, a llama, alpaca or sheep with abundant and long fleece of a good quality.

Suri (Aymara) 'Rhea', the Andean ostrich (Spanish: *avestruz*). In Quechua, it is used as a term for a breed (or variety) of alpaca with long, straight fleece.

T'alla (Aymara) Title of respect for a woman of high status e.g. the wife of a *cacique* (q.v.). In this book it is used to designate a female *uywiri* (q.v.) or *juturi* (q.v.), the sacred sites in the landscape discussed in Chapter 3. It can also have the meaning of 'lady'.

Tambillo or **Tambilliense** (archaeological term) An industrial complex characterized by different types of projectile point (tetragonal, triangular and leaf shapes) and scrapers, knives and perforators (Mostny 1988; Kaltwasser 1963).

Tataqullu (Aymara) Male llamas (literally, 'hill males').

T'awrani (Aymara) 'Woolly' breed (or variety) of llama. It can be distinguished by the abundant fleece round its neck. See *q'ara*.

T''aya pacha (Aymara) The windy season, which corresponds with winter (April to October) in the South-Central Andes.

Tulan (archaeological term) A lithic industry characterized by Le Paige (1971) as consisting of finely retouched tools, often lanceolate points and knives on blades made from silicified rock. See also Núñez and Santoro (1988: 45–8).

Uq''u (Aymara) Wet bottomlands, known in Spanish as *bofedal* (q.v.).

Urk''u (Aymara) The traditional dress still worn by Aymara women in some parts of the Andes. It consists of a large piece of material made from two loom lengths sewn together. In Isluga, the dress is seamed to make a tube which is held in place at the shoulders by a pair of large pins, and at the waist by several firmly woven belts.

Uywiri (Aymara) 'Creator' or 'herder', the spirit of a hill. See *mallku* and *t'alla*.

Visita (Spanish) An official inspection carried out on behalf of the Spanish Crown during the colonial era.

Wakayu (Aymara and Quechua) In Isluga, an alpaca–llama hybrid animal. Among Quechua speakers, *wakayu* or *wakaya* is used to refer to the most widespread breed (or variety) of alpaca with curly fleece. See *suri*.

Waña (Aymara) The large expanses of dry pasture consisting of shrubs and bunch grasses at high altitude.

Warp selvedge (textile term) The edge of a woven fabric formed by the turns of the warp in a web which is set up with a continuous warp. This type of selvedge forms at the top and bottom of a woven length of fabric, as viewed when the textile is still on the loom. It is typical of fabrics produced by women on the Andean loom, in which the warp ends are not cut. In contrast, the warp ends are cut and passed through heddles on the European-style loom formerly used by the men in Isluga to weave *bayeta* cloth.

Wayñu (Aymara and Quechua) The name by which the ceremony during which llamas and alpacas or sheep are marked and ritually 'dressed' is known in Isluga. See Chapter 4.

Weft selvedge (textile term) The edge of a woven fabric which is formed by the turns of the weft at each side of the fabric, as viewed when the cloth is still on the loom.

Wirjin Tayka (Aymara) Literally, 'Virgin Mother', the name by which the Pachamama (q.v.) is revered in Isluga.

Notes

1 Threads through time

1. The effect of Tyler's article is to undermine the authority of the single fieldworker as author. For a fine example of collaboration as co-authors between fieldworkers (an anthropologist and a linguist) and one who answers to fieldworkers, see Arnold, Jiménez and Yapita (1991).
2. I wish to thank Peter Mitchell for drawing my attention to this work.
3. This and all subsequent translations are my own.
4. Hodder, apparently unaware of Tuan's work, also follows a similar semantic trail (Hodder 1990: 45).
5. Kinahan, in turn, has been criticized by Lita Webley (1997: 175) for imposing assumptions concerning gender relations derived from recent times on to the past.
6. For a study of Chukchi herding see Leeds (1965).
7. The total llama and alpaca populations reported for Chile by Novoa and Wheeler seem to be underestimated, especially the alpaca numbers. Guerrero presents higher figures for llamas (60,000), alpacas (40,000) and vicuñas (8,000).
8. Neville Dyson-Hudson (1972: 16) cites Owen Lattimore's observation that 'pure' pastoralism has been acknowledged to be an unlikely phenomenon since 1896. Lattimore warned against using rigid terms for herders and agriculturalists in Inner Mongolia; he pointed out that some herders owned fields, 'while some of the cultivators acquired herds and perhaps, in the next period of unrest, turned nomad altogether' (Lattimore 1941: 130).
9. Ingold comments that 'the *lack* of investment in, or commitment towards, particular resources or persons' among the Hadza as reported by Woodburn is exceptional in anthropological literature (Ingold 1994: 9, original emphasis). However, Alan Barnard maintains that in general terms foragers value sharing and distrust accumulation (Barnard 1992: 54–5; n.d.).
10. For a similar argument, see Asad (1978: 58).
11. The site of Inca Cueva cave 4 in north-western Argentina has yielded yarns of which the majority are said to be made of camelid fibre and a series of C-14 assays ranging from the eleventh to tenth millennium BP (Aschero 1984: 68; Figini *et al.* 1990: 205).
12. For an alternative view see Gade (1999: Chap. 5). Unfortunately, Gade labours under the misapprehension that nomadic pastoralism did not develop 'as an Andean way of life' (ibid.: 108).
13. I wish to thank Howard Morphy for helping me to clarify my thinking on this important point.

2 Camelids, land and water in the South-Central Andes

1 Another classification describes all four forms as subspecies of *Lama glama*, and varieties of *L.g. glama*, using *forma domestica* for the llama and alpaca (Pires Ferreira 1981/2). Novoa and Wheeler state that this use of *forma domestica* is incorrect (Novoa and Wheeler 1984: 116).
2 Other authors who have used osteometric measurements combined with statistical methods to study camelid faunal samples are Miller (1979), Hesse (1982a), Kent (1982) and Yacobaccio *et al.* (1997/8).
3 See Franklin (1982: 464) for an overview of these schemes.
4 This hypothetical pattern overlooks the present-day trading activities undertaken by Andean women, especially Aymara women, but gender roles might have differed in the past.
5 The term affordance was coined by the psychologist James J. Gibson in reference to what he called the 'offerings of nature', both beneficial and harmful in character (Gibson 1979: 18–19). He explained that 'the affordances of the environment are what it offers the animal', but he stressed that the physical properties of the environment (whether horizontal, flat, extended or rigid) 'have to be measured *relative to the animal*' (ibid.: 127).
6 See discussion in Bender (1998: 22, 35). Marcel Mauss used the term *habitus* to refer to the social nature of learned bodily techniques (Mauss 1979 [1935]: 101).
7 Lautaro Núñez (personal communication) suggested that houses constructed by human groups of the Late Archaic and Early Formative periods in the Atacama have a ground plan that resembles the wallowing grounds of the camelids. This is a very suggestive comparison, but unfortunately it cannot be ascertained whether the resemblance is fortuitous or was intended by the builders themselves.
8 At present, it is not possible to correlate proposed changes in environmental conditions in the Atacama with changes observed in the archaeological record with any degree of accuracy. One possibility is that slightly moister conditions prevailed at the time when herding communities began to make their imprint in the archaeological record of the Atacama. The evidence is discussed in Chapter 7.
9 *Bofedal* is the word usually employed in the literature on the subject. Isluga herders prefer the word *potrero*, or *champial* when speaking Spanish, or *uq"u* in Aymara. In the San Pedro de Atacama area, a *potrero* refers to an irrigated and cultivated plot of land.
10 This was the village where I spent most time when doing my ethnographic fieldwork. In Patricia Provoste's census, it was said to have a population of 150 inhabitants (Provoste 1978: Appendix 2). However, many of the families owning houses there do not live there permanently and it is difficult to be sure of the exact population.
11 According to the people of Isluga themselves, there are approximately twelve main settlements in Isluga, plus a series of hamlets and isolated homesteads. Inevitably, there has been some flux through time in the concentration and dispersal of people at these places of residence, which in some measure responds to herd mobility. For a discussion on the classification of 'village' and 'hamlet' see Ortega Perrier (1998: 78–85).
12 *Irpa* means 'canal', 'guide' or 'leader' in Aymara.

3 Caring for herd animals in Isluga

1 For instance, Don Paulino Castro Challapa of Enquelga told me on 31 October 1989 that you have to give the souls food to eat at this time of the year, but he and his family would not be doing so, as they were of 'another religion'.
2 Don Marcos Castro Challapa made the point clear to me by when he explained that the Christian heaven is in 'this world' (18 January 1997). For a discussion of the conjunction of local and exogenous religious knowledge in the Andes, see Harris (1995).

3 Significantly for a way of life that is so involved with herding, in Isluga it is pasture rather than wild flowers that is emphasized in ceremonies such as Carnival and the marking of the herd animals.

4 It was also a custom of the Inkas, for Cristóbal de Molina, El Cuzqueño, relates that during the December celebrations in Cuzco for Camay Quilla, members of Hanancuzco (the upper moiety) fought members of Hurincuzco (the lower moiety), and that the participants were armed with slings and cactus fruit called *coco* (Molina 1943 [*ca* 1575]: 61).

5 The transformation of people into stone is a recurrent theme in the Andes (see, for example, Santa Cruz Pachacuti 1968 [1613]: 284, 293–4; Taylor 1987: 145, n. 3, 195, n. 12). Alternatively, the transformation of stones into warriors also occurs (Santa Cruz Pachacuti 1968 [1613]: 297). However, in this case, the story appears to have a Biblical guise because of its resemblance to the events in Chapter 19 of Genesis. In the Biblical version, Lot's wife, on looking back at the destruction of Sodom, was turned into a pillar of salt (Genesis 19: 26).

6 Tunupa's character and wanderings have been described in further detail by Gisbert (1980: 35–51) and Rostworowski (1986: 24–30).

7 The main river of Isluga drains into the Salar de Coipasa, feeding into 'the dorsal spine'. Bouysse-Cassagne considers the characteristics of Bartholomew to be more appropriate to the environment of the South-Central Andes, and she discusses his Christian associations with volcanoes (1997: 169–77). However, Isluga people do not talk about St Bartholomew.

8 This is the local spelling in Uyuni.

9 Strictly speaking, Octava is not a 'person' like the other saints. Apparently, it is the octave, or the space of eight days beginning with a festival.

10 This information was provided by Doña Natividad and Don Marcos.

11 These identifications are those of the people of Enquelga. Outsiders may be tempted to comment that Saint Philip ought to be paired with his brother St James the Less, but Enquelga people would not agree.

12 This ceremony was co-ordinated by the *jilaqata* of Araxsaya and his wife. People described the office of *jilaqata*, or *cacique* in Spanish, as being equivalent to that of mayor, but the husband and wife who were elected only held the office for one year. The last couple to serve was in 1983 or 1986.

13 Marietta Ortega Perrier observes, correctly, that there are four 'families' in Isluga, two with the same name of Castro, which correspond to what she calls 'lineage-like' families (1998: 66 and 78).

14 In Huasquiña, a town in the *precordillera*, San Antonio is portrayed in a mural painting with his traditional attributes (a child in the right hand and lilies in the left), but also with a llama. Similarly, the church of Santiago de Río Grande in the Atacama possesses an effigy of this saint who is accompanied by a llama.

15 Yellow is also the colour of death, when the setting sun on occasion suddenly turns the land and air yellow. Isluga people call the phenomenon *amaya* and they rush inside at such times, especially the young and the sick. Among the meanings in Aymara dictionaries for *amaya* are 'dead' and a 'corpse'; it is also a term given to the funerary monuments known as *chullpa* (Büttner and Condori Cruz 1984: 8).

16 This ceremony is no longer observed in Enquelga, and the concept of *pukara* will not be discussed further here. Martínez gives more detail regarding the functions of this place (Martínez 1976: 279–82).

17 Juan de Dios Yapita suggested that the word derives from the Aymara *t"ixsiña*, 'to fry' (Büttner and Condori Cruz 1984: 220), in which case it refers to the fate suffered by the people of the past (personal communication).

18 The Aymara suffix *–sa* means 'our', and the people of Enquelga revere this hill as their own. In Aymara, *ch'api* means 'spiny', and *lliji* 'brilliancy' or 'shining'; *lliju-lliju* is 'lightning'.

19 Trucks are of limited use during the rainy season, when the roads may become impassable. In bad weather, sometimes truck-owning families resort to transporting goods by donkey. The maintenance of vehicles may also prove extremely problematic. For an account of the different fates of truck-owning families in the community of Tiraque, Bolivia, in an area that has seen the development of motorized transport for much longer than Isluga, see Lagos (1994: 45–7).

20 The warden of the CONAF shelter in Enquelga in 1997 was a non-Aymara Chilean. He attended a *wayñu* ceremony on 1 February (see Chapter 4), and some of the men present joked that they were herders of llamas, while he was herder of vicuñas. They assigned to him a somewhat devilish role, and in jest made a subtle comment on the contrast between Aymara and non-Aymara people.

21 I consulted widely with the people of Enquelga regarding camelid nomenclature, but I particularly wish to thank Doña Soria Mamani Challapa for her help with this section.

22 Quechua classifications in the Department of Cuzco in Peru for camelids include bird names, but Flores Ochoa does not, in his article of 1978, enlarge on the metamorphological association between birds and alpacas. Dillon and Abercrombie (1988: 56–7) refer to a conception held by the Aymara people of K'ulta in Bolivia that links water birds and camelids. They maintain that the special significance possessed by the water birds is to be found in their association with 'original time', and with the origin, continuing into the present, of domesticated animals. They claim that the majority of llama and alpaca colour and pattern terms derive from water birds, but they do not provide further details (ibid.: 74, n.5).

4 Flowers of the herds

1 This view differs from other published interpretations. Some authors have argued, somewhat vaguely, that people 'identify with' their animals during the marking ceremony (Quispe 1969: 90–2; Bustinza and Lajo 1985: 141), while Flores Ochoa (1977: 236) states that human beings consider their alpacas as persons with whom they have a relationship.

2 Variations of the ceremony have been reported in the ethnographic literature from many parts of the Andes. Published accounts include Boman (1908; from Susques, northwestern Argentina); Spahni (1962; from the Atacama, Chile); Nachtigall (1964; from the Puna de Moquegua, southern Peru); Quispe (1969; from the Department of Ayacucho, Peru) and Palacios Rios (1981; from Chichillapi, southern Peru). These accounts demonstrate that this elaborate ceremony is performed to ensure the fertility of the herds. Often they are descriptive, or only set the ceremony in its seasonal context in the yearly herding cycle. There are exceptions, however. Quispe (1969) discusses the ceremony in the context of social organization and the role of certain affinal relatives, while Arnold (1988: Chap. 7) presents a detailed kinship analysis of the ritual marking of herd animals in Qaqachaka. Aranguren Paz (1975) and Sallnow (1987: 131–6) consider the ritual context of the ceremony as part of a religious cycle. Zorn (1987) examines animal fertility and the symbolic expression of such fertility in the textiles used during the *señalakuy* in Macusani, southern Peru. Mamani Mamani (1990; 1996) discusses the role of music and the marking ceremony in Guallatire, north of Isluga, in northern Chile. In a deeply textured analysis, Arnold and Yapita (1998: Chapter 3) demonstrate how women sing *wayñu* songs 'from the heart' to their herd animals.

3 The arrangement of trees either side of the exit from which the ritually enhanced llamas and alpacas will emerge can be seen as visually analogous to a seventeenth-century drawing of genealogical trees placed either side of the rectangular Inka cave of origin in the work of Santa Cruz Pachacuti Yamqui Salcamaygua (Dransart 1997a: 92–3). The marking ceremony seems to incorporate ancient, non-verbal imagery from the pre-Hispanic and colonial past.

4 Offerings in Isluga tend to be made in multiples or fractions of twelve. Six was the number of *ch'uwa* dishes used in most of the marking ceremonies that I attended in Enquelga. However, Gabriel Martínez (1976: 325) reported that twelve dishes for the *mallku uywiri* and twelve dishes for the *t'alla uywiri* was the ideal number, although he commented that various permutations were used in different *estancias*, depending on the personality of the human hosts of the ceremony and their perception of the 'fierceness' or perhaps the voracity of the *uywiris* in their locality. Don Marcos told me that office bearers should place twelve candles for each saint in the church in Islug marka. He took sixty candles on one occasion, but some people used 120 candles (personal communication 23 January 1997). It is of interest that the Chipaya ritual table exhibited in the Museo de Etnografía y Folklore (MUSEF) in La Paz has twelve dishes. Isluga people have a great deal of respect for the ritual powers of Chipaya people.

5 Gabriel Martínez (1976: 271) observed that the Aymara verb *amtaña* ('to remember') was listed in Bertonio's seventeenth-century dictionary as *amutaatha* ('to remember, to recollect') and that it shares a root with *amu* meaning 'flower bud' (Bertonio 1984, Bk II: 17). Martínez regarded 'remembering' as a liturgical act that is full of meaning. The converse, the forgetting of or negating the Wirjin Tayka and the spirits of the hills, is enacted by individuals who decide to become evangelical protestants (Dransart 1997a: 94–6).

6 *Chuyma p"aqartatistanta* (Arnold 1997: 108).

7 This is a theme to which I will return later.

8 Plate 5.10 (on p. 124) shows examples of woven *k'isa*.

9 Cereceda (1978: 1,035N) states that in ritual contexts, white is considered to be gendered as masculine and any other colour as feminine.

10 See, for example, Tschopik (1951), Flores Ochoa (1977) and Zorn (1987).

11 It should also be noted that in Juan Chapa's testimony, it was he who cut the ears rather than María Ticlla (Mills 1997: 65). In Qaqachaka, men also assume responsibility for wielding the knife (Arnold and Yapita 1998: 128–9).

12 Martínez (1976: 299) discusses the relationship between 'health', 'luck', 'production' and 'existence in general' in Isluga, drawing attention to large, white, stuffed eagles which hang from the roof inside some houses, and from the claws of which paper money is hung. He comments that the significance of the eagle is seen in its role as patron of luck in the form of money. Flores Ochoa (1977: 218–9) reports the Quechua herders of Paratía, in Peru, as possessing small stones called *enqaychu* which serve as amulets in that their physical presence is considered to bring good luck. They are said to be the protectors of the herds, but they are also said to stand for wealth in a society whose subsistence economy depends on herding. On another level, Flores Ochoa says that these stones are containers of *enqa*, a generative and vital force which brings good luck, increased herd sizes and well-being.

13 Ingold (1980: 224) agrees that for most animal species, paternity is notoriously difficult to establish and is virtually never a significant factor in pastoral property reckoning. However, he has encountered occasional cases among Lapp pastoralists in which orphaned reindeer calves have become attached to older animals, usually male geldings whose owners have subsequently laid claim to the calves by virtue of the 'fictive' bond of maternity. In the llama herds I observed, there were always sufficient adult female animals without offspring to 'adopt' orphaned animals. All too often, the young animal pines and dies if it loses its mother. The rigorous climatic conditions in Isluga make such orphans especially vulnerable.

14 For an alternative view on Inka kinship, and a commentary, see Zuidema (1980) and Arnold (1988: 33–4). Inka genealogies are further considered in Dransart (1997c). The dual and diarchic structures of Inka and other Andean cultures have been examined by Rostworowski (1986).

15 One of the instances cited by Arnold has to do with the roles of *jilanqu* (or *jilaqata*) and *mama t'alla*, the husband and wife who are elected to the office of chief authority in

each *ayllu* for one year. She explains that the *jilanqu* 'rents' or borrows his staff of office (*wara*) during the course of the year, but the *mama t'alla* is considered to be the real 'owner' (Arnold 1988: 156).

5 The transformation of fleece into yarn

1 Although I am in agreement with this association between fleece and water, I have reservations about Zuidema and Urton's further development of their theme, in which the concept of a creational deluge is equated with the amniotic fluids of childbirth (see Dransart 1992a: 147). I find their argument somewhat speculative because they take the Spanish colloquialism *salir de madre* ('to overflow', or 'to do something to excess') that Bertonio offers as the Spanish equivalent of the Aymara *hauimucutha* (as spelled by Bertonio [1984, Bk. II: 125]) to imply childbirth. I am not convinced that the available evidence supports their interpretation.
2 In Aymara: '*Ancha sumalla uka t'awra*'.
3 The techniques used in Isluga are fairly similar to those reported for Quechua-speaking communities in Peru by Miller (1979: 39–50).
4 To wind the roving round the hand in this manner is *p"illuntaña* in Aymara.
5 The Aymara verb 'to spin' is *qapuña*.
6 I understand this to be a relatively neutral place, since to bury or to burn them would constitute a ritual offering to the earth.
7 Denise Arnold argues that women are responsible for weaving the outer garments worn by men and women in Qaqachaka, Bolivia, and therefore that women are responsible for defining both men's and women's respective powers in visual form (Arnold 1997: 129).
8 In the dictionary compiled by Büttner and Condori Cruz (1984: 33), *chuyma* is listed as (1) lung, (2) (figuratively) heart and (3) centre. According to Bertonio (1984 [1612], Bk II: 94) this word, properly speaking, translates as 'lungs', but it is also applied to other internal organs, or an internal state of mind, or the pips of fruit. In Isluga, when a camelid is butchered, the heart and organs including the lungs are lifted out in a unit called the *chuyma*.
9 Elayne Zorn (1987) also emphasizes the generative symbolism in her discussion of the ritual textiles of the Macusani people of Puno, Peru.
10 Exceptions are a type of blanket with a soft weft and plaited slings, where the cradle section is woven in weft-faced plain weave.
11 I heard reference to an incident that occurred during Carnival in Isluga. A young woman eloped with a young man to his *estancia*. It was said that she went 'naked' but, in fact, the only garment she lacked was the outermost one, her shawl.
12 In this, I differ from Abercrombie (1986: 104–5), who declares that spinning is a social practice of which the meaning is 'far from overt or conscious' for the practitioner.

6 Historical perspectives on herding technology

1 It is significant that in February 1997 people commented that Don Apolinario's vicuña-coloured alpaca emerged from the *tama* ('herd').
2 The Inka *chaku* hunt described in Chapter 2 is discussed further later in this chapter. At this point it should be remembered that the Inkas culled certain animals during the hunt.
3 For a more full discussion, see Flores Ochoa (1970; 1977) and Murra (1975c).
4 Guaman Poma translated *zapci llama* into Spanish as *común llama* ('llamas under the collective ownership of communities'), but he reiterated that such animals should have been kept for the benefit of the poor (Guaman Poma 1980 [1615]: 910–12).
5 'de tres a tres años'.
6 Garcilaso (1966 [1604]: 260) specifically states that herds were separated by colour.

7 C-14 dates for the tombs range from AD 570 ± 60 to 910 ± 50, and thermoluminescence dates from AD 510 ± 150 to 920 ± 120 (Llagostera *et al.* 1988: 65).

8 Rebecca Stone-Miller dates the textile to AD 500–800 (1992: fig. 11).

9 See, for example, Menzel's discussion of the Wari epoch 1B female divinity laden with emblems of maize heads (Menzel 1977: 55).

10 Rivera (1980: 88 and plate 6) says that this textile came from site AZ 121. However, Muñoz (1989: 146, fig. 4a) says that it is from AZ 115.

11 Plate 6.7 is of a fragment of a Wari-style tunic in the Royal Museum of Scotland (accession No. 1954 1428). I examined this textile under low-level magnification using a binocular microscope. The black areas use a weft of camelid yarn, which was perhaps dyed black, since some of the fibres are in fact light coloured, suggesting that the hairy fibres resisted the dye.

7 The emergence of herding societies in the Atacama

1 Purifica is the name used by local residents, while archaeologists refer to it as Puripica.

2 *Vega* is the Spanish term for an open plain or meadow.

3 For further information, see Thompson *et al.* (1984), Thompson *et al.* (1988), Thompson *et al.* (1985) and Seltzer (1990).

4 See Fuentes (1984) for a discussion on the perturbation caused by goats to the drought-deciduous plant communities of the Norte Chico, the semi-arid area south of the Atacama and north of Santiago. Since goats have different foraging preferences to those of the native camelids, the vegetation structure may be modified under low goat-browsing pressures; with higher browsing levels the shrub vegetation is soon degraded and becomes dominated by unpalatable cacti and spiny *Acacia* shrubs. The reduction of existing vegetation results in the erosion of topsoil, thus limiting the chances for future recovery. Fuentes points out that low population levels of goats seem capable of shifting the equilibrium level in plant communities, but that the critical level beyond which the vegetation and soils will not recover to their original state has yet to be determined (ibid.: 47).

5 In this area they are known as *estancia*, which does not imply a village or hamlet as it does in Isluga.

6 The theoretical basis on which the archaeological work reported in Núñez (1988) rests is based on what he calls *movilidad transhumántica* (transhumantic mobility) (Núñez 1980).

7 For a similar observation regarding Susques in north-western Argentina, see Yacobaccio *et al.* (1997/8: 428; 1998: 7–8).

8 The site descriptions presented here are based on my fieldwork notes for the excavations undertaken at sites TA 1 (Tambillo 1), TU 67, TU 52, PU 1, TU 85, TU 54, TU 57, TU 64 and TU 56. Descriptions of other sites discussed here are based on the unpublished reports compiled by Le Paige (MS) and Núñez (1988), and the specialist reports by Popper (1977), Hesse (1982a; 1982b; 1984) and Holden (1990) were also consulted. Figures 7.3 to 7.5 show the location of most of these sites.

9 An earlier version of this account of TU 67 was published in Dransart (1997).

10 This is called *Toba Tulan* or *Toba desvitrificada* in the Spanish literature on the site.

11 Harvest profiles are known as such since they assume that the animals' death was caused by human activity.

12 In a more recent article, Berenguer argues that the Taira panel, like the camelids from Puripica and Kalina, dates from the archaic period (1995: 22). However, it should be noted that the Tulan 60 and Taira camelids were drawn in perspective, with all four legs depicted, rather than in profile with only two legs as at Puripica and Kalina. The Tulan and Taira camelids are less schematic, and their feet are carefully drawn.

13 Black polished pottery is present at TU 57 (1980 ± 80 BP); elsewhere reported dates for contexts where black polished ware is found are 1700 ± 150 BP (Sa 226) for tomb

2532 in the Quitor 6 cemetery, and 1750 ± 80 BP (I-1205) for tomb 3397 in Quitor 5 (Le Paige 1963: 25). (Quitor is one of the oases of San Pedro de Atacama, see figure 6.4 on p. 143). It has also been found in a midden in the site of Las Cuevas in northwest Argentina at 1965 ± 30 BP (GrN-5399) (Cigliano 1970: 104). In general, fine polished black pottery is said to be associated with funerary contexts (Tarragó 1976: 61), and so its early appearance in a house at Tulan is of interest.

14 The four thermoluminescence dates for Chiu Chiu 200 are 2870 ± 300, 2950 ± 300, 2850 ± 260 and 2950 ± 310 BP (Benavente 1982: 82).

8 The yarns and fabrics of Tulan societies

1 An 'assemblage' is the term applied to a collection of artefacts found together at an archaeological site.

2 Three C-14 dates are available for La Capilla: 3670 ± 160 BP (GAK 8778); 3450 ± 90 BP (I 11,642); and 2790 ± 140 BP (GAK 8777), indicating an Archaic period occupation.

3 Two C-14 determinations are available for this cemetery: 1660 ± 90 BP (GAK 2206) and 1590 ± 170 BP (GAK 2893) (Núñez 1976a: 93–4).

4 Elsewhere, narrow woven bands have been reported from Camarones 15, with a C-14 date of 3060 ± 100 BP (GAK 5813) (Rivera *et al.* 1974: 100; Núñez 1976a: 122), and in the Azapa Valley at AZ 71 (Santoro 1981: 41), tomb 79 of which has a C-14 determination of 2940 ± 150 BP (GAK 7404) (cited by Rivera 1985: 53).

5 Father Cobo (1653), the chronicler of *Historia del Nuevo Mundo*, mentions the use of the root of *chapichapi* for dyeing fleece red; *chapichapi* has been identified as *Relbunium microphyllum* in the Cuzco area (Fester and Lexow 1973: 234). The species listed for Chile, *R. hypocarpium*, is said to occur generally between Aconcagua to Llanquihue (Fester and Lexow 1973: 236), which is much further south than the Atacama. *Relbunium* is listed in two dyers' manuals, one from Peru and the other from Bolivia. Zumbühl (1979: 61) lists *antanco* (*R. celiatum*), 'a little plant which grows close to the ground', and *chapi* (*Galium*), a shrub about 30 cm high, which is said to be similar to *antanco*. The sample of sheep wool included in Zumbühl's manual, dyed with the root of *Galium*, is described as 'reddish brown'. Cajías and Fernández (1987: 27) list *chapichapi* (*R. microphyllum*), the root of which is said to produce 'vivid red'. *Ch'api* is the Aymara word for 'spine', which suggests a thorny plant, but the authors do not mention spines as being characteristic of these plants. In an ethnobotanical study carried out in Toconce, northern Chile, *R. croceum* is mentioned as having the local name of *cantoria*, and it is said to be used as a spice and in herbal infusions (Castro *et al.* 1982: 190).

6 C-14 dates for the preceramic at La Galgada range between 3590 ± 75 BP (Uga-4583) and 4110 ± 50 BP (TX-3664) (Grieder *et al.* 1988: 69).

7 A rectangular-shaped bag from the 'black refuse' deposit at Punta Pichalo also has a series of longer yarns between the final row of looping and the cord of the edging, but this bag is of a loop-and-twist structure (Bird 1946: figure 26d).

8 Catalogue Nos 180, 275, 296, 317, 91.489, and 1959.15.1 in King (1965: Appendix A).

9 It should be noted that Pollard (1971: 45) calls the same site RAnL 88. Four samples dated by thermoluminescence suggest that it is contemporary with TU 54 (Benavente 1982: 82).

10 M. Antonia Benavente kindly gave me permission to examine fabrics from Chiu Chiu 200 during a period of political unrest against the Pinochet dictatorship in January 1988, and I was not able to study the entire sample during my visit to Santiago de Chile.

11 Neither fabric structure nor the construction of the constituent yarns are reported in sufficient detail in the literature consulted on sites from the north of Chile to enable a rigorous set of correlations to be made, and there are at least some exceptions to the

pattern proposed here. Dauelsberg (1974) reports the use of camelid fibre in a loop-and-twist bag (No. 6231) from the coastal site of Quiani 7. The use of camelid fibre at this site is worthy of further investigation, as it has been combined with cotton in the same yarns that form the warp in at least one of the twined fabrics.

12 The theme of patterns of association between fibre and structure and between structure and function was explored further in Dransart (1992b). Unfortunately, the article was re-typed immediately prior to publication with neither my authorization nor my knowledge. Many errors were introduced into the text. One of the most grievous errors committed by the typist was to change the standardized S and Z notation for spin directions to a and b.

13 Bird (Bird *et al.* 1985: 106) discusses the lack of recognizable spindles and whorls at Huaca Prieta, Peru. In the nearby town of Catavio, he observed a woman spinning cotton without a drop spindle. She had a mass of prepared cotton, loosely tied to the end of an unworked stick approximately two feet long, which she held as a distaff between the left arm and her body. With her left hand, she drew out and drafted the fibres, while she twisted the fibres by turning an unworked shoot of a shrub with her right hand. Bird says there was no whorl, and that at no time did the spindle hang free or rotate by itself. The spun yarn was wrapped round the shaft. Although Bird says that the yarn produced in this manner resembled the weight and quality of the preceramic yarns of Huaca Prieta, there can be no certainty that this was the method employed in the production of the preceramic yarns. However, he says that its use would leave no recognizable traces in the archaeological record, and that similar pieces of the woody shrub used by the woman to twist the fibre were present in the preceramic deposits of Huaca Prieta (ibid.: 107).

14 For an example in bone from north-western Argentina, see Dransart (1995: 235).

15 Given the probability of disturbed layers in the rock shelter and cave sites (TU 67 and TU 55), one has to recognize that some of the yarns may be of a more recent date.

16 In late pre-Hispanic textiles, weft twining is used only at each end of a cloth as an edging, not to make a complete textile.

17 For an account of *quipus* in colonial times see (Loza 1998).

18 I wish to thank Denise Arnold for discussing her work with me in advance of publication, as well as discussing some of the other texts cited in this coda.

19 The quotation of the phrase 'webs of significance' is taken from Geertz (1973: 5). Rosaleen Howard-Malverde (1997: 7–10) has considered some of the problems involved in the culture-as-text approach that Geertz's work inspired, especially concerning the imposition of inappropriate paradigms in particular case studies. She argues that it is important to 'distinguish the perspective of the analyst from that of the cultural agent', who is the producer of culture, pointing out that Andean peoples are 'prolific producers of text (be it spoken, written, woven, inscribed)'. Howard-Malverde also argues for a concept of text as context, commenting that 'texts and textualising practices are active agents in the shaping of experience and, effectively, the creation of context' (ibid.: 7)

Conclusions: earth, water, fleece and fabric

1 M. Antonia Benavente suggested that llamas were used as beasts of burden about 800 BC at site Chiu Chiu 200 (Benavente 1981). However, it is not clear what evidence is available to support such a suggestion.

2 The Wankarani-sculptured camelid heads in Bolivia may be contemporary with TU 54 (Guerra Gutiérrez 1994; 1995).

3 Carolina Agüero Piwonka (1993: 74) calculated that a yarn wrapping the outer part of a turban from the site of Camarones 15, Arica, was 3 m long. In most cases, it is difficult to assess the length of the yarns used in these headdresses without dismantling them.

4 As an instance, a warp-faced textile in a tunic from tomb 21 at the Coyo Oriente cemetery has a count of 40 to 44 warp and 12 to 13 weft threads per 10 mm, in the collections of the Museo R.P. Gustavo Le Paige, S.J., San Pedro de Atacama.

5 Gilmore (1950: plate 43, bottom) reproduced a photograph by courtesy of the New York Zoological Society of an alpaca with fleece reaching to the ground.

6 María del Carmen Reigadas has begun a programme of research attempting the notoriously difficult task of distinguishing between camelid species in archaeological samples of fleece (Reigadas 1992). There are enormous problems in interpreting the findings of such work.

7 Gayton observed that two-toned yarns were used with any frequency only in Paracas embroideries in order to extend the colour range of the embroidery threads rather than to enhance the variety of yarn characteristics (1967: 287).

8 'Que el huso no agarre la mano y no deje así hilar.'

9 Dransart (1993) has discussed ceramic spindle whorls incised with zigzag motifs from pre-Hispanic workbaskets from the Central Andes.

10 Members of the Pentecostal sect do not ritually sprinkle blood on the ground. They are learning to understand the sacrifice of the herd animal in terms of the sacrifice of Christ, and they speak of 'the blood of Christ' when the animal's lifeblood drains from the jugular.

Bibliography

Abercrombie, T.A. (1986), 'The politics of sacrifice: an Aymara cosmology in action', unpublished doctoral thesis, Chicago, Illinois: University of Chicago.

Adovasio, J.M. and Lynch, T.F. (1973), 'Preceramic textiles and cordage from Guitarrero Cave, Peru', *American Antiquity* 38, 84–90.

Adovasio, J.M., and Maslowski, R.F. (1980), 'Cordage, basketry and textiles', in Lynch, T.F., ed., 253–90.

Agüero Piwonka, C. (1993), 'Análisis técnicos', in *Identidad y prestigio en los Andes. Gorros, turbantes y diademas*, Museo Chileno de Arte Precolombino, 74–5, Santiago: Museo Chileno de Arte Precolombino.

Agüero Piwonka, C. (1995), 'Indicadores textiles de grupos formativos: proposición de una tipología de turbantes', *Hombre y Desierto. Una Perspectiva Cultural* 9, 97–114.

Aguerre, A.M., Fernandez Distel, A.A. and Aschero, C.A. (1973), 'Hallazgo de un sitio acerámico en la Quebrada de Inca Cueva (Provincia de Jujuy)', *Relaciones* (N.S.) VII, 197–231.

Aguerre, A.M., Fernandez Distel, A.A. and Aschero, C.A. (1975), 'Comentarios sobre nuevas fechas en la cronología arqueológica precerámica de la Provincia de Jujuy', *Relaciones* (N.S.) IX, 211–14.

Aguilar Meza, T. (1988), 'Las trampas para cazar camélidos', in *Llamichos y paqocheros. Pastores de llamas y alpacas*, Flores Ochoa, J., ed., 59–65, Cuzco: Consejo Nacional de Ciencia y Tecnología.

Albeck, M.E., ed. (1994), *Taller. De costa a selva. Producción e intercambio entre los pueblos agroalfareros de los Andes Centro Sur* (6–11 April 1992), Buenos Aires: Instituto Interdisciplinario Tilcara, Facultad de Filosofía y Letras, Universidad de Buenos Aires.

Aldenderfer, M. (1990), 'Cronología y definición de fases arcaicas en Asana, Sur del Perú', *Chungará* 24/5, 13–35.

Aldunate del S., C., Berenguer R., J. and Castro R., V., eds (1985), *Estudios en arte rupestre*, Santiago: Museo Chileno de Arte Precolombino.

Altieri, R.A. (1941), 'Sobre 11 kipus peruanos', *Notas del Instituto de Antropología* II(1), Universidad Nacional de Tucumán.

Ampuero Brito, G. (1986), *Diaguitas. Pueblos del Norte Verde*, Santiago: Museo Chileno de Arte Precolombino.

Anders, M. (1988), progress report in *Willay* 28, 5–7.

Anonymous (1987), 'Peligro de desecación de la Vega de Turi', *Creces* 8 (1/2), 10–11.

Appleyard, H.M. (1978), *Guide to the Identification of Animal Fibres*, Leeds: WIRA.

Aranguren Paz, A. (1975), 'Las creencias y ritos mágicos religiosos de los pastores puneños', *Allpanchis Phuturinqa* 8, 103–32.

Arnold, D.Y. (1988), 'Matrilineal practice in a patrilineal setting: rituals and metaphors of kinship in an Andean ayllu', unpublished doctoral thesis, London: University of London.

Arnold, D.Y. (1997), 'Making men in her own image: gender, text and textile in Qaqachaka', in Howard-Malverde, R., ed., 99–131.

Arnold, D.Y., Jiménez, D. and Yapita, J. de D. (1991), 'Scattering the seeds: shared thoughts on some songs to the food crops', *Amerindia* 16, 105–78.

Arnold, D.Y. and Yapita, J. de D. (1998), *Río de vellón, río de canto. Cantar a los animales, una poética andina de la creación*, La Paz: ILCA and Hisbol.

Arnold, D.Y. Yapita, J. de D. *et al.* (2000), *El rincón de las cabezas: luchas textuales, educación y tierras en los Andes*, La Paz: Universidad Mayor San Andrés, Facultad de Humanidades y Ciencias de la Educación and ILCA, Colección Academia 9.

Asad, T. (1978), 'Equality in nomadic social systems?', *Critique of Anthropology* 11, 57–65.

Asad, T. (1986), 'The concept of cultural translation in British Social Anthropology', in Clifford, J., and Marcus, G.E., eds, 141–64.

Ascher, M. (1991), *Ethnomathematics: A Multicultural View of Mathematical Ideas*, New York: Chapman and Hall.

Ascher, M. and Ascher, R. (1981), 'El quipu como lenguaje visible', in Lechtman, H. and Soldi, A.M., eds, 407–32.

Aschero, C. A. (1984), 'El sitio ICC-4: un asentamiento precerámico en la Quebrada de Inca Cueva (Jujuy, Argentina)', *Estudios Atacameños* 7, 62–72.

Astorga, R. (1986), 'Secamos el altiplano', *Revista del Domingo*, 9 February 1986, 8–11, Santiago.

Avila Felipe, M.F. (1979), 'Relación peso vivo/peso vellón en alpacas huacaya y suri del Centro de Producción La Raya', unpublished undergraduate dissertation, Puno: Universidad Nacional Técnica del Altiplano.

Ayala Loayza, J.L. (1988), *Diccionario Español–Aymara Aymara–Español*, Lima: Editorial Juan Mejía Baca.

Barnard, A. (1992), *Hunters and Herders of Southern Africa*, Cambridge: Cambridge University Press.

Barnard, A. (n.d.), 'The foraging mode of thought', paper presented at the Eighth International Conference on Hunting and Gathering Societies, Osaka, October 1998.

Barnes, R. and Eicher, J.B. (1992), 'Introduction', in Barnes, R. and Eicher, J.B., eds, 1–7.

Barnes, R. and Eicher, J.B. (eds) (1992), *Dress and Gender: Making and Meaning in Cultural Contexts*, New York and Oxford: Berg.

Barros Valenzuela, R. (1963), 'Anotaciones sobre los Lamidos en Chile', *Revista Universitaria* XLVIII (=Anales de La Academia Chilena de Ciencias Naturales 26), 57–67.

Barth, F. (1961), *Nomads of South Persia*, Oslo: Universitetsforlaget.

Baumann, M.P., ed. (1996), *Cosmología y música en los Andes*, Frankfurt am Main and Madrid: Vervuert Verlag and Iberoamericana.

Benavente Aninat, M.A. (1981), 'Chiu-Chiu 200: un campamento de pastores', unpublished undergraduate dissertation, Santiago: Universidad de Chile.

Benavente Aninat, M.A. (1982), 'Chiu-Chiu 200. Una comunidad pastora temprana en la Provincia del Loa (II Región)', *Actas del IX Congreso Nacional de Arqueología*, 75–94, La Serena.

Benavente Aninat, M.A. (1985), 'Reflexiones en torno al proceso de domesticación de camélidos en los valles del centro y sur de Chile', *Boletín del Museo Regional Araucanía (Temuco)* 2, 37–52.

Bender, B. (1998), *Stonehenge: Making Space*, Oxford: Berg.

Bennett, W.C. (1936), 'Excavations in Bolivia', *Anthropological Papers of the American Museum of Natural History* 35, 331–505.

Bennett, W.C. and Bird, J.B. (1949), *Andean Culture History*, New York: American Museum of Natural History, Handbook Series 15.

Berenguer R., J. (1995), 'El arte rupestre de Taira dentro de los problemas de la arqueología atacameña', *Chungara* 27(1), 7–43.

Berenguer R., J., Castro R., V., Aldunate del S., C., Sinclaire A., C. and Cornejo B., L. (1985), 'Secuencia del arte rupestre en el alto Loa: una hipótesis de trabajo', in Aldunate del S., Berenguer, J., and Castro, V., eds, 87–108.

Berenguer R., J. and Martinez C., J.L. (1986), 'El río Loa, el arte rupestre de Taira y el mito de Yakana', *Boletín del Museo Chileno de Arte Precolombino* 1, 79–99.

Bernhardson, W. (1985), 'Tierra, trabajo y ganadería indígena en la economía regional de Arica', *Chungara* 15, 151–67.

Bertonio, L. (1984 [1612]), *Vocabulario de la lengua Aymara*, La Paz: CERES, IFEA and MUSEF.

Betanzos, J. de (1968 [1551]), 'Suma y narración de los Incas', in *Biblioteca de Autores Españoles* 209, Esteve Barba, F., ed., 1–56, Madrid: Ediciones Atlas.

Bird, J.B. (1946), 'Excavations in Northern Chile', *Anthropological Papers of the American Museum of Natural History New York* 38(4), 171–316.

Bird, J.B. (1954), *Paracas Fabrics and Nazca Needlework, 3rd century BC–3rd century AD*, Washington, D.C.: The Textile Museum (catalogue raisonné with textile analysis by Bellinger, L.),

Bird, J.B., Hyslop, J., and Dimitrijevic Skinner, M. (1985), 'The preceramic excavations at the Huaca Prieta Chicama Valley, Peru', *Anthropological Papers of the American Museum of Natural History New York* 62 (1),

Blanco Velasquez, M.R. (1980), 'Peso vivo, peso vellón y rendimiento de vellón de alpacas en la C.A.P. Huaycho Ltda No 44', unpublished undergraduate dissertation, Puno: Universidad Nacional Técnica del Altiplano.

Boisset, G., Llagostera, A. and Salas, E. (1969), 'Excavaciones arqueológicas en Caleta Abtao, Antofagasta', *Actas del V Congreso Nacional de Arqueología*, 75–112, La Serena.

Bökönyi, S. (1969), 'Archaeological problems and methods of recognizing animal domestication', in Ucko, P.J. and Dimbleby, G., eds, 219–29, London: Duckworth.

Bökönyi, S. (1989), 'Definitions of animal domestication', in Clutton-Brock, J., ed., 22–7.

Bolton, R. and Mayer, E., eds (1977), *Andean Kinship and Marriage*, Washington, D.C.: American Anthropological Association, Special Publication 7.

Boman, E. (1908), *Antiquités de la région andine de la République Argentine et du Desert d'Atacama*, Paris: Imprimerie Nationale.

Bourdieu, P. (1977), *Outline of a Theory of Practice*, Cambridge: Cambridge University Press.

Bouysse-Cassagne, T., ed. (1997), *Saberes y memorias en los Andes. In memoriam Thierry Saignes*, Paris and Lima: Institut des Hautes Etudes de l'Amérique Latine and Institut Français d'Etudes Andines.

Bouysse-Cassagne, T. (1997), 'De Empédocles a Tunupa: evangelización, hagiografía y mitos', in Bouysse-Cassagne, T., ed., 157–212.

Bowman, I. (1924), *Desert Trails of the Atacama*, New York: American Geographical Society, Special Publication 5.

Brothwell, D.R. (1975), 'Salvaging the term "domestication" for certain types of man–animal relationship: the possible value of an eight-point scoring system', *Journal of Archaeological Science* 2, 397–400.

Browman, D.L. (1974), 'Pastoral nomadism in the Andes', *Current Anthropology* 15, 188–96.

Bruhns, K. O. (1988–9), 'Prehispanic weaving and spinning implements from southern Ecuador', *The Textile Museum Journal*, 70–7.

Bustinza Choque, V. (1985), *Razas de alpacas del altiplano: suri y wacaya*, Puno: Universidad Nacional del Altiplano.

Bustinza Choque, V. and Avila Cazorla, E., eds (1985), *Proyecto: piel de alpaca. Informe final*, Puno: Universidad Nacional del Altiplano.

Bustinza, V. and Lajo, V.H. (1985), 'Aspectos socio-culturales de comunidades alpaqueras', in Bustinza Choque, V. and Avila Cazorla, E., eds, 130–42.

Bustinza Choque, V. and Sapana Valdivia, R. (1985), 'Cualidades de la fibra de alpaca en comunidades campesinas', in Bustinza Choque, V. and Avila Cazorla, E., eds, 99–114.

Büttner, T. and Condori Cruz, D. (1984), *Diccionario Aymara–Castellano. Arunakan liwru Aymara–Kastillanu*, Puno: Proyecto Experimental de Educación Bilingüe.

Cajías, M. and Fernández, B. (1987), *Manual de tintes naturales*, La Paz: SEMTA (=Manuales Técnicos 12),

Campbell, L.G. (1985), 'Women and the great rebellion in Peru, 1780–1783', *The Americas* 42(2), 163–96.

Cañete y Dominguez, P.V. (1974 [1791]), 'Documentos Números 1 y 2', *Norte Grande* 1(2), 233–51.

Carcelén, X. (1992), 'The good shepherdess', in *America Bride of the Sun. 500 years Latin America and the Low Countries*, Royal Museum of Fine Arts, Antwerp, p. 469, Gent: Ministry of the Flemish Community Administration of External Relations and Imschoot.

Cardozo, A. (1975), *Orígen y filogenía de los camélidos sudamericanos*, La Paz: Academia Nacional de Ciencias de Bolivia.

Cardozo, A. and Martínez, Z. (1981), 'Producción anual de fibra y peso vivo de llamas a diferentes edades', *Estación experimental de Ulla Ulla. Informe de investigaciones agropecuarias*, La Paz: Instituto Nacional de Fomento Lanero (=Serie: Estudios Especializados EE 31), 89–95.

Caro, D. (1985), *Those Who Divide Us: Resistance and Change among Pastoral Ayllus in Ulla Ulla, Bolivia*, Ann Arbor, Michigan: University Microfilms International.

Castillo G., G. (1989), 'Agricultores y pescadores del Norte Chico: el complejo Las Animas', in Hidalgo *et al.*, eds, 265–76.

Castro, M., Villagran, C. and Arroyo, M. (1982), 'Estudio etnobotánico en la Precordillera y Altiplano de los Andes del Norte de Chile (18–19° S)', in *Hombre y los Ecosystemas, Vol. II: Biología humana y aspectos de la antropología social en el transecto Arica – Lago Chungará*, Man and the Biosphere 6, 133–99.

Cereceda, V. (1978), 'Sémiologie des tissus andins: les *talegas* d'Isluga', *Annales: Economies, Sociétés et Civilisations* 33(5–6), 1017–35.

Cereceda, V. (1987), 'Aproximaciones a una estética andina: de la belleza al *tinku*', in *Tres reflexiones sobre el pensamiento andino*, Medina, J., ed., 133–231, La Paz: Hisbol (= Biblioteca Andina 1),

Cigliano, E.M. (1970), 'Problemas referentes al sitio arqueológico de Las Cuevas, Departamento de Rosario de Lerma, Provincia de Salta', *Relaciones* (N.S.) 5(1), 99–104.

Clapperton, C.M. (1983), 'The glaciation of the Andes', *Quaternary Science Reviews* 2, 83–155.

Clapperton, C.M. (1990), 'Quaternary glaciations in the southern hemisphere: an overview', *Quaternary Science Reviews* 9, 299–304.

Clapperton, C.M. and Sugden, D.E. (1988), 'Holocene glacier fluctuations in South America and Antarctica', *Quaternary Science Reviews* 7, 185–98.

Clifford, J. and Marcus, G.E., eds (1986), *Writing Culture: The Poetics and Politics of Ethnography*, Berkeley, Los Angeles, and London: University of California Press.

Clutton-Brock, J., ed. (1989), *The Walking Larder: Patterns of Domestication, Pastoralism, and Predation*, London: Unwin Hyman.

Clutton-Brock, J. and Grigson, C., eds (1984), *Animals and Archaeology: 3 Early Herders and their Flocks*, Oxford: BAR International Series 202.

Cobo, B. (1956 [1653]), 'Obras II', in *Biblioteca de Autores Españoles* 92, Mateos, F., ed., 6–275, Madrid: Atlas.

CONAF (1982), *Evaluación y catastro de recursos del Parque Nacional Volcán Isluga. Informe final*, Santiago: Corporación Nacional Forestal, Ministerio de Agricultura, República de Chile.

Conklin, W.J. (1985), 'Pucara and Tiahuanaco tapestry: time and style in a sierra weaving tradition', *Ñawpa Pacha* 21, 1–43.

Conklin, W.J. (1990), 'El sistema informativo de los kipus del horizonte medio', in Mackey, C., *et al.*, 21–38.

Conklin, W. and Moseley, M.E. (1988), 'The patterns of art and power in the Early Intermediate Period', in Keatinge, R.W., ed., 145–63.

Cook, A. G. (1992), 'The stone ancestors: idioms of imperial attire and rank among Huari figurines', *Latin American Antiquity* 3(4), 341–64.

Costa Junqueira, M.A. (1988), 'Reconstitución física y cultural de la población tardía del cementerio de Quitor-6 (San Pedro de Atacama)', *Estudios Atacameños* 9, 99–126.

Craig, A.K. (1985), 'Cis-Andean environmental transects: late Quaternary ecology of northern and southern Peru', in Masuda *et al.*, 23–44.

Crapanzano, V. (1985), *Waiting: The Whites of South Africa*, New York: Random House.

Custred, G. (1979), 'Hunting technologies in Andean culture', *Journal de la Société des Américanistes* 66, 7–19.

Dahl, G. and Hjort, A. (1976), *Having Herds: Pastoral Herd Growth and Household Economy*, Stockholm: University of Stockholm, Stockholm Studies in Social Anthropology 2.

Dauelsberg, P. (1974), 'Excavaciones arqueológicas en Quiani', *Chungará* 4, 7–38.

Dedenbach-Salazar Sáenz, S. (1990), *Inka pachaq llamanpa willaynin. Uso y crianza de los camélidos en la epoca incaica*, Bonn: Bonner Amerikanistische Studien 16.

Desrosiers, S. (1997), 'Lógicas textiles y lógicas culturales en los Andes', in Bouysse-Cassagne, T., ed., 325–49.

Deza Galindo, J.F. (1989), *Diccionario Aymara–Castellano Castellano–Aymara*, Lima: Consejo Nacional de Ciencias y Tecnología.

Diez de San Miguel, G. (1964 [1567]), *Visita hecha a la provincia de Chucuito*, Espinoza Soriano, W., ed., Lima: Ediciones de la Casa de Cultura.

Dillehay, T.D., Netherly, P.J. and Rosen, J. (1989), 'Middle preceramic public and residential sites on the forested slope of the western Andes, northern Peru', *American Antiquity* 54(4), 733–59.

Dillon, M. and Abercrombie, T. (1988), 'The destroying Christ: an Aymara myth of the conquest', in *Rethinking History and Myth*, Hill, J.D., ed., 50–77, Urbana and Chicago: University of Illinois Press.

Donkin, R.A. (1977), 'Spanish red: an ethnographic study of cochineal and opuntia cactus', *Transactions of the American Philosophical Society* 67(5).

Donnan, C.B. (1978), *Moche Art of Peru: Pre-Columbian Symbolic Communication*, Los Angeles, California: Museum of Cultural History, University of California.

Doyon-Bernard, S.J. (1990), 'From twining to triple cloth: experimentation and innovation in ancient Peruvian weaving (*ca* 5000–400 BC)', *American Antiquity* 55(1), 68–87.

Dransart, P. (1988), 'Continuidad y cambio en la producción textil tradicional aymara', *Hombre y Desierto: una Perspectiva Cultural* 2, 41–57.

Dransart, P. (1991a), 'Llamas, herders and the exploitation of raw materials in the Atacama Desert', *World Archaeology* 22(3), 304–19.

Dransart, P. (1991b), 'Fibre to fabric: the role of fibre in camelid economies in Prehispanic and Contemporary Chile', unpublished doctoral thesis, University of Oxford and Linacre College.

Dransart, P. (1992a), 'Pachamama: the Inka earth mother of the long sweeping dress', in Barnes, R. and Eicher, B., eds, 145–63.

Dransart, P. (1992b), 'Fabrics of the Atacama: a study of fibre and structure', in *Contributions to New World Archaeology*, Saunders, N., ed., 256–71, Oxford: Oxbow Books.

Dransart, P. (1992c), 'Form and function in the basketry of the Central and South Central Andes: an overview from prehispanic times to the present', in *Basketmakers: Meaning and Form in Native American Baskets*, Mowat, L., Morphy H. and Dransart, P., eds, 130–45, Oxford: Pitt Rivers Museum.

Dransart, P. (1993), 'Pre-Columbian work baskets', *Journal of Museum Ethnography* 4, 123–42.

Dransart, P., ed. (1995a), *Andean Art: Visual Expression and its Relation to Andean Beliefs and Values*, Aldershot, Hampshire: Avebury (Worldwide Archaeology Series).

Dransart, P. (1995b), 'Inner worlds and the event of a thread in Isluga, northern Chile', in Dransart, P., ed., 228–42.

Dransart, P. (1996), 'Las flores de los rebaños en Isluga: la vida cultural de los ganaderos y camélidos en el norte de Chile', *Nuevo Texto Crítico* 9(18), 29–39.

Dransart, P. (1997a), 'Cultural transpositions: writing about rites in the llama corral', in Howard-Malverde, R., ed., 85–98.

Dransart, P. (1997b), 'Rockshelters and ritual activities in the Atacama Desert of Northern Chile', in *The Human Use of Caves*, Bonsall, C. and Tolan-Smith, C., eds, 207–16, Oxford: BAR International Series 667.

Dransart, P. (1997c), 'Afinidad, descendencia, y la política de las representaciones de género: ¿quién fue la *Quya* de Ataw Wallpa?', in *Más allá del silencio: las fronteras de género en los Andes*, Arnold, D., ed., 475–90, La Paz: Ediciones ILCA, and St Andrews: Centre for Indigenous American Studies and Exchange, University of St Andrews.

Dransart, P. (1998), 'A short history of rosaries in the Andes,' in *Beads and Bead Makers: Gender, Material Culture and Meaning*, Sciama, L.D., ed., 129–46, Oxford: Berg.

Dransart, P. (2000), 'Cultural tourism in an interconnected world: tensions and aspirations in Latin America', in *Indigenous People in an Interconnected World*, Smith, C. and Ward, G. eds, 145–64; 199, St Leonards, Australia: Unwin Hyman.

Dransart, P. (2002), 'Concepts of spiritual nourishment in the Andes and Europe: a study of rosaries in cross-cultural contexts', *Journal of the Royal Anthropological Institute* 8(1).

Druss, M. (1978), *Environment, Subsistence Economy, and Settlement Patterns of the ChiuChiu Complex (ca 2700 to 1600 BC) of the Atacama Desert, Northern Chile*, Ann Arbor: U.M.I.

Ducos, P. (1978), '"Domestication" defined and methodological approaches to its recogniton in faunal assemblages', in *Approaches to faunal analysis in the Middle East*, Meadow, R.H. and Zeder, M.A., eds, 53–6, Harvard Museum: Peabody Museum Bulletin 2.

Ducos, P. (1989), 'Defining domestication: a clarification', in Clutton-Brock, J., ed., 28–30.

Duga, L. (1985), 'Características más importantes de las fibras de los camélidos', in *Estado actual de las investigaciones sobre camélidos en la República argentina*, Cajal, J.L. and Amaya, J.N., eds, 129–32, Buenos Aires: Ministerio de Educación y Justicia, Secretaria de Ciencia y Técnica, Programa Nacional de Recursos Naturales Renovables.

Dwyer, E.B. (1979), 'Early Horizon tapestry from south coastal Peru', in *The Junius B. Bird Pre-Columbian Textile Conference*, Rowe, A.P., Benson, E. and Schaffer, A.-L., eds, 61–75, Washington D.C.: The Textile Museum and Dumbarton Oaks.

Dyson-Hudson, N. (1972), 'The study of nomads', in *Perspectives on Nomadism*, Irons, W. and Dyson-Hudson, N., eds, Leiden: E.J. Brill, 2–29.

Emery, I. (1966), *The Primary Structure of Fabrics*, Washington, D.C.: The Textile Museum.

Engel, F. (1963), 'A preceramic settlement on the Central Coast of Peru: Asia, Unit 1', *Transactions of the American Philosophical Society* 53.

Fallet, M. (1961), 'Vergleichende Untersuchungen zur Wollbildung südamerikanischer Tylopoden', *Zeitschrift für Tierzüchtung und Züchtungsbiologie* 75(1), 34–56.

Fernandez Baca, S. (1971), *La alpaca. Reproducción y crianza*, Lima: IVITA Boletín de Divulgación 7.

Fernandez Distel, A.A. (1980), 'Elementos indicadores de pautas de vida migratoria en etnias del borde de la puna jujeña (Argentina)', *Actas del V Congreso Nacional de Arqueología Argentina* 1, 99–101, San Juan.

Fester, G.A. and Lexow, S.G. (1973), 'Las raices del genero "Rebulnium" [sic] en la tintorería americana', *Anales de la Sociedad Científica Argentina* 136, 233–270.

Figini, A.J., Carbonari, J.E. and Huarte, R.A. (1990), 'Museo de la Plata radiocarbon measurements II', *Radiocarbon* 32(2), 197–208.

Flannery, K.V., Marcus, J. and Reynolds, R.G. (1989), *The Flocks of the Wamani: A Study of Llama Herders on the Punas of Ayacucho, Peru*, San Diego, California: Academic Press, Inc.

Flores Ochoa, J.A. (1968), *Los pastores de Paratía. Una introducción a su estudio*, Mexico: Instituto Indigenista Interamericana (=Serie Antropología Social [10]),

Flores Ochoa, J.A. (1970), 'Notas sobre rebaños en la visita de Gutiérrez Flores', *Historia y Cultura* 4, 163–96, Lima: Museo Nacional de Historia.

Flores Ochoa, J.A., ed. (1977), *Pastores de puna. Uywamichiq punarunakuna*, Lima: Instituto de Estudios Peruanos.

Flores Ochoa, J. (1978), 'Classification et denomination des camélides sud-americains', *Annales: Economies, Sociétés et Civilisations* 33 (5–6), 1006–16.

Flores Ochoa, J. (1979a), *Pastoralists of the Andes. The Alpaca Herders of Paratía*, trans. Bolton, R., Philadelphia: Institute for the Study of Human Issues.

Flores Ochoa, J. (1979b), 'Distorciones en el uso del ecosistema de la Puna alta y los programas de cooperación técnica', *Antropología Andina* 3, 14–20.

Flores Ochoa, J. (1980), 'Causas que originaron la actual distribución de las alpacas y llamas', in *El hombre y su ambiente en los Andes Centrales*, Millones, L. and Tomoeda, H., eds, 63–92, Osaka: Senri Ethnological Studies 10.

Flores Ochoa, J. (1983), 'Pastores de llamas y alpacas en los Andes – balance bibliográfico', *Revista Andina* 1 (1), 175–218.

Focacci, G. and Erices, S. (1972–3), 'Excavación en túmulos de San Miguel de Azapa (Arica, Chile)', *Actas del VI Congreso de Arqueología Chilena 1971*, 47–62, Santiago (=*Boletín de Prehistoria*, special number),

Forde, C.D. (1934), *Habitat, Economy and Society*, London: Methuen.

Fortich Baca, V.I. (1980), 'Cabellos humanos de enterratorios prehistóricos de la Quebrada de Humahuaca', *Actas del V Congreso Nacional de Arqueología Argentina* 1, 129–36, San Juan.

Frame, M. (1986), 'The visual images of fabric structures in ancient Peruvian art', in Rowe, A.P., ed., 47–80.

Frame, M. (1990), *Andean Four-Cornered Hats: Ancient Volumes*, New York: The Metropolitan Museum of Art.

Franklin, W.L. (1975), 'Guanacos in Peru', *Oryx* 13, 191–202.

Franklin, W.L. (1982), 'Biology, ecology and relationship to man of the South American camelids', in *Mammalian Biology in South America*, Mares, M.A., and Genoways, H., eds, Pittsburgh: Pymatuning Laboratory of Ecology, Special Publication Series 6.

Franquemont, E.M. (1986), 'Threads of time: Andean cloth and costume', in *Costume as Communication*, Schevill, M.B., ed., Bristol RI: Haffenreffer Museum of Anthropology and Seattle: University of Washington Press.

Fuentes, E.R. (1984), 'Human impacts and ecosystem resilience in the southern Andes', *Mountain Research and Development* 4(1), 45–9.

Fuentes, J. (1965), *Tejidos prehispánicos de Chile. Colección Max Uhle, Museo Histórico Nacional*, Santiago: Editorial Andrés Bello.

Gade, D.W. (1977), 'Llama, alpaca y vicuña: ficción y realidad', in Flores Ochoa, J.A., ed., 113–20.

Gade, D.W. (1999), *Nature and Culture in the Andes*, Madison: University of Wisconsin Press.

Garcilaso de La Vega, El Inga (1966 [1609, 1616]), *Royal commentaries of the Incas and general history of Peru*, trans. by Livermore, H.V., Austin: University of Texas Press.

Gavilán, V. (1985), *Mujer aymara y producción textil. El altiplano de Tarapacá*, Santiago: Centro de Estudios de la Mujer.

Gayton, A.H. (1967), 'The cultural significance of Peruvian textiles: production, function and aesthetics', in Rowe, J.H. and Menzel, D., eds, 275–92.

Geertz, C. (1973), *The Interpretation of Cultures: Selected Essays by Clifford Geertz*, New York: Basic Books.

Geyh, M.A. (1967), 'Hannover radiocarbon measurements IV', *Radiocarbon* 9, 198–217.

Gibbon, L. Grassic (1978 [1946]), *A Scots Quair*, London: Hutchinson and Co Ltd.

Gibson, J.J. (1979), *The Ecological Approach to Visual Perception*, Boston: Houghton Mifflin Company.

Gilmore, R.M. (1950), 'Fauna and ethnozoology of South America', in *Handbook of South American Indians*, Steward, J.H., ed., Washington D.C.: Smithsonian Institution, Bureau of American Ethnology Bulletin 143, Vol. 6, 354–464.

Gisbert, T. (1980), *Iconografía y mitos indígenas en el arte*, La Paz: Editorial Gisbert.

Gomez Para, D.H. (1975), 'Toconce. Estudio de una comunidad andina', unpublished undergraduate dissertation, Antofagasta: Departamento de Ciencias Sociales, Universidad del Norte.

González Holguín, D. (1952 [1608]), *Vocabulario de la lengua general del todo el Perú llamada lengua qquichua o del Inca*, Lima: Universidad Nacional Mayor de San Marcos.

Goodell, G. (1968), 'A study of Andean spinning in the Cuzco region', *Textile Museum Journal* 2(3), 2–8.

Goody, J. (1976), *Production and Reproduction: a Comparative Study of the Domestic Domain*, Cambridge: University Press.

Gose, P. (1994), *Deathly Waters and Hungry Mountains: Agrarian Ritual and Class Formation in an Andean town*, Toronto: University of Toronto Press.

Graffam, G. (1992), 'Beyond state collapse: rural history, raised fields, and pastoralism in the South Andes', *American Anthropologist* 94(4), 882–904.

Grau, J. (1984), 'Chinchilla', in *Evolution of Domesticated Animals*, Mason, I.L., ed., 260–3, London: Longman.

Grebe Vicuña, M.E. (1983), 'En torno a los ritos terapéuticos astrales de Isluga', *Chungará* 10, 155–64.

Grebe Vicuña, M.E. (1984), 'Etnozoología andina : concepciones e interacciones del hombre andino con la fauna altiplánica', *Estudios Atacameños* 7, 455–72.

Grieder, T. (1986), 'Preceramic and Initial Period textiles from La Galgada, Peru', in Rowe, A.P. ed., 19–29.

Grieder, T., Bueno Mendoza, A., Smith Jr, C.E., and Malina, R.M. (1988), *La Galagada, Peru: a Preceramic Culture in Transition*, Austin: University of Texas Press.

Guaman Poma de Ayala, F. (1980 [1615]), *El primer nueva corónica y buen gobierno*, Murra, J.V. and Adorno, R., eds, Mexico City: Siglo Veintiuno.

Guerra G., L. (1994), *El arte en la prehistoria orureña*, Oruro, Bolivia: Centro Diocesano de Pastoral Social, Serie 'Nosotros' 1.

Guerra G., L. (1995), *Preambulo a la arqueología de Wankarani*, Oruro, Bolivia: Centro Diocesano de Pastoral Social, Serie 'Nosotros' 2.

Guerrero Lara, R. (1986), 'Los camélidos sudamericanos y su significado para el hombre de la puna', *Diálogo Andino* 5, 7–89.

Gutiérrez Flores, P. (1970 [1572]), 'Resultado de la visita secreta lega que hizieron en la provincia de Chucuito . . .', *Historia y Cultura* 4, 5–48, Lima: Museo Nacional de Historia.

Harcourt, R. d' (1974), *Textiles of Ancient Peru and their Techniques*, Denny, G.G. and Osborne, C.M., eds, Seattle: University of Washington Press (paperback edition),

Hardman-de-Bautista, M. (1988), 'Andean ethnography: the role of language structure in observer bias', *Semiotica* 71(3/4), 339–72.

Harris, O. (1983), 'Los muertos y los diablos entre los Laymi de Bolivia', *Chungará* 11, 135–52.

Harris, O. (1987), 'De la fin du monde. Notes depuis le Nord-Potosi', *Cahiers des Amériques Latines* 6, 93–117.

Harris, O. (1995), 'Knowing the past: plural identities and the antinomies of loss in Highland Bolivia', in *Counterworks: Managing the Diversity of Knowledge*, Fardon, R. ed., London: Routledge, 105–23.

Hastorf, C. (1993), *Agriculture and the Onset of Political Inequality Before the Inka*, Cambridge: University Press.

Hedges, R.E.M., Housley, R.A., Law, I.A., and Bronk, C.R. (1989), 'Radiocarbon dates from the Oxford AMS system: *Archaeometry* datelist 9', *Archaeometry* 31(2), 207–34.

Herre, W. (1963), 'The science and history of domestic animals', in *Science in Archaeology*, Brothwell, D. and Higgs, E., eds, 235–49, Bristol: Thames and Hudson.

Hesse, B. (1982a), 'Animal domestication and oscillating climates', *Journal of Ethnobiology* 2(1), 1–15.

Hesse, B. (1982b), 'Archaeological evidence for camelid exploitation in the Chilean Andes', *Saugetierkandeliche Mitteilungen* 30(3), 201–11.

Hesse, B. (1984), 'Archaic exploitation of small mammals and birds in northern Chile', *Estudios Atacameños* 7, 42–61.

Hesse, B. and Wapnish, P. (1985), *Animal Bone Archaeology*, Washington: Taraxacum (Manuals on Archaeology 5).

Heusser, C.J. (1974), 'Vegetation and climate of the Southern Chilean Lake District during and since the Last Interglaciation', *Quaternary Research* 4, 290–315.

Heusser, C.J. (1981), 'Palynology of the Last Interglacial-Glacial Cycle in midlatitudes of Southern Chile', *Quaternary Research* 16, 293–321.

Heusser, C.J. and Streeter, S.S. (1980), 'A temperature and precipitation record of the past 16,000 years in southern Chile', *Science* 210, 1,345–7.

Hidalgo L., J., Schiappacasse F., V., Niemeyer F., H., Aldunate del S., C. and Solimano R., I. eds (1989), *Culturas de Chile. Prehistoria desde sus origenes hasta los albores de la conquista,* Santiago: Editorial Andrés Bello.

Hilari Mendizabel, H.D. (1985), 'Densidad folicular y diámetro de fibra de alpacas Huacaya de la Comunidad de Umasuyo Bajo-Ayaviri', unpublished undergraduate dissertation, Puno: Universidad Nacional del Altiplano.

Hodder, I. (1990), *The Domestication of Europe: Structure and Contingency in Neolithic Societies*, Oxford: Blackwell.

Holden, T.G. (1990), 'Taphonomic and methodological problems in reconstructing diet from ancient human gut and faecal remains', unpublished doctoral thesis: Institute of Archaeology, University of London.

Holden, T.G. (1991), 'Evidence of prehistoric diet from northern Chile: coprolites, gut contents and flotation samples from the Tulán Quebrada', *World Archaeology* 22(3), 320–31.

Holy, L., ed. (1987), *Comparative Anthropology*, Oxford: Basil Blackwell.

Howard-Malverde, R. (1997), 'Introduction. Between text and context in the evocation of cultures', in Howard-Malverde, R., ed., 3–18.

Howard-Malverde, R., ed. (1997), *Creating Context in Andean Cultures*, New York, Oxford: Oxford University Press (Oxford Studies in Anthropological Linguistics).

Ingold, T. (1984), 'Time, social relationships and the exploitation of animals: anthropological reflections on prehistory', in Clutton-Brock, J. and Grigson, C., eds, 3–12.

Ingold, T. (1986), *Evolution and Social Life*, Cambridge: Cambridge University Press.

Ingold, T. (1980/1988), *Hunters, Pastoralists and Ranchers*, Cambridge: Cambridge University Press.

Ingold, T., ed. (1988), *What is an Animal?*, London: Unwin Hyman.

Ingold, T. (1994), 'From trust to domination: an alternative history of human–animal relations', in *Animals and Human Society: Changing Perspectives*, Manning, A. and Serpell, J., eds, 1–22. London: Routledge.

Jennings, J.D., ed. (1983), *Ancient South Americans*, San Francisco: W.H. Freeman and Co.

Kaltwasser P., J. (1963), 'Descripción de artefactos líticos de Tambillo', *Anales de la Universidad del Norte* 2, 135–45.

Keatinge, R.W., ed. (1988), *Peruvian Prehistory*, Cambridge: Cambridge University Press.

Kent, J. (1982), *The Domestication and Exploitation of the South American Camelids: Methods of Analysis and their Application to Circum-Lacustrine Archaeological Sites in Bolivia and Peru*, Ann Arbor, Michigan: U.M.I.

Kessel, J.J.M.M. Van (1980), *Holocausto al progreso: los aymara de Tarapacá*, Amsterdam: Cedla Incidentele Publication 16.

Kessel B., J. van (1988), 'Tecnología aymara: una perspectiva cultural', *Hombre y Desierto: una Perspectiva Cultural* 2, 58–88.

Kinahan, J. (1991), *Pastoral Nomads of the Central Namib Desert: The People History Forgot*, Windhoek: Namibia Archaeological Trust and New Namibia Books.

King, M.E. (1965), *Textiles and Basketry of the Paracas Period, Ica Valley, Peru*, Ann Arbor, Michigan: U.M.I.

King, M.E. (1978), 'Analytical methods and prehistoric textiles', *American Antiquity* 43(1), 89–96.

Klein, C. (1995), 'Wild Woman in Colonial Mexico: an encounter of European and Aztec concepts of the other', in *Reframing the Renaissance: Visual Culture in Europe and Latin America 1450–1650*, Farago, C. ed., 244–63, New Haven and London: Yale University Press.

Koford, C.B. (1957), 'The vicuña and the puna', *Ecological Monographs* 27(2), 153–219, Durham, N.C.: Duke University Press.

Kolata, A.L. (1983), 'The South Andes', in Jennings, J.D., ed., 241–85.

Koster, H.A. and Chang, C. (1994), 'Introduction', in *Pastoralists at the Periphery: Herders in a Capitalist World*, Chang, C. and Koster, H.A., eds, 1–21, Tucson: University of Arizona Press.

Kuznar, L.A. (1990), 'Pastoralismo temprano en la sierra alta del Departamento de Moquegua, Perú', *Chungará* 24/25, 54–68.

Lagos, M.L. (1994), *Autonomy and Power: The Dynamics of Class and Culture in Rural Bolivia*, Philadelphia: University of Pennsylvania Press.

Lange Topic, T., McGreevy, T.H. and Topic, J.R. (1987), 'A comment on the breeding and herding of llamas and alpacas on the north coast of Peru', *American Antiquity* 52(4), 832–5.

Lanino Rozas, I. (1977), *Antecedentes de las explotaciones ganaderas en Isluga, altiplano de la Provincia de Iquique*, Iquique: Universidad del Norte, Sede Iquique, Proyecto de Diagnóstico del Altiplano Convenio SERPLAC.

Latcham, R.E. (1938), *Arqueología de la región atacameña*, Santiago: Prensas de la Universidad de Chile.

La Tour, G.D. de (1954), 'The guanaco', *Oryx* 2(5), 273–9.

Lattimore, O. (1941), *Mongol Journeys*, London: Jonathan Cape.

Laurencich Minelli, L. (1996), *La scrittura dell'antico Perù : un mondo da scoprire*, Bologna: CLUEB.

Laurencich Minelli, L. (2000), 'Los textiles peruanos del documento "Exsul immeritus Blas Valera populo suo": una propuesta de lectura', in Solanilla Demestre, V., ed., *Actas de la I Jornada Internacional sobre Textiles Precolombinos*, Barcelona: Universitat Autónoma de Barcelona, Department d'Art.

Lavallée, D., Julien, M., Wheeler, J. and Karlin, C. (sous la direction de D. Lavallée) (1985), *Telarmachay. Chasseurs et pasteurs préhistoriques des Andes*, Paris: Institut Français des Etudies Andines, Editions Recherches sur les Civilisations, Synthèse (20).

Lechtman, H. (1981), 'Introducción', in Lechtman, H. and Soldi, A.M., eds, 11–22.

Lechtman, H. and Soldi, A.M., eds (1981), *La tecnología en el mundo andino. Runakunap kawsayninkupaq rurasqankunaqa*, 1, Mexico City: Instituto de Investigaciones Antropológicas, Universidad Nacional Autónoma de México (=Serie Antropológica 36),

Leeds, A. (1965), 'Reindeer herding and Chukchi social institutions', in *Man, Culture and Animals*, Leeds, A., and Vayda, A.P., eds, 87–128, Washington, D.C.: American Association for the Advancement of Science (78).

Lemonnier, P. (1986), 'The study of material culture today: toward an anthropology of technical systems', *Journal of Anthropological Archaeology* 5, 147–86.

León, J.A. (1939), 'Algunos consideraciones sobre los camélidos de los Andes', *Boletín del Museo de Historia Natural "Javier Prado"* 11, 95–105.

Le Paige, G. (1963), 'Continuidad o discontinuidad de la Cultura Atacameña', *Anales de la Universidad del Norte* 2, 7–25.

Le Paige, G. (1965), 'San Pedro de Atacama y su zona (14 temas)', *Anales de la Universidad del Norte* 4.

Le Paige, G. (1971), *Industrias líticas de San Pedro de Atacama*, Santiago: Editorial Orbe.

Le Paige, G. (MS), manuscripts on file, housed in the library of the Instituto de Investigaciones Arqueológicas R.P. Gustavo Le Paige S.J., San Pedro de Atacama.

Lévi-Strauss, C. (1970), *The Raw and the Cooked: Introduction to a Science of Mythology: 1*, London: Jonathan Cape Ltd.

Lindberg, I. (1967), 'Técnicas en tejidos del área andina de la Provincia de Antofagasta', *Revista de la Universidad del Norte* 2, 1–16.

Llagostera Martínez, A. (1979), '9,700 years of maritime subsistence on the Pacific: an analysis by means of bioindicators in the north of Chile', *American Antiquity* 44(2), 309–24.

Llagostera M., A., Baron P., A.M. and Bravo V., L. (1984), 'Investigaciones arqueológicas en Tulor-1', *Estudios Atacameños* 7, 133–51.

Llagostera Martínez, A. and Costa Junqueira, M.A. (1984), *Museo Arqueológico R.P. Gustavo Le Paige S.J.*, Santiago: Departamento de Extensión Cultural del Ministerio de Educación (Serie Patrimonio Cultural Chileno, Colección Museos Chilenos),

Llagostera M., A., Torres, C.M. and Costa J., M.A. (1988), 'El complejo psicotrópico en Solcor-3 (San Pedro de Atacama)', *Estudios Atacameños* 9, 61–98.

Locke, L.L. (1923), *The Ancient Quipu, or Peruvian Knot Record*, New York: The American Museum of Natural History.

Lounsbury, F.G. (1978), 'Aspects du système de parenté inca', *Annales: Economies, Sociétés et Civilisations* 33(5–6), 991–1005.

Lownie, H. (1978), 'Research problems regarding the domestication of South American mammals', in *Research Problems in Zooarchaeology*, Brothwell, D.R., Thomas, K.D. and Clutton-Brock, J., eds, 113–20, London: Institute of Archaeology Occasional Publication 3.

Loza, C.B. (1998), 'Du bon usage del quipus face à l'administration coloniale espagnole (1500–1600)', *Population* 1–2, 139–60.

Lumbreras, L.G. (1974), *The Peoples and Cultures of Ancient Peru*, Washington, D.C.: Smithsonian.

Lumbreras, L.G. (1981), *Arqueología de la América andina*, Lima: Editorial Milla Batres.

Lumbreras, L.G. (1990), *Visión arqueológica del Perú milenario*, Lima: Editorial Milla Batres.

Lynch, T.F. (1975a), 'Algunos problemas básicos del estadio de caza-recolección andina: trashumancia', *Estudios Atacameños* 3, 7–10.

Lynch, T.F. (1975b), 'La cosecha inoportuna, trashumancia y el proceso de domesticación', *Estudios Atacameños* 3, 75–80.

Lynch, T.F., ed. (1980), *Guitarrero Cave: Early Man in the Andes*, New York: Academic Press.

McCracken, R.D. (1971), 'Lactase deficiency: an example of dietary evolution', *Current Anthropology* 12, 479–517.

MacKenzie, M.A. (1991), *Androgynous Objects: String Bags and Gender in Central New Guinea*, Reading: Harwood Academic Publishers.

Mackey, C., Peyra, H., Radicati, C., Rodriguez, H. and Valverde, H. (1990), *Quipu y yupana. Colección de escritos*, Lima: CONCYTEC.

Macías de Arias, A. (n.d.), 'La textilería y artesanías de cuero. Técnicas de teñido indígena', paper presented at the *Segundo Encuentro Latinoamericano de Artesanía Tradicional*, Universidad de Antofagasta, 19–22 January, 1988.

Mamani Mamani, M. (1985), *Ensayo de topónimos más comunes de la Primera Región de Tarapacá*, Arica: Universidad de Tarapacá.

Mamani Mamani, M. (1990), 'Myth and music in the livestock marking ritual of the Chilean Andes', *Latinamericanist* 25(2), 1–7.

Mamani Mamani, M. (1996), 'El simbolismo, la reproducción y la música en el ritual. Marca y floreo de ganado en el altiplano chileno', in Baumann, M.P., ed., 221–45.

Markgraf, V. (1989), 'Palaeoclimates in Central and South America since 18,000 BP based on pollen and lake-level records', *Quaternary Science Reviews* 8, 1–24.

Martínez, G. (1975), 'Características de orden antropológico y socio-económico de la comunidad de Isluga (I Región)', *Norte Grande* 1(3–4), 403–26.

Martínez, G. (1976), 'El sistema de los uywiris en Isluga', *Anales de la Universidad del Norte (Homenaje al Dr Gustavo Le Paige S.J.)* 10, 255–327.

Martínez, G. (1980), 'Paisaje y pensamiento para una semántica del espacio Aymara', unpublished dissertation, Lima: Pontificia Universidad Católica del Perú.

Martínez, G. (1989), 'Estructuras binarias y ternarias en pueblo Isluga', in *Espacio y pensamiento. 1: Andes Meridionales*, 109–48, La Paz: Hisbol (=Biblioteca Andina 6, Medina, J., ed.),

Masuda, S., Shimada, I. and Morris, C., eds (1985), *Andean Ecology and Civilization*, Tokyo: University Press.

Mauss, M. (1979 [1935]), 'Body Techniques', in *Sociology and Psychology: Essays*, trans. Brewster, B. 95–123, London: Routledge and Kegan Paul.

Mayer, E. and Bolton, R., eds (1980), *Parentesco y matrimonio en los Andes*, Lima: Pontificia Universidad Católica del Perú.

Meighan, C.W. (1980), 'Archaeology of Guatacondo, Chile', in *Prehistoric Trails of Atacama: Archaeology of Northern Chile*, Meighan, C.W. and True, D.L., eds, 99–126, Los Angeles: Institute of Archaeology, University of California.

Meigs, P. (1966), 'The Atacama Desert', *Geography of Coastal Deserts: Arid Zone Research Series* 28, 110–13, Liège: UNESCO.

Meisch, L.A. (1986a), 'Spinning in Bolivia', *Spin-Off*, 10(1), 25–9.

Meisch, L.A. (1986b), 'Weaving styles in Tarabuco, Bolivia', in Rowe, A.P. ed., 243–74.

Mengoni Goñalons, G.L. (1997/8), 'Comentario', *Relaciones* 22/23, 423–6.

Menzel (1964), 'Style and time in the Middle Horizon', *Ñawpa Pacha* 2, 1–105.

Menzel, D. (1977), *The Archaeology of Ancient Peru and the Work of Max Uhle*, Berkeley: R.H. Lowie Museum of Anthropology, University of California.

Merlino, R.J. and Rabey, M.A. (1978), 'El ciclo agrario-ritual en la Puna argentina', *Relaciones* (N.S.) 12, 47–70.

Merlino, R.J. and Rabey, M.A. (1985), 'Ecología cultural de la Puna argentina II: la estructura de los ecosistemas' *Actas de la IV Convención Internacional sobre Camélidos Sudamericanos*, 219–68, Punta Arenas: La Universidad de Magallanes.

Michieli, C.T. (1979), 'Relaciones entre las técnicas textiles de la Región de Cuyo, Norte de Chile y Costa peruana', *Actas: Jornadas de Arqueología del Noroeste Argentino* (=*Antiquitas* 2), 1–14, Buenos Aires.

Miller, G.R. (1979), *An Introduction to the Ethnoarchaeology of the Andean Camelids*, Ann Arbor, Michigan: U.M.I.

Miller, G.R. and Burger, R.L. (1995), 'Our father the cayman, our dinner the llama: animal utilization at Chavín de Huántar, Peru', *American Antiquity* 60(3), 421–48.

Miller, G.R. and Gill, A.L. (1990), 'Zooarchaeology at Pirincay, a Formative Period site in highland Ecuador', *Journal of Field Archaeology* 17(1), 49–68.

Mills, K. (1997), *Idolatry and its Enemies: Colonial Andean Religion and Extirpation, 1640–1750*, Princeton, University of Princeton Press.
Mimica, J. (1992), *Intimations of Infinity: The Cultural Meanings of the Iqwaye Counting and Number System*, Oxford: Berg.
Molina, C. de ('El Cuzqueño') (1943 [*ca* 1575]), *Fábulas y ritos de los Incas*, Lima: Los Pequeños Grandes Libros de Historia Americana, Series I, IV.
Molina, C. de (1947 [*ca* 1574/5]), *Ritos y fábulas de los Incas*, Buenos Aires: Editorial Futuro.
Monast, J.E. (1972), *Los indios aimaraes. ¿Evangelizados o solamente bautizados?*, Buenos Aires: Cuadernos Latinoamericanos Ediciones Carlos Lohlé.
Morgan, L.H. (1877), *Ancient Society*, New York: World Publishing.
Morris, C. (1988), 'Progress and prospect in the archaeology of the Inca', in Keatinge, R.W., ed., 233–56.
Morris, C. and Von Hagen, A. (1993), *The Inka Empire and its Andean Origins*, New York: American Museum of Natural History and Abbeville Press.
Morúa, M. de (1946 [1590]), *Los orígenes de los Inkas*, Porras Barrenechea, ed., Lima: Los Pequeños Grandes Libros de Historia Americana, Series I, XI.
Mostny, G. (1988), *Prehistoria de Chile*, Santiago de Chile: Editorial Universitaria (eighth edition).
Mostny, G., Jeldes, F., Gonzalez, R. and Oberhauser, F. (1954), *Peine, un pueblo atacameño*, Santiago: Instituto de Geografía, Universidad de Chile, Publication 4.
Mujica, E. (1985), 'Altiplano–coast relationships in the South-Central Andes: from indirect to direct complementarity', in Masuda *et al.*, eds, 103–40.
Muñoz Ovalle, I. (1982), 'Las sociedades costeras en el litoral de Arica durante el periodo arcaico tardío y sus vinculaciones con la costa peruana', *Chungará* 9, 124–51.
Muñoz Ovalle, I. (1989), 'El período formativo en el Norte Grande (1000 a.C. a 500 d.C)', in Hidalgo L., J. *et al.*, eds, 107–28.
Muñoz Ovalle, I., and Chacama Rodriguez, J. (1982), 'Investigaciones arqueológicas en las poblaciones precerámicas de la costa de Arica', in *Documentos de Trabajo No 2*, Muñoz O. I., and Córdova G. J., eds, 3–93, Arica: Universidad de Tarapacá.
Murra, J.V. (1975a), 'Rebaños y pastores en la economía del Tawantinsuyu', in *Formaciones económicas y políticas del mundo andino*, 117–144, Lima: Instituto de Estudios Peruanos.
Murra, J.V. (1975b), 'La función del tejido en varios contextos sociales y políticos', in *Formaciones económicas y políticas del mundo andino*, 145–170, Lima: Instituto de Estudios Peruanos.
Murra, J.V. (1975c), 'Un reino ayamara en 1567', in *Formaciones económicas y políticas del mundo andino*, 193–223, Lima: Instituto de Estudios Peruanos.
Murra, J.V. (1975d), 'Las etno-categorías de un *khipu* estatal', in *Formaciones económicas y políticas del mundo andino*, 242–54, Lima: Instituto de Estudios Peruanos.
Murra, J.V.,and Morris, C. (1976), 'Dynastic oral tradition, administrative records and archaeology in the Andes', *World Archaeology* 7(3), 268–79.
Nachtigall, H. (1966), 'Ofrendas de llamas en la vida ceremonial de los pastores de la Puna de Moquegua (Perú) y de la Puna de Atacama (Argentina), y consideraciones histórico-culturales sobre la ganadería indígena', *Actas y Memorias del XXXVI Congreso Internacional de Americanistas* 3, 193–8, Seville.
Niemeyer, H., and Schiappacasse, V. (1963), 'Investigaciones arqueológicas en las terrazas de Conanoxa, Valle de Camarones (Provincia de Tarapacá)', *Revista Universitaria* Year 48 (*Anales de La Academia Chilena de Ciencias Naturales* 26), 101–66.
Niemeyer F., H. and Schiappacasse F., V. (1968), 'Tres industrias líticas de la Puna de

Atacama, Parte I. El yacimiento Miscanti Sur', *Revista Universitaria* Year 53 (=*Anales de La Academia Chilena de Ciencias Naturales* 31), 133–44.

Niemeyer, H. and Schiappacasse, V. (1976), 'Los yacimientos arqueológicos de la Laguna Meniques', *Anales de la Universidad del Norte (Homenaje al Dr Gustavo Le Paige, S.J.)* 10, 31–57.

Niemeyer, H. and Schiappacasse, V. (1977), 'Investigación de un sitio temprano de cazadores-recolectores arcaicos en la desembocadura del valle de Camarones', *Actas del VII Congreso de Arqueología de Chile* 1, 115–18, Altos de Vilches.

Niemeyer, H., Schiappacasse, V. and Solimano, I. (1972–3), 'Padrones de poblamiento en la Quebrada de Camarones (Prov. de Tarapacá) (Estudio preliminar que comprende el sector medio y superior del valle)', *Actas del VI Congreso de Arqueología Chilena 1971*, 115–37, Santiago (=*Boletín de Prehistoria*, special number),

Novoa, C. and Wheeler, J.C. (1984), 'Lama and alpaca', in *Evolution of Domesticated Animals*, Mason, I.L., ed., 116–28, London: Longman.

Núñez A., L. (1975), 'Dinámica de grupos precerámicos en el perfil costa-altiplano, Norte de Chile', *Estudios Atacameños* 3, 59–74.

Núñez A., L. (1976a), 'Registro de fechas radiocarbónicas del Norte de Chile', *Estudios Atacameños* 4, 74–123.

Núñez Atencio, L. (1976b), 'Informe de actividades del Programa Paleoindio', *Estudios Atacameños* 4, 145–9.

Núñez A., L. (1980), 'Hipótesis de movilidad transhumántica en la Puna de Atacama: Quebrada de Tulán (nota preliminar)', *Actas del V Congreso de Arquelogía Argentina* 1, 19–56, San Juan.

Núñez, L. (1981), 'Asentamiento de cazadores-recolectores tardíos de la Puna de Atacama: hacia el sedentarismo', *Chungará* 8, 137–68.

Núñez A., L. (1982), 'Temprana emergencia de sedentarismo en el desierto chileno: Proyecto Caserones', *Chungará* 9, 80–122.

Núñez, L. (1983), 'Paleoindian and Archaic cultural periods in the arid and semiarid regions of northern Chile', in *Advances in World Archaeology* 2, Wendorf, F. and Close, A.E., eds, 161–203.

Núñez Atencio, L. (1988), 'Análisis multidisciplinario de domesticación y crianza inicial de camélidos del Norte de Chile', unpublished report for Fondecyt Project 1,017–86.

Núñez A., L. (1989), 'Hacia la producción de alimentos y la vida sedentaria (5.000 a.C. a 900 d.C.)', in Hidalgo L.J. *et al.*, eds, 81–105.

Núñez, L. (1994), 'Emergencia de complejidad y arquitectura jerarquizada en la Puna de Atacama: las evidencias del sitio Tulan-54', in Albeck, M.E., ed., 85–108.

Núñez A., L. and Dillehay, T.D. (1979), *Movilidad giratoria, armonia social y desarrollo en los Andes Meridionales: patrones de tráfico e interacción económica (ensayo)*, Antofagasta: Universidad del Norte.

Núñez, L. and Santoro, C.M. (1988), 'Cazadores de la puna seca y salada del área centro-sur andina (Norte de Chile)', *Estudios Atacameños* 9, 11–60.

Núñez H., P. and Zlatar M., V. (1980), 'Coexistencia de comunidades recolectoras-cazadoras', *Actas del V Congreso Nacional de Arqueología Argentina* 1, 79–92, San Juan.

O'Neale, L.M. (1937), 'Archaeological explorations in Peru, Part III: textiles of the Early Nazca period', *Anthropology, Memoirs of the Field Museum of Natural History* 2(3), 119–218, Chicago.

O'Neale, L. (1942), 'Textile periods in ancient Peru II: Paracas caverns and the Grand

Necropolis', *University Publications in American Archaeology and Ethnology* 39(2), 143–201.

Oakland, A. (1986), 'Tiahuanaco tapestry tunics and mantles from San Pedro de Atacama, Chile', in Rowe, A.P., ed., 101–21.

Olmos Figueroa, O. G. (1980), 'Análisis cuantitativo y comparativo de dos tests estratigráficos de los campamentos Puripica-1 y Tulán 52, San Pedro de Atacama', Chile, unpublished undergraduate dissertation, Antofagasta: Universidad del Norte.

Orlove, B.S. (1982), 'Native Andean pastoralists: traditional adaptations and recent changes', in Salzman, P.C., ed., 95–152.

Ortega Perrier, M. (1998), 'By reason or by force: Isligueño identity and Chilean nationalism', unpublished doctoral thesis, Cambridge: University of Cambridge.

Palacios Ríos, F. (1977), 'Pastizales de regadío para alpacas', in Flores Ochoa, J.A., ed., 155–70.

Palacios Ríos, F. (1981), 'Tecnología del pastoreo', in Lechtman, H. and Soldi, A.M., eds, 217–32.

Palacios Ríos, F. (1984), 'Bilateralidad y propiedad en una comunidad de pastores', *Boletín del Instituto de Estudios Aymaras* Series 2, 16, 47–62.

Pärssinen, M. (1992), *Tawantinsuyu: the Inca State and its Political Organization*, Helsinki: Societas Finlandiae, Studia Historica 43.

Paskoff, R.P. (1977), 'Quaternary of Chile: the state of research', *Quaternary Research* 8, 2–31.

Payne, S. (1968), 'The origins of domestic sheep and goats: a reconsideration in the light of the fossil evidence', *Proceedings of the Prehistoric Society* (N.S.) 34, 368–84.

Pires Ferreira, E. (1981/2), 'Nomenclatura y nueva clasificación de los camélidos sudamericanos', *Revista do Museu Paulista* (N.S.) 28, 203–19.

Platt, T. (1975), 'Experiencia y experimentación: los asentamientos andinos en las cabeceras del valle de Azapa', *Chungará* 5, September 1975, 33–60.

Pollard, G.C. (1971), 'Cultural change and adaptation in the central Atacama Desert of Northern Chile', *Ñawpa Pacha* 9, 41–64.

Pollard, G.C. and Drew, I.M. (1975), 'Llama herding and settlement in prehispanic northern Chile: application of an analysis for determining domestication', *American Antiquity* 40(3), 296–305.

Popper, V. S. (1977), 'Prehistoric cultivation in the Puna de Atacama, Chile', unpublished undergraduate dissertation, Cambridge, Massachusetts: Harvard University.

Posnansky, A. (1945), *Tihuanacu. La cuna del hombre americano*, La Paz: Ministerio de Educación.

Provoste Fernandez, P. (1978), *Consideraciones para promover organizaciones colectivas en Isluga*, Iquique: Universidad del Norte, Centro Isluga de Investigaciones Andinas (mimeographed),

Quispe M.U. (1969), *La herranza en Choque Huarcaya y Huancasancos, Ayacucho*, Lima: Instituto Indigenista Peruano, Serie Monográfica 20.

Rabassa, J. and Clapperton, C.M. (1990), 'Quaternary glaciations of the Southern Andes', *Quaternary Science Reviews* 9, 153–74.

Rabey, M.A. (1989), 'Are llama-herders in the South central Andes true pastoralists?', in Clutton-Brock, J., ed., 269–76.

Radicati Di Primeglio, C., ed. (1979), *El sistema contable de los Incas*, Lima: Libreria Studium S.A.

Radicati Di Primeglio, C. (1984), 'El secreto de la quilca', *Revista de Indias*, XLIV(173), 11–60, plates I–IV.

Randall, R. (1982), 'Qoyllur Rit'i, an Inca fiesta of the Pleidades: reflections on time and space in the Andean world', *Bulletin de l'Institut Français d'Etudes Andines* 11(1–2), 37–81.

Reigadas, M. del C. (1992), 'La punta del ovillo: determinación de domesticación y pastoreo a partir del análisis microscopico de fibras y folículos de camélidos', *Arqueología* 2, 9–52.

Rivera, M. (1980), 'Algunos fenómenos de complementaridad económica a través de los datos arqueológicos en el área centro sur andina: la fase Alto Ramírez reformulada', *Estudios Arqueológicos*, Special Number, 71–103.

Rivera, M.A. (1985), 'Alto Ramírez y Tiwanaku, un caso de interpretación simbólica a través de datos arqueológicos en el área de los valles occidentales S. del Perú y N. de Chile', *Diálogo Andino* 4, 39–58.

Rivera, M.A., Soto R., P., Ulloa T., L. and Kushner L., D. (1974), 'Aspectos sobre el desarrollo tecnológico en el proceso de agriculturización en el norte prehispano, especialmente Arica (Chile)', *Chungará* 3, 79–107.

Rivière, G. (1983), 'Quadripartition et ideologie dans les communautés aymaras de Carangas (Bolivie)', *Bulletin de l'Institut Français d'Etudes Andines* 12 (3–4), 41–62.

Rodriguez V., W. (1985), 'Huywa ch'uwaña en Wallatani', in Bustinza Choque, V. and Avila Cazorla, E., eds, 148–51.

Rose, D.B. (1996), *Nourishing Terrains: Australian Aboriginal Views of Landscape and Wilderness*, Canberra: Australian Heritage Commission.

Rostworowski de Diez Canseco, M. (1986), *Estructuras andinas del poder*, Lima: Instituto de Estudios Andinos (second edition),

Rowe, A.P. (1977), *Warp-Patterned Weaves of the Andes*, Washington, D.C.: The Textile Musem.

Rowe, A.P., ed. (1986), *The Junius B. Bird Conference on Andean Textiles*, Washington D.C.: The Textile Museum.

Rowe, J.H. (1967) 'Stages and periods in archaeological interpretation', in Rowe, J.H. and Menzel, D., eds, 1967, 1–15.

Rowe, J.H. (1979), 'An account of the shrines of ancient Cuzco', *Ñawpa Pacha* 17, 1–80.

Rowe, J.H. (1985), 'Probanza de los Incas nietos de conquistadores', *Histórica* IX(2), 193–245.

Rowe, J.H. and Menzel, D., eds (1967), *Peruvian Archaeology: Selected Readings*, Palo Alto, California: Peek Publications.

Rydén, S. (1936), *Archaeological Researches in the Department of La Candelaria (Prov. Salta, Argentina)*, Göteborg: Elanders Boktryckeri Aktiebolag.

Ryder, M.L. (1969), 'Changes in the fleece of sheep following domestication (with a note on the coat of cattle)', in *The Domestication and Exploitation of Plants and Animals*, Ucko, P.J. and Dimbleby, G.W., eds, London: Duckworth, 495–521.

Ryder, M.L. (1978), *Sheep and Wool for Handicraft Workers*, Edinburgh: published privately.

Ryder, M.L. and Gabra-Sanders, T. (1985), 'The application of microscopy to textile history', *Textile History* 16(2), 123–40.

Sallnow, M.J. (1987), *Pilgrims of the Andes: Regional Cults and Cusco*, Washington, D.C.: Smithsonian Institution Press.

Salzman, P.C., ed. (1982), *Contemporary and Nomadic Peoples: Africa and Latin America*, Williamsburg, Virginia: Department of Anthropology, College of William and Mary, Studies in Third World Societies, Publication 17.

Santa Cruz Pachacuti Yamqui Salcamaygua, J. de (1879 [1613]), 'Relación de antigüedades deste reyno del Pirú', in *Tres Relaciones de Antigüedades Peruanos*, Jimenez de la Espada, ed., 229–328, Madrid.

Santa Cruz Pachacuti Yamqui Salcamaygua, J. de (1968 [1613]), 'Relación de antigüedades deste reyno del Perú', in *Biblioteca de Autores Españoles* 209, Esteve Barba, F., ed., 279–319, Madrid: Ediciones Atlas.

Santillán, H. de (1968 [1563]), 'Relación del origen, descendencia, política y gobierno de los Incas', in *Biblioteca de Autores Españoles* 209, Esteve Barba, F., ed., 97–149, Madrid: Ediciones Atlas.

Santoro Vargas, C. (1981), 'Formativo temprano en el extremo norte de Chile', *Chungará* 8, 33–77.

Sarmiento de Gamboa, P. (1942 [1572]), 'Historia Indica', in *Don Francisco de Toledo, supremo organizador del Perú* 3, Levillier, R., ed., 1–159, Buenos Aires: Espasa Calpe SA.

Sawyer, A.R. (1962), 'A group of early Nazca sculptures in the Whyte collection', *Archaeology* 15(3), 152–9.

Sawyer, A.R. (1967), 'Tiahuanaco tapestry design', in Rowe, J.H. and Menzel, D., eds, 165–76.

Schneider, J. and Weiner, A.B. (1986), 'Cloth and the organization of human experience', *Current Anthropology* 27, 178–84.

Seltzer, G.O. (1990), 'Recent glacial history and paleoclimate of the Peruvian–Bolivian Andes', *Quaternary Science Reviews* 9, 137–52.

Serracino, G. and Stehberg L., R. (1975), 'Vida pastoril en la precordillera andina (Guatín [sic], San Pedro de Atacama, Chile)', *Estudios Atacameños* 3, 81–99.

Sherbondy, J.E. (1988), 'Mallki: ancestros y cultivo de árboles en los Andes', in *Sociedad andina pasado y presente. Contribuciones en homenaje a la memoria de César Fonseca Martel*, Matos Mendieta, R., ed., 101–35, Lima: Fomciencias.

Sherratt, A. (1981), 'Plough and pastoralism: aspects of the secondary products revolution', in *Pattern of the Past. Studies in Honour of David Clarke*, Hodder, I., Isaac, G. and Hammond, N., eds, 261–305, Cambridge: Cambridge University Press.

Shimada, M. and Shimada, I. (1985), 'Prehistoric llama breeding and herding on the north coast of Peru', *American Antiquity* 50, 3–26.

Shimada, M., and Shimada, I. (1987), 'Comment on the functions and husbandry of alpaca', *American Antiquity* 52(4), 836–9.

Sikkink, L. (1997), 'Water and exchange: the ritual of *yaku cambio* as communal and competetive encounter', *American Ethnologist* 24(1), 170–98.

Sikkink, L. (1998), 'Mountains and modern landscape: folk stories about Azanaques and Thunapa', information and abstracts, in *Kay Pacha: Earth, Land, Water and Culture in the Andes*, Dransart, P. and Sillar, B., eds, Lampeter: University of Wales.

Simpson, D.P. (1973), *Cassell's Latin–English Dictionary*, London: Cassell, fifth edition, fourth impression.

Simpson Vuilleumier, B. (1971), 'Pleistocene changes in the fauna and flora of South America', *Science* 173 (3,999), 771–780.

Spahni, J.C. (1962), 'L' "enfloramiento" ou le culte du lama chez les indiens du désert d'Atacama (Chili)', *Bulletin Société Suisse des Américanistes* 24, 26–36.

Stahl, P.W. (1988), 'Prehistoric camelids in the lowlands of Western Ecuador', *Journal of Archaeological Science* 15: 355–65.

Stahl, P.W. and Norton, P. (1987), 'Precolumbian animal domesticates from Salango, Ecuador', *American Antiquity* 52(2), 382–91.

Standen, V. and Núñez, L. (1984), 'Indicadores antropológicos-físicos y culturales del cementerio precerámico Tiliviche-2 (Norte de Chile)', *Chungará* 12, 135–54.

Stiles, D. (1977), 'Ethnoarchaeology: a discussion of methods and applications', *Man* (N.S.) 12, 87–103.

Stine, S. and Stine, M. (1990), 'A record from lake Cardiel of climate change in southern South America', *Nature* 345, 705–7.

Stobart, H. (1996), '*Tara* and *q'iwa* – worlds of sound and meaning', in Baumann, M.P., ed., 67–81.

Stone, R.R. (1986), 'Color patterning and the Huari artist: the "Lima Tapestry" revisited', in Rowe, A.P., ed., 137–49.

Stone-Miller, R. (1992), 'Camelids and chaos in Huari and Tiwanaku textiles', in *The Ancient Americas: Art from Sacred Landscapes*, Townsend, R.F., ed., 334–45, Munich: Prestel Verlag and the Art Insitute of Chicago.

Stone-Miller, R. (1994), 'To weave for the sun: an introduction to fiber arts of the ancient Andes', in *To Weave for the Sun: Ancient Andean Textiles in the Museum of Fine Arts, Boston*, Stone-Miller, R., Paul, A., Niles, S.A. and Young-Sánchez, M., 11–24, London: Thames and Hudson.

Strathern, M. (1991), *Partial Connections*, Savage, Maryland: Rowman and Littlefield.

Tapper, R. (1988), 'Animality, humanity, morality, society', in Ingold, T., ed., 47–62.

Tarrago de Font, M.N. (1976), 'Alfarería típica de San Pedro de Atacama (Norte de Chile)', *Estudios Atacameños* 4, 37–64.

Taussig, M.T. (1980), *The Devil and Commodity Fetishism in South America*, Chapel Hill, North Carolina: The University of North Carolina Press.

Taylor, G., (1987), *Ritos y tradiciones de Huarochirí del siglo XVII*, Lima: Instituto de Estudios Peruanos.

Thompson, L.G., Davis, M.E., Mosley-Thompson, E., and Liu, K-B. (1988), 'Pre-Incan agricultural activity recorded in dust layers in two tropical ice cores', *Nature* 336, 763–5.

Thompson, L.G., Davis, M.E., Mosley-Thompson, E., Sowers, T.A., Henderson, K.A., Zagorodonov, V.S., Lin, P.-N., Mikhalenko, V.N., Campen, R.K., Bolzan, J.F., Cole-Dai, J. and Francou, B. (1998), 'A 25,000-year tropical climate history from Bolivian ice cores', *Science* 282(5,395), 1,858–64.

Thompson, L.G., Mosley-Thompson, E., Bolzan, J.F., and Koci, B.R. (1985), 'A 1,500-year record of tropical precipitation in ice cores from the Quelccaya ice cap, Peru', *Science* 229, 971–3.

Thompson, L.G., Mosley-Thompson, E., Dansgaard, W. and Grootes, P.M. (1986), 'The Little Ice Age as recorded in the stratigraphy of the tropical Quelccaya ice cap', *Science* 234, 361–4.

Thompson, L.G., Mosley-Thompson, E. and Morales Arnao, B. (1984), 'El Niño Southern Oscillation events recorded in the stratigraphy of the tropical Quelccaya ice cap, Peru', *Science* 226, 50–3.

Tomoeda, H. (1985), 'The llama is my *chacra*: metaphor of Andean pastoralists', in Masuda, S. *et al.*, eds, 277–99.

Torres, C.M. (1987), 'The iconography of the prehispanic snuff trays from San Pedro de Atacama, northern Chile', *Andean Past* 1, 191–246.

Torres, C.M. and Conklin, W.J. (1995), 'Exploring the San Pedro de Atacama/Tiwanaku relationship', in Dransart, P., ed., 78–108.

Torrico, C. (MS), 'Living weavings: the symbolism of Bolivian herders' sacks'.

Torrico Prado, B. (1971), *Indígenas en el corazón de America*, La Paz and Cochabamba: Editorial 'Los Amigos del Libro'.

Troll, C. (1958), 'Las culturas superiores andinas y el medio geográfico', *Revista del Instituto de Geografía* 5, Lima.

Troll, C. (1968), 'The cordilleras of the tropical Andes. Aspects of climatic phyto-geographical and agrarian ecology', in *Geo-Ecology of the Mountainous Regions of the*

Tropical Americas, Troll, C., ed., 15–56, Bonn: UNESCO Colloquium Geographicum Band 9.

Tschopik J.R.H. (1951), 'The Aymara of Chucuito, Peru. 1. Magic', *Anthropological Papers of the American Museum of Natural History New York* 44(2).

Tuan, Yi-Fu (1984), *Dominance and Affection: the Making of Pets*, New Haven and London: Yale University Press.

Tyler, S.A. (1986), 'Post-modern ethnography: from document of the occult to occult document', in Clifford, J. and Marcus, G.E., eds, 122–40.

Ulloa, L. (1981), 'Evolución de la industria textil prehispánica en la zona de Arica', *Chungará* 8, 97–108.

Ulloa T., L. (1985), 'Vestimientas y adornos prehispánicos en Arica', in *Arica, diez mil años*, 15–23, Santiago: Museo Chileno de Arte Precolombino.

Urton, G. (1988), *At the Crossroads of the Earth and Sky: an Andean Cosmology*, Texas: University of Texas Press (paperback edition).

Urton, G., with Nina Llanos, P. (1997), *The Social Life of Numbers: A Quechua Ontology of Numbers and Philosophy of Arithemetic*, Austin: University of Texas Press.

Valdes, X., Montecino, S., Leon. K. de and Mack, M. (1983), *Historias testimoniales de mujeres del campo*, Santiago: PEMCI.

Valdivia, P. de (1953 [1550–1]), *Cartas de Pedro de Valdivia que tratan del descubrimiento de Chile*, Medina, J.T., ed., Santiago de Chile: Fondo Histórico y Bibliográfico de José Toribio Medina.

Villalba O.M. (1988), *Cotocollao: una aldea formativo del valle de Quito*, Quito: Museos del Banco Central del Ecuador, Miscelánea Antropológica Ecuatoriana, Serie Monográfica 2.

Vita Finzi, C. (1959), 'A pluvial age in the Puna de Atacama', *The Geographical Journal* 125(3–4), 401–3.

Vreeland, J.M. (1974), 'Procedimiento para la evaluación y classificación del material textil andino', *Arqueológicas* 15, 70–96.

Webley, L. (1997), 'Wives and sisters: changing gender relations among Khoe pastoralists in Mamqualand', in *Our Gendered Past: Archaeological Studies of Gender in Southern Africa*, Wadley, L., ed., 167–208, Johannesburg: Witwatersrand University Press.

Wheeler, J.C. (1984), 'On the origin and early development of camelid pastoralism in the Andes', in Clutton-Brock, J. and Grigson, C., eds, 395–410.

Wheeler Pires-Ferreira, J., Pires-Ferreira, E. and Kaulicke, P. (1976), 'Preceramic animal utilization in the central Peruvian Andes', *Science* 194 (4,264), 483–90.

Wheeler, J.C., Russel, A.J.F., and Redden, H. (1995), 'Llamas and alpacas: pre-conquest breeds and post-conquest hybrids', *Journal of Archaeological Science* 22, 833–40.

Wildman, A.B. (1954), *The Microscopy of Animal Textile Fibres*, Leeds: WIRA.

Willey, G.R. (1946), 'The culture of La Candelaria', in *Handbook of South American Indians* 2, Steward, J.H., ed., 661–72.

Wing, E.S. (1972), 'Appendix IV: Utilization of animal resources in the Peruvian Andes,' in *Excavations at Kotosh, Peru. A report on the third and fourth expeditions*, Izumi, S. and Terada, K., eds, Tokyo: University Press, 327–51.

Wing, E.S. (1977), 'Animal domestication in the Andes', in *Origins of Agriculture*, Reed, C.A., ed., The Hague: Mouton, 837–59.

Wing, E.S. (1988), 'Use of animals by the Inca as seen at Huanuco Pampa', in *Economic Prehistory of the Central Andes*, Wing, E.S. and Wheeler, J., eds, 167–79, Oxford: BAR International Series 427.

Wirrman, D. and Mourguiart, P. (1995), 'Late Quaternary spatio-temporal limnological variations in the altiplano of Bolivia and Peru', *Quaternary Research* 43, 344–354.

Woodburn, J. (1982), 'Egalitarian societies', *Man* (N.S.) 17, 431–51.

Wouters, J. and Verhecken, A. (1989a), 'The coccid insect dyes: HPLC and computerized diode-array analysis of dyed yarns', *Studies in Conservation* 34, 189–200.

Wouters, J. and Verhecken, A. (1989b), 'The scale insect dyes (*Homoptera: Coccidea*), Species recognition by HPLC and diode-array analysis of the dyestuffs', *Annales de la Société Entomologique de France* 25(4), 393–410.

Yacobaccio, H.D. and Madero, C.M. (1994), 'Etnoarqueología de pastores surandinos: una herramienta para conocer el registro arqueológico', *Jornadas de Arqueología e Interdisciplinas*, 203–36, Buenos Aires: Consejo Nacional de Investigaciones Científicas y Técnicas, Programa de Estudios Prehistóricos.

Yacobaccio, H.D., Madero, C.M., Malmierca, M.P. (1998), *Etnoarqueología de pastores surandinos*, Buenos Aires: Grupo Zooarqueología de Camélidos.

Yacobaccio, H.D., Madero, C.M., Malmierca, M.P. and Reigadas, M. del C. (1997/8), 'Caza, domesticación y pastoreo de camélidos en la puna argentina', *Relaciones* 22/23, 389–429.

Zárate, A. de (1862 [1555]), 'Historia del descubrimiento y conquista del Perú', in *Biblioteca de Autores Españoles* 26, Vedia, E. de, ed., 459–574, Madrid: M. Rivadeneyra.

Zeuner, F.E. (1963), *A History of Domesticated Animals*, London: Hutchison.

Zlatar, V. (1983), 'Replanteamiento sobre el problema Caleta Huelén 42', *Chungará* 10, 21–8.

Zorn, E. (1987), 'Un análisis de los tejidos en los atados rituales de los pastores', *Revista Andina* 10, 489–526.

Zuidema, R.T. (1980), 'El sistema de parentesco incaico: una nueva visión teórica', in Mayer, E. and Bolton, R., eds, 57–113.

Zuidema, R.T. (1990), *Inca Civilization in Cusco*, trans. Decoster, J.-J., Austin: University of Texas Press.

Zuidema, R.T. and Urton, G. (1976), 'La constelación de la llama en los Andes peruanos', *Allpanchis Phuturinqa* 9, 59–119.

Zumbühl, H. (1979), *Tintes naturales*, Huancayo: Kamaq Maki.

Index

Page references for plates and figures are shown in italics